ENCYCLOPAEDIA OF
NANO BIOLOGY

ENCYCLOPAEDIA OF
NANO BIOLOGY

Robert William

ANMOL PUBLICATIONS PVT. LTD.
NEW DELHI - 110 002 (INDIA)

ANMOL PUBLICATIONS PVT. LTD.
H.O.: 4374/4B, Ansari Road, Darya Ganj,
New Delhi-110 002 (India)
Ph.: 23278000, 23261597
B.O.: No. 1015, Ist Main Road, BSK IIIrd Stage
IIIrd Phase, IIIrd Block,
Bangalore - 560 085 (India)
Visit us at: www.anmolpublications.com

Encyclopaedia of Nano Biology

First Published, 2006
ISBN 81-261-2985-9

PRINTED IN INDIA

Printed at Mehra Offset Press, Delhi.

Contents

Preface vii

1. Nano-technology, Nano-biology & Bio-materials: An Overview 1

2. Quantum Biology, Nanomedicine and Molecular Nanotechnology 23

3, Nanosafety, Nanolabs & Nanobodies 76

4. Nanobiology & Nanomedicine: Tools & Techniques 135

5. Impact of Nanotechnology on Food Agriculture, Fishing, Veterinary and Environment 188

Glossary 198

Bibliography 240

Index 249

Preface

This book entitled "Encyclopaedia of Nano Biology" deals primarily with related areas in nanotechnology and their relationship with nanobiology and biomaterials. An introduction overview of quantum biology, nanomedicine and molecular nanotechnology is provided. The rather important issues of nanosafety, nanolabs and nanobodies are discussed in detail. The tools and techniques of nanobiology and nanomedicine are summarized and their future impacts on food, agriculture, fishing, veterinary & environment are discussed briefly. Crucial issues in nanobiology and nanomedicine are listed and explained.

No doubt, nanotechnology overlaps with a wide range of the existing sciences. So, whichever field of Physics, Chemistry, Computer Science or Biology (Biomedical) one specializes in, one will be able to find some degree of involvement. Subjects that seem at present to have the most direct and constructive involvement with Nano Biology include molecular biology and biomedical engineering (through genetics and microbiology). The applications of nano biology are near limitless, but the main focus seems to be on its use in medical applications. That seems to be what's driving the research, since it would give us the opportunity to cure most diseases, even possibly eliminating the concept of natural death. There are so many other applications, however, ranging from mundane to incredibly profound. Other applications include advanced biological imaging, stem-cell engineering, and medical diagnostic tools. Nature provides the inspiration for this new biotechnology – from time immemorial, nanomachines have been at work in all organic cells, building our world of plants, animals and people from a store of chemical elements. Now several nanosciences include nanosurgery, cell pharmacology and others.

Nanosurgery is a generic term including molecular repair and cell surgery.

Cell surgery is modifying cellular structures using medical nanomachines.

Cell pharmacology is delivery of drugs by medical nanomachines to exact locations in the body. There is already a wide range of products in the nanotechnology sector – the world market was estimated at almost US$100 billion in 2004. These products are used primarily in the areas of material sciences, medicine, environmental technology, and computer hardware. Major areas where Nanotech products will invite some kind of regulation by government agencies include: Biological products (vaccines, blood products, tissues); Cosmetics; Devices; Foods (for humans and animal feed, though generally not meat and poultry); Dietary supplements; Drugs (human and animal); Radiation Emitting Electronic Products; Color additives used in food, drugs, cosmetics, devices; Combination products (i.e., drug-device, drug-biologic, and device-biologic products).

—Editor

1

Nano-technology, Nano-biology & Bio-materials: An Introductory Overview

Nanotechnology: Building Blocks

Although it seems at first that Nature has provided a limited number of basic building blocks—amino acids, lipids, and nucleic acids—the chemical diversity of these molecules and the different ways they can be polymerized or assembled provide an enormous range of possible structures. Furthermore, advances in chemical synthesis and biotechnology enable one to combine these building blocks, almost at will, to produce new materials and structures that have not yet been made in Nature. These self-assembled materials often have enhanced properties as well as unique applications. The selected examples below show ways in which clever synthetic methodologies are being harnessed to provide novel biological building blocks for nanotechnology. The protein polymers produced by Tirrell and coworkers (1994) are examples of this new methodology. In one set of experiments, proteins were Top: a 36-mer protein polymer with the repeat sequence (ulanine-glycine)3 – glutamic acid – glycine. Bottom: idealized folding of this protein polymer, where the glutamic acid sidechains (+) are on the surface of the folds. designed from first principles to have folds in specific locations and surfacereactive groups in other places (Krejchi et al. 1994; 1997). One of the target sequences was -((AG)3EG)- 36. The hypothesis was that the AG regions would form hydrogen-bonded networks of beta sheets and that the glutamic acid would provide a functional group for surface modification. Synthetic DNAs coding for these proteins were produced, inserted into an *E. coli* expression system, and the desired proteins were produced and harvested. These biopolymers formed chain-folded lamellar crystals with the anticipated folds. In addition to serving as a source of totally new materials, this type of research also enables us to test our understanding of amino acid interactions and our ability to predict chain folding. Biopolymers produced via biotechnology are monodisperse; that is, they have precisely defined and controlled chain lengths; on the other hand, it is virtually impossible to produce a monodisperse synthetic polymer. It has recently been shown that polymers with well-defined chain lengths can have unusual liquid crystalline properties.

For example, Yu et al. (1997) have shown that bacterial methods for polymer synthesis can be used to produce poly(gamma-benzyl alpha L-glutamate) that exhibits smectic ordering in solution and in films. The distribution in chain length normally found for synthetic polymers makes it unusual to find them in smectic phases. This work is important in that it suggests that we now have a route to new smectic phases whose layer spacings can be controlled on the scale of tens of nanometers. The biotechnology-based synthetic approaches described above generally require that the final product be made from the natural, or L-amino acids. Progress is now being made so that biological machinery (e.g., *E. coli*), can be co-opted to incorporate non-natural amino acids such as b-alanine or dehydroproline or fluorotyrosine, or ones with alkene or alkyne functionality (Deming et al. 1997). Research along these lines opens new avenues for producing controlled-length polymers with controllable surface properties, as well as biosynthetic polymers that demonstrate electrical phenomena like conductivity. Such molecules could be used in nanotechnology applications. Novel chemical synthesis methods are also being developed to produce "chimeric" molecules that contain organic turn units and hydrogen-bonding networks of amino acids (Winningham and Sogah 1997). Another approach includes incorporating all tools of chemistry into the synthesis of proteins, making it possible to produce, for example, mirror-image proteins. These proteins, by virtue of their D-amino acid composition, resist biodegradation and could have important pharmaceutical applications (Muir et al. 1997). Arnold and coworkers are using a totally different approach to produce proteins with enhanced properties such as catalytic activity or binding affinity. Called "directed evolution," this method uses random mutagenesis and multiple generations to produce new proteins with enhanced properties. Directed evolution, which involves DNA shuffling, has been used to obtain esterases with five- to six-fold enhanced activity against *p*-nitrobenzyl esters (Moore et al. 1997).

The ability of biological molecules to undergo highly controlled and hierarchical assembly makes them ideal for applications in nanotechnology. The self-assembly hierarchy of biological materials begins with monomer molecules (e.g., nucleotides and nucleosides, amino acids, lipids), which form polymers (e.g., DNA, RNA, proteins, polysaccharides), then assemblies (e.g., membranes, organelles), and finally cells, organs, organisms, and even populations (Rousseau and Jelinski 1991, 571-608). Consequently, biological materials assembly on a very broad range of organizational length scales, and in both hierarchical and nested manners (Aksay et al. 1996; Aksay 1998). Research frontiers that exploit the capacity of biomolecules and cellular systems to undergo self-assembly have been identified in two recent National Research Council reports (NRC 1994 and 1996). Examples of self-assembled systems include monolayers and multilayers, biocompatible layers, decorated membranes, organized structures such as microtubules and biomineralization, and the intracellular assembly of CdSe semiconductors and chains of magnetite. A number of researchers have been exploiting the predictable basepairing of DNA to build molecular-sized, complex, three-dimensional objects. For example, Seeman and coworkers (Seeman 1998) have been investigating these properties of DNA molecules with the goal of forming complex 2-D and 3-D periodic structures with defined topologies.

DNA is ideal for building molecular nanotechnology objects, as it offers synthetic control, predictability of interactions, and well-controlled "sticky ends" that assemble in highly specific fashion. Furthermore, the existence of stable branched DNA molecules permits complex and interlocking shapes to be formed. Using such technology, a number of topologies have been prepared, including cubes (Chen and Seeman 1991), truncated octahedra (Zhang and Seeman 1994), and Borromean rings (Mao et al. 1997). Other researchers are using the capacity of DNA to self-organize to develop photonic array devices and other molecular photonic components (Sosnowski et al. 1997).

This approach uses DNA-derived structures and a microelectronic template device that produces controlled electric fields. The electric fields regulate transport, hybridization, and denaturation of oligonucleotides. Because these electric fields direct the assembly and transport of the devices on the template surface, this method offers a versatile way to control assembly. There is a large body of literature on the self-assembly on monolayers of lipid and lipid-like molecules (Allara 1996, 97-102; Bishop and Nuzzo 1996). Devices using self-assembled monolayers are now available for analyzing the binding of biological molecules, as well as for spatially tailoring. Idealized truncated octahedron assembled from DNA. This view is down the four-fold axis of the squares. Each edge of the octahedron contains two double-helical turns of DNA surface activity. The technology to make self-assembled monolayers (SAMs) is now so well developed that it should be possible to use them for complex electronic structures and molecular-scale devices.

Research stemming from the study of SAMs (e.g., alkylthiols and other biomembrane mimics on gold) led to the discovery of "stamping" (Kumar and Whitesides 1993). This method, in which an elastomeric stamp is used for rapid pattern transfer, has now been driven to < 50 nanometer scales and extended to nonflat surfaces. It is also called "soft lithography" and offers exciting possibilities for producing devices with unusual shapes or geometries. Self-assembled organic materials such as proteins and/or lipids can be used to form the scaffolding for the deposition of inorganic material to form ceramics such as hydroxyapatite, calcium carbonate, silicon dioxide, and iron oxide. Although the formation of special ceramics is bio-inspired, the organic material need not be of biological origin. An example is production of template-assisted nanostructured ceramic thin films (Aksay et al. 1996). A particularly interesting example of bio-inspired self-assembly has been described in a recent article by Stupp and coworkers (Stupp et al. 1997). This work, in which organic "rod-coil" molecules were induced to selfassemble, is significant in that the molecules orient themselves and selfassemble over a wide range of length scales, including mushroom-shaped clusters; sheets of the clusters packed side-by-side; and thick films, where the sheets pack in a head-to-tail fashion. The interplay between hydrophobic and hydrophilic forces is thought to be partially responsible for the controlled assembly.

Dispersion and Coatings

Nanoparticles and nanospheres have considerable utility as controlled drug delivery systems (Hanes et al. 1997). When suitably encapsulated, a pharmaceutical can be delivered to the appropriate site, its concentration can be maintained at proper levels for long periods of time, and it can be prevented from undergoing premature degradation. Nanoparticles (as opposed to micron-sized particles) have the advantage that they are small enough that they can be injected into the circulatory system. Highly porous materials are also ideal candidates for controlled drug delivery (Schnur et al. 1994) and for tissue engineering (Hubbell and Langer 1995). An example of a controlled drug delivery system comes from the area of microtubules. Phospholipid bilayers can self-assemble into long cylindrical tubes with diameters usually below a micron and lengths up to hundreds of microns (Schnur 1993).

During synthesis, drugs can be entrained original patterned silicon *elastomeric stamp elastomeric stamp* substrate. An elastomeric stamp (top left) is made from an original master (bottom left). The stamp is dipped into the biological material (top right) and the pattern is transferred to the substrate (bottom right). Mushroom-shaped clusters formed from self-assembly of rod-coil molecules; these clusters can undergo further packing to form sheets. in these nanotubes, and the final product can be used as a controlled delivery system. Tubules prepared from phospholipid bilayers are ideal for such applications because they are biocompatible. Dendrimers can be prepared so they are of discrete size and contain specific functional groups (Karak and Maiti 1997). They can be functionalized and used in biomedicine. Examples include gene transfer agents for gene therapy, made to carry and control the relaxivity of paramagnetic MRI (magnetic resonance imaging) contrast agents (Toth et al. 1996), and to deliver drugs on a controlled release basis.

Role of Nanoparticles in Health and Pollution

Although beyond the scope of this review, it is important to keep in mind the potential role of atmospheric nanoparticles in photocatalytic and thermal production of atmospheric pollutants. Atmospheric aerosols in heavily polluted areas have the potential to accelerate ozone formation reactions. Furthermore, because they are respirable, they could represent a health hazard. Two controversial studies (the Harvard University six-city study and the American Cancer Society study) have linked the presence of fine particular matter to premature mortality (*Chemical and Engineering News* 1997). Atmospheric aerosols generally contain two major components: one is composed of amorphous carbon that has fullerene-like particles dispersed in it;

the second is inorganic and consists of oxides and sulfides supported on clay minerals. In particular, the iron oxide, manganese oxide, and iron sulfide nanoparticles have band-gaps that could enhance the photocatalytic adsorption of solar radiation. In addition, these materials are acidic and may be coated with water, which would enhance their catalytic ability to crack hydrocarbons and create free radicals (Chianelli 1998). At present this is an underexplored area of research that bears scrutiny. It is interesting to note that some microorganisms produce and sequester CdSe and CdS nanoparticles in response to high levels of toxic Cd++ in their environment (Brus 1996). A large number of organisms also have the ability to precipitate ferrimagnetic minerals Fe_3O_4 and Fe_3S_4.

High Surface Area Materisl

Supported polymeric membranes can be used to remove low molecular weight organics from aqueous solution (Ho et al. 1996). They work by filling the pores of microfiltration or ultrafiltration membranes with functional, polymeric, or oligomeric liquids that have an affinity for the compound of interest. These ideas can be extended to microemulsions and used for separations of large biological molecules. For example, non-ionic microemulsions (i.e., oil-water systems) have been used for hemoglobin extraction. These results augur well for the separation of other biological materials (Qutubuddin et al. 1994). Molecular imprinting can also be used to produce high surface area adsorbents that are enantioselective for amino acids and other biological molecules (Vidyasankar et al. 1997).

Crystalline bacterial cell surface layers (S-layers) are composed of repeating protein units. These layers self-assemble and have a high binding capacity. They have been explored as patterning elements for molecular nanotechnology. For example, they have been used to pattern cadmium sulfide superlattices (Shenton et al. 1997).

Functional Nanostructures

The smallest possible computer would ideally be able to perform computations on a molecular scale. Even though the computation may be carried out on that scale, the issue becomes one of having enough molecules to obtain sufficient signal-to-noise ratio to read out the answer. Consequently, at least at present, these computations must be carried out in bulk or with extremes in temperature. There has been a recent development along this line. Adleman (1994) has shown how the rules of DNA selfassembly, coupled with polymerase chain reaction (PCR) amplification of DNA, can lead to a molecular computer of sorts. The system was used to solve the Hamiltonian path problem, a classic and difficult (NP complete) mathematical problem that involves finding a path between vertices of a graph. The starting and ending vertices of each edge of the graph were encoded as the first and second halves of a strand of DNA. A solution to the problem (what is the path between two specific vertices?) was obtained by using PCR primers for the two vertices. Others have extended these ideas and shown that it is possible to make DNA add (Guarnieri et al. 1996). Although these are exciting demonstrations, at present this technology has a number of drawbacks, including the labor and time it takes to analyze the results of the computation, and the uncertainties associated with wet chemistry.

Biological molecules and assemblies, such as the photochemical reaction center, are capable of capturing light with good quantum efficiency and transforming it into chemical energy. If properly exploited, such assemblies have potential applications as biomolecule information processing units. Bacteriorhodopsin, from the purple membrane bacterium *Halobacterium halobium*, is one such system that has been studied extensively and has been commercialized into optical holographic memories (Birge 1995). In the bacterium, the protein bacteriorhodopsin self-assembles into ordered lipid patches. The protein absorbs light and undergoes a cycle involving a complex series of intermediates, resulting in a proton being pumped across the membrane. It was information developed from understanding the basic science behind the way that bacteriorhodopsin works that led to the use of bacteriorhodopsin as a biomolecule information processing unit. To be used for information storage, the protein is placed under nonbiological conditions It is dispersed and immobilized into a matrix (e.g., collagen or another polymeric substance) and held at liquid nitrogen temperatures. At this low temperature, the protein acts as an optically-driven bistable switch. One form, the light-adapted form of bacteriorhodopsin, absorbs light at 570 nm. When irradiated with green light at 77 K, it switches to a different stable form that absorbs at 630 nm. Using light of different colors enables one to read and write images onto these memories. By subtracting one memory from the other, these memories are especially useful for realtime, rapid comparison of images.

The molecular motors found in biology provide for bacterial locomotion, as well as for the active transport and delivery of molecules. For example, the bacterial flagellar motor is about 20 nm in diameter, and is comprised of a complex assembly of more than 10 different proteins (Imae and Atsumi 1989). The role of the motor is to rotate the helical flagella of the bacterium so that it is able to swim. Chemical energy (in this case protons or sodium) is transduced into mechanical energy. Other examples of molecular motors include RNA polymerase (Yin et al. 1995), F1-ATPase (Noji et al. 1997), myosin, and kinesin (Seventh Biophysical Discussions 1995). The fuel that powers these motors is ATP (adenosine triphosphate). A number of researchers have proposed schemes by which such motors could be used to deliver molecules, one at a time, for the purpose of the ground-up assembly of nanoscale devices (NRC 1996). It is envisioned that the highways could be actin or tubulin, which would need to be immobilized onto a surface. Myosin or kinesin, which naturally travel along these highways, could be used to deliver molecular "packages" to a specific assembly site.

In the human body, the function of the high density lipoproteins is to transport cholesterol. The ~ 7.5 nm discoidal lipoprotein assemblies are sandwiched between discs of phospholipids and stacked, poker-chip style.

The lipoproteins stabilize the cholesterol particles and assist in their transport. Current research involves manipulating and fusing the assemblies and particles with an atomic force microscope (ATM) tip (Sligar 1998).

There has been considerable research activity in molecular electronics and bioelectronics, particularly in Japan (Aizawa 1994). Although this area of nanotechnology is still not as well developed as others in this report, it bears watching. For example, Shionoya and coworkers at the Institute for Molecular Science of the Okazaki National Research Institutes in Japan have proposed that novel combinations of DNA, metal ligands, DNA templating, and proteins could produce molecular wires; molecular hoops through which DNA could be threaded; and doublestranded peptides whose helix pitch could be controlled by an entrained metal. The active site containing the metal could be induced to go from Cu(I)tetrahedral to CU(II)square planar, perhaps by electrons delivered by a scanning tunneling microscope (STM) tip.

Consolidated Materials

Biomolecule/inorganic interactions can be used to produce ceramics with increased toughness. Fundamental studies of biomineralization, in which an organic substance (usually protein or peptide or lipid) interacts with an inorganic phase (e.g., calcium carbonate or hydroxyapatite) have led to the bioinspired synthesis of composite materials. Novel combinations of DNA, metal ligands, DNA templating, and proteins are being investigated for molecular wires, inductors, and switches (photo courtesy of Shionoya and coworkers, Inst. for Molecular Science).

The structure and porosity of the inorganic phase can be controlled by templating with an organic surfactant, vesicular arrays, or liquid crystalline materials. Micelle-templated synthesis can produce ceramics with 20-100 Å pore dimensions (Ying 1998). These tailored pores can be used as catalysts and absorbents, and for gas/liquid separations and thermal and acoustic insulation. Their selectivity makes them very useful for biochemical and pharmaceutical separations. Bioceramics can also be made that are more compatible with teeth and bone. An interesting example of an organic/inorganic composite is the new packaging material that has been developed to replace the polystyrene "clam-shell" for fast food products. Composed of potato starch and calcium carbonate, this foam combines the advantages of good thermal insulation properties and light weight with biodegradability (Stucky 1978).

It has been well documented that a very large number of organisms have the ability to precipitate ferrimagnetic minerals such as Fe3O4 and Fe3S4. In addition, linear chains of membrane-bound crystals of magnetite, called magnetosomes, have been found in microorganisms and fish (Kirschvink et al. 1992). For example, the Fe3O4 domain size in the organism *A. magnetotactum* is about 500 Å, and a chain of 22 of those domains has a magnetic moment of 1.3 x 10-15. It is not immediately clear how these particles could be exploited for nonbiological applications.

Opportunities and Challenges

This section of the review describes current limits on the biological aspects of nanotechnology. It outlines selected broad areas of research opportunity, and it sets forth challenges for the intermediate and long-term future.

We owe much of the recent progress in the biological aspects of nanotechnology and nanoparticles to important enabling technologies. Many of these enabling technologies are described in other parts of this report. Those that have had a significant impact on biological measurements include techniques and instrumentation such as optical traps, laser tweezers, and "nano-pokers" to measure femtoNewton forces (Svoboda and Block 1994), and AFM and STM (Wildoer et al. 1998).

There is also a need to move detection away from the ensemble, to the single molecule scale. This includes driving instrumentation toward singlemolecule spectroscopy; single-spin detection (Rugar et al. 1994); chemical analysis of nanoliter volumes (Hietpas et al. 1996, 139-158); nuclear magnetic resonance microcoils; nanoscale electrode arrays and chemical sensing and detecting technology (McConnell 1996, 97-102); separations technology; chemical analysis of single cells (Hietpas et al. 1996); new biological transformations; and chemical probes of nanostructures. Enhanced computational infrastructure, both hardware and algorithm development, will also be required to investigate supramolecular assemblies and to understand interactions that occur over a wide variety of time and length scales. At present, we are fairly well equipped to handle quantum mechanical calculations. At the next level of computational complexity, the molecular dynamics force fields and atomic charges can presently handle up to about a million particles and still retain their accuracy. The next scale of complexity, mesoscale modeling, currently requires the use of pseudo atoms. Much more research is required to improve the accuracy of finite element calculations and the modeling of materials applications (Goddard 1998).

Much of the progress in the biological aspects of nanotechnology has come from research on surface interactions. It will be important to continue to develop a better understanding of the interface between biomolecules and surfaces. Of particular interest are questions about whether surfaces cause biomolecules to denature, the optimum length of linker groups, and ways to communicate from the biomolecule, through the linker, to the substrate. Another challenge is to produce nanometer ultrathin films that have stable order (Jaworek et al. 1998).

Although much progress has been made on ways to confer additional robustness on biomolecules, it will be necessary to continue to improve the stability and reliability of biomolecular assemblies. Such research may be along the lines of understanding the thermal stability of biomolecules, perhaps by examining extremophiles and molecules produced by directed evolution. It will also be necessary to produce molecules that can work in the absence of water, or to devise ways in which aqueous solutions can be used reliably.

The hierarchical self-assembly of biomolecules is often touted as one of their key attributes for nanotechnology applications. However, there remains much to be learned about how to make periodic arrays of biomolecular assemblies, how to use biological templating in an efficient manner, how to mimic biological self-assembly with nonbiological molecules, and how to exploit differences between biological and nonbiological self-assembly.

Combinatorial chemistry has had an enormous impact on drug discovery and development in the past several years. Ability to develop molecularly sensitive screening techniques is one of the limiting factors of combinatorial chemistry. It would be intriguing to discover screening methods that are particularly sensitive for picking up new phenomena exhibited by biological nanoparticles.

The largest impact of biological sciences on nanotechnology may prove to be through the lessons one can learn from biology, rather than from directly employing the biological molecules themselves. To this end, it is important to support research on various aspects of biomimetics, including activities in which synthetic and biologically-based molecules are combined in chimeric fashion.

Nanotechnology for Bio Materials and Tissues

Historically, synthetic materials have not served as sufficient implants.For example, the current average lifetime of an orthopedic implant, such as hip, knee, ankle, etc., is only 15 years. Similarly, for the vascular community, small diameter vascular grafts are only functional 25% of time past 5 years of use. The same limited lifetimes for a number of failing organs could be stated for cartilage, bladder, and the peripheral and central nervous systems. Clearly, conventional materials, or those materials with constituent dimensions greater than 1 micron have not invoked proper cellular responses to regenerate tissue that allows for these devices to be successful for long periods of times.

In contrast, due to their ability to mimic the dimensions of constituent components of natural tissues, like proteins, nanophase materials may be an exciting successful alternative. Nanophase materials are defined as materials with constituent dimension less than 100 nm in at least one direction. Materials investigated to date include nanophase ceramics, metals, polymers, and composites. Nanophase materials have revolutionized many traditional engineering fields such as those involving catalysis, electronics, optics, magnetics, etc. For example, in the catalytic field, higher surface reactivity of nanophase materials allows for faster reaction times.

To date,there has also been the emergence of data that nanophase materials may be optimal materials for tissue engineering applications. This is not only due to their ability to simulate dimensions of proteins that comprise tissues, but also because of their higher reactivity for interactions of proteins that control cell adhesion and, thus, the ability to regenerate tissues.

There are numerous reasons to consider using nanophase materials as better tissue engineering materials.While most of the promise is conjecture, limited scientific evidence does exist and is generating wide support.

The first objective of this course is to define current problems with implants. You will be familiar with issues related to implant failure particular for orthopedic,dental, cartilage,vascular,bladder,and the nervous systems.

The second objective is to understand how cells interact with implanted materials to generate a biological response.We will describe how proteins interact with materials to control cell function leading to either a desirable or undesirable biological response to an implanted material.

The third objective is to understand how nanophase materials could be used as better implants. What are their promising properties and what experimental evidence exists? We will sort through the experimental data and separate promise from "hype". The discussion will make you be able to indicate what are the promising properties of nanophase materials and how to make them to control initial protein interactions leading to successful regeneration of tissue.

The last objective is to understand what further information will be needed to elucidate if nanophase materials will make better biomaterials for these organ systems. While there is no golden standard chemistry for various implant applications, this course will discuss the possibility of a standard topography that may be useful to regenerate tissue for a wide range of materials including ceramics, metals, and polymers. You will be able to identify future experiments needed for the further consideration of nanophase materials to regenerate tissue

This course will begin with a general overview of current problems with orthopedic, dental, cartilage, vascular, bladder, and central as well as peripheral nervous system devices. These organs will be chosen since they are the ones in which the most implantable devices have been used to date. One day will be spent describing situations in which these devices are needed and why we have not been able to develop sufficient materials to successfully regenerate these tissues. To do this, we will cover the basic anatomy, physiology, and cellular structures of these organs. This will help us understand the complexity of our organs and consequently to design materials that match the most important properties of tissue to promote tissue regeneration.

Next,we will cover how the body responds to implanted materials by first adsorbing a thin layer of proteins and then through select cellular responses. Topics such as how proteins adsorb to surfaces,how cells recognize proteins, how cells adhere and function on materials, and how tissues may or may not regrow will be the focus of this day.The biological response section will be complemented with results from experiments that have delineated fundamental material properties that control such responses. By understanding how the body originally responds to implanted materials, we can begin to develop materials to obtain a desirable response, such a tissue regrowth.

We will then interject a new classification of materials that are beginning to be considered to serve as better biomaterials: nanophase materials. One day will be spent defining these materials and highlighting their special properties pertinent for implants. This includes their mechanical, optical, electrical, and surface properties. We will also discuss various routes to synthesize nanophase materials. Although, to date, the special properties of nanophase materials have been used in non-biological applications, we will emphasize their role in biological applications. We will constantly compare theory will experimental evidence in this section. The day will end with the most promising results obtained thus far in terms of increased tissue regeneration on nanophase materials.

Lastly, we will discuss the most promising future directions for the use of nanophase materials in biological applications. This day will focus on key unanswered questions if nanophase materials are to be used in the body. Central to this question is the immunological response to nanoparticles, which will comprise a majority of the day.

To enhance learning as well as stimulate discussion, the course will be interactive with participation from the learners throughout.

Current Implant Failures

- Current Materials Used
- Current Modes of Failure: Poor Bonding to Surrounding Bone – Generation of Wear Debris – Stress and Strain Imbalances at the Tissue Implant Interface
- Current Statistics on Orthopedic Implant Failure

Cartilage

- Current Treatment Methods and Materials Used
- Current Modes of Failure: Poor Regeneration of Cartilage Tissue – Generation of Wear Debris – Stress and Strain Imbalances at the Tissue Implant Interface
- Current Statistics on Cartilage Implant Failure

Dental

- Current Materials Used
- Current Modes of Failure: Poor Bonding to Surrounding Bone – Generation of Wear Debris – Stress and Strain Imbalances at the Tissue Implant Interface

- Current Statistics on Dental Implant Failure

Vascular

- Current Materials Used
- Current Modes of Failure: Poor Bonding to Surrounding Vascular – Endothelium Removal Due to Shear Stress – Generation of Stenosis - Stress and Strain Imbalances at the Tissue Implant Interface
- Current Statistics on Vascular Implant Failure

Bladder

Central and Peripheral Nervous System

- Current Materials Used
- Current Modes of Failure: Poor Bonding to Surrounding Tissue – Build-up of Glial Scar Tissue – Poor Maintenance of Electrical Properties – Stress and Strain Imbalances at the Tissue Implant Interface
- Current Statistics on Central and Peripheral Nervous System Implant Failure

- Current Materials Used
- Current Modes of Failure: Poor Bonding to Surrounding Tissue – Stress and Strain Imbalances at the Tissue Implant Interface
- Current Statistics on Bladder Implant Failure

Biological Response of Implanted Materials

- Properties of Proteins
- Adsorption
- Conformation/Bioactivity

Cellular Recognition of Proteins Adsorbed on Materials Surfaces

- Adhesion
- Migration
- Differentiation

Cellular Extracellular Matrix Deposition Leading to Tissue Regeneration

- Bone
- Dental Tissue
- Cartilage

- Vascular
- Bladder
- Central and Peripheral Nervous System.

REFERENCES

Averback, R.S., J. Bernholc, and D.L. Nelson. 1991. *MRS Symposium Proceedings* Vol. 206.

Baird, D. & Shew, A.: 2004, 'Probing the History of Scanning Tunneling Microscopy', in: D. Baird, A. Nordmann & J. Schummer (eds.), *Discovering the Nanoscale*, Amsterdam: IOS Press, pp. 145-156.

Baird, D., Nordmann, A. & Schummer, J. (eds.): 2004, *Discovering the Nanoscale*, Amsterdam: IOS Press.

Baker, R.T.K. Synthesis, properties and applications of graphite nanofibers. In *R&D status and trends*, ed. Siegel et al.

Bate, R., Frazier, G., Frensley, W., Reed., M., "An Overview of Nanoelectronics," *Texas Instruments Technical Journal*, July-August 1989, pp. 13-20.

Beck, J.S., J.C. Vartuli, W.J. Roth, M.E. Leonowicz, C.T. Kresge, K.D. Schmitt, C.T.-W. Chu, D.H. Olsen, E.W. Shepard, S.B. McCullen, J.B. Higgins, and J.L. Schlenker. 1992. *J. Am. Chem. Soc.* 114:10834.

Bensaude-Vincent, B.: 2001, 'The Construction of a Discipline: Materials Science in the United States', *Historical Studies in the Physical and Biological Sciences*, 31, 223-248.

Bensaude-Vincent, B.: 2004, 'Two Cultures of Nanotechnology?', *Hyle: International Journal for Philosophy of Chemistry*, 10(2), 65-82.

Berube, D.M.: 2004, 'The Rhetoric of Nanotechnology', in: D. Baird, A. Nordmann & J. Schummer (eds.), *Discovering the Nanoscale*, Amsterdam: IOS Press, pp. 173-192.

Bowes, C.L., A. Malek, and G.A. Ozin. 1996. *Chem. Vap. Deposition* 2:97.

Braun, P.V., P. Osenar, and S.I. Stupp. 1996. *Nature*. 368: 2.

Braun, T.; Schubert, A. & Zsindely, S.: 1997, 'Nanoscience and nanotechnology on the balance', *Scientometrics*, 38, 321-325.

Brinker, C.J. 1996. *Curr. Opin. Solid State Mater. Sci.* 1:798.

Brown, N.: 2004, 'Needle on the Real: Technoscience and Poetry at the Limits of Fabrication', in: N.K. Hayles (ed.), *Nanoculture: Implications of the New Technoscience*, Bristol, UK: Intellect Books, pp. 173-190.

Bueno, O.: 2004, 'The Drexler-Smalley Debated on Nanotechnology: Incommensurability at Work?', *Hyle: International Journal for Philosophy of Chemistry*, 10(2), 83-98.

Bueno, O.: 2004, 'Von Neumann, Self-Reproduction and the Constitution of Nanophenomena', in: D. Baird, A. Nordmann & J. Schummer (eds.), *Discovering the Nanoscale*, Amsterdam: IOS Press, pp. 101-115.

Carroll, J.S.: 2001, 'Social Science Research Methods for Assessing Societal Implications of Nanotechnology', in: M.C. Roco and W.S. Bainbridge (eds.), *Societal Implications of Nanoscience and Nanotechnology*, Dordrecht: Kluwer, pp. 188-192.

Chen, C-Y., S.L. Burkett, H.-X. Li, and M.E. Davis. 1993. *Microporous Mater*. 2:27.

Chianelli, R.R. 1998. Synthesis, fundamental properties and applications of nanocrystals, sheets, and fullerenes based on layered transition metal chalcogenides. In *R&D status and trends*, ed. Siegel et al.

Chianelli, R.R., M. Daage, and M.J. Ledoux. 1994. *Advances in Catalysis* 40:177.

Claeson, T., and Likharev, K.," Single Electronics," *Scientific American*, June 1992, pp. 80-85.

Coenen, C.: 2004, 'Nanofuturismus: Anmerkungen zu seiner Relevanz, Analyse und Bewertung', *Technikfolgenabschätzung ' Theorie und Praxis*, 13 (2), 78-85.

Crandall, B.C. and Lewis, J., *Nanotechnology: Research and Perspectives*, MIT Press, Cambridge, 1992.

Crow, M. & Sarewitz, D.: 2001, 'Nanotechnology and Societal Transformation', in: M.C. Roco and W.S. Bainbridge (eds.), *Societal Implications of Nanoscience and Nanotechnology*, Dordrecht: Kluwer, pp. 45-54.

de Souza e Silva, A.: 2004, 'The Invisible Imaginary: Museum Spaces, Hybrid Reality and Nanotechnology', in: N.K. Hayles (ed.), *Nanoculture: Implications of the New Technoscience*, Bristol, UK: Intellect Books, pp. 27-82.

Decker, M.; Fiedeler, U.; Fleischer, Th.: 2004, 'Ich sehe was, was Du nicht siehst... zur Definition von Nanotechnologie', *Technikfolgenabschätzung ' Theorie und Praxis*, 13 (2), 10-16.

Dresselhaus, M., and G. Dresselhaus. 1995. *Ann. Rev. Mat. Sci*. 25:487.

Dresselhaus, M.S. 1998. Carbon-based nanostructures. In *R&D status and trends*, ed. Siegel et al.

Dresselhaus, M.S., G. Dresselhaus, and P. Eklund. 1996. *Science of fullerenes and carbon nanotubes*. San Diego: Academic Press.

Drexler, K.E.: 2004, 'Nanotechnology: From Feynman to Funding', *Bulletin of Science, Technology & Society*, 24, 21-27.

Dupuy, J.-P.: 2004, 'Complexity and Uncertainty: A Prudential Approach To Nanotechnology', in: European Commission (Community Health and Consumer Protection): *Nanotechnologies: A Preliminary Risk Analysis on the Basis of a Workshop, Brussels, 1-2 March 2004*, pp. 71-93 (www.europa.eu.int/comm/health/ ph_risk/documents/ev_20040301_en.pdf).

Dupuy, J.-P.: 2004, 'Pour une évaluation normative du programme nanotechnologique', *Annales des Mines*, (February 2004), 27-32.

Einsiedel, E.F.; Goldenberg, L.: 2004, 'Dwarfing the Social? Nanotechnology Lessons from the Biotechnology Front', *Bulletin of Science, Technology & Society*, 24, 28-33.

ETC: 2003a, *The Big Down: Atom Tech ' Technologies Converging at the Atomic Scale*. Winnipeg, Canada: Action Group on Erosion, Technology and Concentration [www.etcgroup.org/documents/TheBigDown.pdf].

ETC: 2003b, 'No Small Matter II: The Case for a Global Moratorium ' Size Matters!' *Occasional Paper Series* 7(1) [www.etcgroup.org/documents/Occ.Paper_Nanosafety.pdf].

Etzkowitz, H., 2001, 'Nano-Science and Society: Finding a Social Basis for Science Policy', in: M.C. Roco and W.S. Bainbridge (eds.), *Societal Implications of Nanoscience and Nanotechnology*, Dordrecht: Kluwer, pp. 121-128.

European Comission: *European Workshop on Social and Economic Research on Nanotechnologies and Nanosciences, Brussels, 14-15 April 2004* [www.stage-research.net/STAGE/PAGES/Nano.html].

European Commission (Community Health and Consumer Protection): 2004, *Nanotechnologies: A Preliminary Risk Analysis on theBasis of a Workshop, Brussels, 1-2 March 2004* [www.europa.eu.int/comm/health/ ph_risk/documents/ev_20040301_en.pdf].

European Commission: 2004, *Converging Technologies: Shaping the Future of European Societies (Report of the High Level Expert Group "Foresighting the New Technology Wave")*, Brussels: European Commission Research [http://europa.eu.int/comm/research/conferences/2004/ntw/index_en.html].

F. Buot, "Mesoscopic Physics and Nanoelectronics: Nanoscience and Nanotechnology," *Physics Reports*, pp.73-174, 1993.

F. Capasso and S. Datta, "Quantum Electron Devices," *Physics Today*, pp.74-82, February 1990.

Feynman, R., "There's Plenty of Room at the Bottom: An invitation to Enter a New Field of Physics," Talk at the Annual Meeting of the American Physical Society, 29 December 1959. Reprinted in Appendix B of Crandall and Lewis. See citation above.

Fielder, F.A.; Reynolds, G.H.: 1994, 'Legal Problems of Nanotechnology: An Overview', *Southern California Interdisciplinary Law Journal*, 3, 593-629.

Fleischer, Th.; Decker, M. & Fiedeler, U. (eds.): 2004, *Große Aufmerksamkeit für kleine Welten ' Nanotechnologie und ihre Folgen*, special issue of *Technikfolgenabschätzung ' Theorie und Praxis*, 13 (2), 5-85 [www.itas.fzk.de/tatup/042/inhalt.htm].

Fogelberg, H. & Glimell, H.: 2003, 'Molecular Matters: In Search of the Real Stuff', in: H. Fogelberg & H. Glimell, *Bringing Visibility to the Invisible: Towards A Social Understanding of Nanotechnology*, Göteborg: Göteborg University, pp. 5-32 [www.sts.gu.se/publications/STS_report_6.pdf].

Fogelberg, H. & Glimell, H.: 2003, *Bringing Visibility to the Invisible: Towards A Social Understanding of Nanotechnology*, Göteborg: Göteborg University [www.sts.gu.se/publications/STS_report_6.pdf].

Fogelberg, H.: 2003, 'The Grand Politics of Technoscience: Contextualizing Nanotechnology', in: H. Fogelberg & H. Glimell, *Bringing Visibility to the Invisible: Towards A Social Understanding of Nanotechnology*, Göteborg: Göteborg University, pp. 33-54 [www.sts.gu.se/publications/STS_report_6.pdf].

Fogelberg, H.: 2003, 'The Material Culture of Nanotechnology', in: H. Fogelberg & H. Glimell, *Bringing Visibility to the Invisible: Towards A Social Understanding of Nanotechnology*, Göteborg: Göteborg University, pp. 99-114.

Frazier, G., "An Ideology For Nanoelectronics," in *Concurrent Computations: Algorithms, Architecture, and Technology*, Plenum Press, New York, 1988.

Glimell, H.: 2001, 'Challenging Limits ' Excerpts from an Emerging Ethnography of Nano Physicists', in: H. Glimell & O. Johlin (eds.), *The Social Production of Technology: On the Everyday Life with Things*, Göteburg, SE: BAS Publisher, chapter 7, pp. 111-131 (reprinted in: H. Fogelberg & H. Glimell, *Bringing Visibility to the Invisible: Towards A Social Understanding of Nanotechnology*, Göteborg: Göteborg University, pp. 115-139.

Glimell, H.: 2001, 'Dynamics of the Emerging Field of Nanoscience', in: M.C. Roco and W.S. Bainbridge (eds.), *Societal Implications of Nanoscience and Nanotechnology*, Dordrecht: Kluwer, pp. 156-160.

Glimell, H.: 2003, 'A Nano Narrative: The Mircopolitics of a New Generic Enabling Technology', in: H. Fogelberg & H. Glimell, *Bringing Visibility to the Invisible: Towards A Social Understanding of Nanotechnology*, Göteborg: Göteborg University, pp. 55-77 [www.sts.gu.se/publications/STS_report_6.pdf].

Glimell, H.: 2003, 'Dynamics of the Emerging Field of Nanoscience', in: H. Fogelberg & H. Glimell, *Bringing Visibility to the Invisible: Towards A Social Understanding of Nanotechnology*, Göteborg: Göteborg University, pp. 79-85 [www.sts.gu.se/publications/STS_report_6.pdf].

Glimell, H.: 2004, 'Grand Visions and Lilliput Politics: Staging the Exploration of the 'Endless Frontier'', in: D. Baird, A. Nordmann & J. Schummer (eds.), *Discovering the Nanoscale*, Amsterdam: IOS Press, pp. 231-246.

Goldhaber-Gordon, D., Montemerlo, M.S., Love, J.C., Opiteck, G.J., and Ellenbogen, J.C., "Overview of Nanoelectronic Devices," submitted to the *Proceedings of the IEEE*, February 1997. For more information, please send e-mail to nanotech@mitre.org.

Gorman, M.E.; Groves, J.F. & Shrager, J.: 2004, 'Societal Dimensions of Nanotechnology as a Trading Zone: Results from a Pilot Project', in: D. Baird, A. Nordmann & J. Schummer (eds.), *Discovering the Nanoscale*, Amsterdam: IOS Press, pp. 63-73.

Grinbaum, A. & Dupuy, J.-P.: 2004, 'Living with Uncertainty: Toward the Ongoing Normative Assessment of Nanotechnology', *Techne: Research in Philosophy and Technology*, 8(3) (forthcoming).

Grinbaum, A.: 2004, 'La condition de l'homme moderne et les nanotechnologies', in: G. Nivat (ed.), *Les limites de l´humain. 39èmes Rencontres Internationales de Genève*, L´Age d´Homme, Genève, p. 141.

Grunwald, A.: 2004, 'Ethische Aspekte der Nanotechnologie. Eine Felderkundung', *Technikfolgenabschätzung ' Theorie und Praxis*, 13 (2), 71-78.

Gupta, V.K. & Pangannaya, N.B.: 2000, 'Carbon nanotubes: bibliometric analysis of patents', *World Patent Information*, 22, 185-189.

Hansson, S.O.: 2004, 'Great Uncertainty about Small Things', *Techne: Research in Philosophy and Technology*, 8(3) (forthcoming).

Haruta, M. 1997. *Catalyst surveys of Japan* 1:61 and references therein.

Hennig, J.: 2004, 'Changes in the Design of Scanning Tunneling Microscopic Images from 1980 to 1990', *Techne: Research in Philosophy and Technology*, 8(3) (forthcoming).

Hessenbruch, A.: 2004, 'Nanotechnology and the Negotiation of Novelty', in: D. Baird, A. Nordmann & J. Schummer (eds.), *Discovering the Nanoscale*, Amsterdam: IOS Press, pp. 135-144.

Hullmann, A. & Meyer, M.: 2003, 'Publications and Patents in Nanotechnology. An overview of previous studies and the state of the art', *Scientometrics*, 58 (3) 507-527.

Huo, Q., D.I. Margolese, U. Ciesla, P. Feng, T.E. Gier, P. Sieger, R. Leon, P.M. Petroff, F. Schuth, and G.D. Stucky. 1994. *Nature* 368:317.

Jena, P., S.N. Khanna, and B.K. Rao. 1996. In *Science and technology of atomically engineered materials*, ed. P. Jena. River Edge, NJ: World Scientific.

Johansson, M.: 2003, 'Plenty of room at the bottom: Towards an anthropology of nanoscience', *Anthropology Today*, 19 (No. 6), 3-6.

Johnson, A.: 2004, 'The End of Pure Science: Science Policy from Bayh-Dole to the NNI', in: D. Baird, A. Nordmann & J. Schummer (eds.), *Discovering the Nanoscale*, Amsterdam: IOS Press, pp. 217-230.

Jounet, C., W.K. Maser, P. Bernier, A. Loiseau, M. Lamy de la Chapelle, S. Lefrant, P. Deniard, R. Lee, and J.E. Fischer. 1997. *Nature* 388:756.

K.E. Drexler, *Nanosystems: Molecular Machinery, Manufacturing, and Computation,* Wiley, New York, 1992.

Karger, J., and D.M. Ruthven. 1992. *Diffusion in zeolites.* New York: J. Wiley.

Keiper, A.: 2003, 'The Nanotechnology Revolution', *The New Atlantis*, 2, 17-34.

Khushf, G.: 2004, 'A Hierarchical Architecture for Nano-scale Science and Technology: Taking Stock of the Claims About Science Made By Advocates of NBIC Convergence', in: D. Baird, A. Nordmann & J. Schummer (eds.), *Discovering the Nanoscale*, Amsterdam: IOS Press, pp. 21-33.

Khushf, G.: 2004, 'Systems Theory and the Ethics of Human Enhancement: A Framework for NBIC Convergence', *Annals of the New York Academy of Sciences*, 1013, 124-149.

Krätschmer, W., L.D. Lamb, K. Fostiropoulos, and D.R. Huffman. 1990. *Nature* 347:354.

Kresge, C.T., M.E. Leonowicz, W.J. Roth, J.C. Vartuli, and J.S. Beck. 1992. *Nature* 359:710.

Kupperman, A., S. Nadimi, S. Oliver, G. Ozin, J. Garcés, and M. Olken. 1993. *Nature* 365:239.

Kuusi, O.; Meyer, M.: 2002, 'Technological generalizations and leitbilder ' the anticipation of technological opportunities', *Technological Forecasting & Social Change*, 69, 625-639.

Landon, B.: 2004, 'Less is More: Much Less is Much More: The Insistent Allure of Nanotechnology Narratives in Science Fiction', in: N.K. Hayles (ed.), *Nanoculture: Implications of the New Technoscience*, Bristol, UK: Intellect Books, pp. 131-146.

Laszlo, P.: 2004, 'Is There Life After Partington?', *Hyle: International Journal for Philosophy of Chemistry*, 10(2), 169-178.

Lenhard, J.: 2004, 'Nanoscience and the Janus-Faced Character of Simulations', in: D. Baird, A. Nordmann & J. Schummer (eds.), *Discovering the Nanoscale*, Amsterdam: IOS Press, pp. 93-100.

Lent, C.S., Tougaw, P.D., Porod, W., Bernstein, G.H., "Quantum Cellular Automata," *Nanotechnology*, Vol. 4, p. 49, 1993.

Lewak, S.E.: 2004, 'What's the Buzz? Tell Me What's A-Happening: Wonder, Nanotechnology, and Alice's Adventures in Wonderland', in: N.K. Hayles (ed.), *Nanoculture: Implications of the New Technoscience*, Bristol, UK: Intellect Books, pp. 201-.

Lide, D.R., ed. 1993-1994. *CRC Handbook of Chemistry and Physics*, 74th ed.

Lin-Easton, P.C.: 2001, 'It's Time for Environmentalists to Think Small ' Real Small: A Call for the Involvement of Environmental Lawyers in Developing Precautionary Policies Molecular Nanotechnology', *Georgetown International Law Review*, 14, 106-134.

López, J.: 2004, 'Bridging the Gaps: Science Fiction in Nanotechnology', *Hyle: International Journal for Philosophy of Chemistry*, 10(2), 129-152.

Lösch, A.: 2004, 'Nanomedicine and Space: Discursive Orders of Mediating Innovations', in: D. Baird, A. Nordmann & J. Schummer (eds.), *Discovering the Nanoscale*, Amsterdam: IOS Press, pp. 193-202.

Marshall, K.: 2004, 'Atomizing the Risk Technology', in: N.K. Hayles (ed.), *Nanoculture: Implications of the New Technoscience*, Bristol, UK: Intellect Books, pp. 147-160.

Martin, T.P., N. Malinowski, U. Zimmerman, U. Naher, and H. Schaber. 1993. *J. Chem. Phys*. 99:4210.

Martin, T.P., U. Naher, H. Schaber, U. Zimmerman. 1993. *Phys. Rev. Lett*. 70:3079.

Mayer, S.: 2002, 'From genetic modification to nanotechnology: the dangers of 'sound science", in: T. Gilland (ed.), *Science: Can We Trust the Experts?*, London: Hodder and Stoughton, pp. 1-15.

Mehta, M.D.: 2002, 'Nanoscience and Nanotechnology: Assessing the Nature of Innovation in These Fields', *Bulletin of Science, Technology & Society*, 22(4), 269-273.

Mehta, M.D.: 2004, 'From Biotechnology to Nanotechnology: What Can We Learn from Earlier Technologies?, *Bulletin of Science, Technology & Society*, 24, 34-39.

Meyer, M. & Kuusi, O.: 2004, 'Nanotechnology: Generalizations in an Interdisciplinary Field of Science and Technology', *Hyle: International Journal for Philosophy of Chemistry*, 10(2), 153-168.

Meyer, M.: 2000a, 'Does science push technology? Patents citing scientific literature', *Research Policy*, 29, 409-434.

Meyer, M.: 2000b, 'Patent citations in a novel field of technology: What can they tell about interactions of emerging communities of science and technology?', *Scientometrics*, 48, 151-178.

Meyer, M.: 2001a, 'Patent citations in a novel field of technology: An exploration of nano-science and nano-technology', *Scientometrics*, 51, 163-183.

Meyer, M.: 2001b, 'Socio-economic Research on Nanoscale Science and Technology: A European Overview and Illustration', in: M.C. Roco and W.S. Bainbridge (eds.), *Societal Implications of Nanoscience and Nanotechnology*, Dordrecht: Kluwer, pp. 217-241.

Meyer, M.; Persson, O.: 1998, 'Nanotechnology ' interdisciplinarity, patterns of collaboration and differences in application', *Scientometrics*, 47, 195'205.

Milburn, C.: 2002, 'Nanotechnology in the age of post-human engineering: science fiction as science', *Configurations*, 10, 261-295 [reprinted in: N.K. Hayles (ed.), *Nanoculture: Implications of the New Technoscience*, Bristol, UK: Intellect Books, 2004, pp. 109-130].

Milburn, C.: 2004, 'Nano/Splatter: Disintegrating the Postbiological Body', *New Literary History*, 35 (in print).

Mnyusiwalla, A.; Abdaliah, S.D.; Singer, P.A.: 2003, 'Mind the gap: science and ethics in nanotechnology', *Nanotechnology*, 14, R9-R13.

Mody, C.C.M.: 2004, 'How Probe Microscopists Became Nanotechnologists', in: D. Baird, A. Nordmann & J. Schummer (eds.), *Discovering the Nanoscale*, Amsterdam: IOS Press, pp. 119-133.

Mody, C.C.M.: 2004, 'Instruments in Training: The Growth of American Probe Microscopy in the 1980s', in: D. Kaiser (ed.), *Pedagogy and the Practice of Science: Producing Physical Scientists, 1800-2000*, Cambridge, MA: MIT Press (forthcoming).

Mody, C.C.M.: 2004, 'Small, but Determined: Technological Determinism in Nanoscience', *Hyle: International Journal for Philosophy of Chemistry*, 10(2), 99-128.

Mody, C.C.M.: 2004, *Crafting the Tools of Knowledge: The Invention, Spread, and Commercialization of Probe Microscopy, 1960-2000*, Ph.D. dissertation, Cornell University.

Montemerlo, M.S., Love, J.C., Opiteck, G.J., Goldhaber, D. J., and Ellenbogen, J.C., "Technologies and Designs for Electronic Nanocomputers," MITRE Technical Report 96W0000044, The MITRE Corporation, McLean, VA, July 1996. For more information, send e-mail to nanotech@mitre.org.

Moor, J.H. & Weckert, J.: 2004, 'Nanoethics: Assessing the Nanoscale From an Ethical Point of View', in: D. Baird, A. Nordmann & J. Schummer (eds.), *Discovering the Nanoscale*, Amsterdam: IOS Press, pp. 301-310.

Munn Sanchez, E.: 2004, 'The Expert's Role in Nanoscience and Technology', in: D. Baird, A. Nordmann & J. Schummer (eds.), *Discovering the Nanoscale*, Amsterdam: IOS Press, pp. 257-266.

Nordmann, A.: 2003, 'Shaping the World Atom by Atom: Eine nanowissenschaftliche WeltBildanalyse', in: A. Grunwald (ed.), *Technikgestaltung zwischen Wunsch und Wirklichkeit*, Berlin: Springer, pp. 191-199.

Nordmann, A.: 2004, 'Molecular Disjunctions: Staking Claims at the Nanoscale', in: D. Baird, A. Nordmann & J. Schummer (eds.), *Discovering the Nanoscale*, Amsterdam: IOS Press, pp. 51-62.

Nordmann, A.: 2004, 'Nanotechnology: Convergence and Integration', Presentation at *EuroNanoForum, Trieste, December 10, 2003* (Proceedings in preparation).

Nordmann, A.: 2004, 'Nanotechnology's WorldView: New Space for Old Cosmologies', *IEEE Technology and Society Magazine*, 23 (forthcoming).

Nordmann, A.: 2004, 'Social Imagination for Nanotechnology', in: European Commission (Community Health and Consumer Protection): *Nanotechnologies: A Preliminary Risk Analysis on the Basis of a Workshop, Brussels, 1-2 March 2004*, pp. 111-113 [www.europa.eu.int/comm/health/ph_risk/documents/ev_20040301_en.pdf].

Nordmann, A.: 2004, 'Was ist TechnoWissenschaft? ' Zum Wandel der Wissenschaftskultur am Beispiel von Nanoforschung und Bionik', in: T. Rossmann & C. Tropea (eds.), *Bionik ' Neue Forschungsergebnisse aus Natur-, Ingenieur- und Geisteswissenschaften*, Berlin: Springer, 2004 (forthcoming)

Paschen, H.; Coenen, C.; Fleischer, T.; Grünwald, R.; Oertel, D. & Revermann, C.: 2004, *Nanotechnologie: Forschung, Entwicklung, Anwendung*, Berlin: Springer (*Nanotechnologie, TAB-Arbeitsbericht 92*, Berlin: Büro für Technikfolgen-Abschätzung beim Deutschen Bundestag).

Pitt, J.C.: 2004, 'The Epistemology of the Very Small', in: D. Baird, A. Nordmann & J. Schummer (eds.), *Discovering the Nanoscale*, Amsterdam: IOS Press, pp. 157-163.

Pressman, J.: 2004, 'Nano Narrative: A Parable from Electronic Literature', in: N.K. Hayles (ed.), *Nanoculture: Implications of the New Technoscience*, Bristol, UK: Intellect Books, pp. 191-200.

Prigogine, I., and S. Rice. 1988. *Advances in chemical physics*, Vol. 70, Parts 1 & 2. New York: J. Wiley.

R. T. Bate, "Nanoelectronics," *Nanotechnology*, Vol. 1, pp. 1-7, 1990.

Rademann, K., B. Kaiser, U. Even, F. Hensel. 1987. *Phys. Rev. Lett.* 59:2319.

Rao, C.N.R., B.C. Satishkumar, and A. Govindaraj. 1997. *Chem. Commun.* 1581.

Rao, M.B., and S. Sircar. 1993. *Gas Separation and Purification* 7:279.

Reed, M.A., "Quantum Dots," *Scientific American,* January 1993, pp. 118-123.

Roberts, J.A.: 2004, 'Deciding the Future of Nanotechnologies: Legal Perspectives on Issues of Democracy and Technology', in: D. Baird, A. Nordmann & J. Schummer (eds.), *Discovering the Nanoscale*, Amsterdam: IOS Press, pp. 247-255.

Robinson, C.: 2004, 'Images in NanoScience/Technology', in: D. Baird, A. Nordmann & J. Schummer (eds.), *Discovering the Nanoscale*, Amsterdam: IOS Press, pp. 165-169.

Robison, W.L.: 2004, 'Nano-Ethics', in: D. Baird, A. Nordmann & J. Schummer (eds.), *Discovering the Nanoscale*, Amsterdam: IOS Press, pp. 285-300.

Roco, M.C. & Bainbridge, W.S. (eds.): 2001, *Societal implications of nanoscience and nanotechnology*, (Proceedings of a workshop organized by the National Science Foundation, September 28-29, 2000), Kluwer: Dordrecht [available online at http://itri.loyola.edu/nano/societalimpact/nanosi.pdf]

Roco, M.C. & Tomellini, R. (eds.): 2002, *Nanotechnology: Revolutionary Opportunities and Societal Implications, Workshop, Lecce (Italy), 31 January - 1 February 2002*, Luxemburg: Office for Official Publications of the European Communities, [200 pp.]

Roco, M.C.; Bainbridge, W.S. (eds.): 2002, *Converging Technologies for Improving Human Performance: Nanotechnology, Biotechnology, Information Technology and the Cognitive Science*, Arlington, VA: National Science Foundation.

Roher, H. 1993. *Jpn. J. Appl. Phys.* 32:1335.

Rohlfing, E.A., D.M. Cox, and A. Kaldor. 1984. *J. Chem. Phys.* 81:3846.

Rosen, A. 1998. A periodic table in three dimensions: A sightseeing tour in the nanometer world. In *Advances in quantum chemistry*. In press.

Ruthven, D.M., S. Farooq, K.S. Knaebel. 1994. *Pressure swing adsorption*. New York: VCH Publishers.

Sarewitz, D.; Woodhouse, E.: 2003, 'Small is Powerful', in: A. Lightman, D. Sarewitz & Chr. Desser, (eds.), *Living with the Genie: Essays on Technology and the Quest for Human Mastery*, Washington, DC: Island Press, pp. 63-83.

Schiemann, G.: 2004, 'Dissolution of the Nature-Technology Dichotomy? Perspectives on Nanotechnology from an Everyday Understanding of Nature', in: D. Baird, A. Nordmann & J. Schummer (eds.), *Discovering the Nanoscale*, Amsterdam: IOS Press, pp. 209-213.

Schmidt, J.C.: 2004, 'Unbounded Technologies: Working Through Technological Reductionism of Nanotechnology', in: D. Baird, A. Nordmann & J. Schummer (eds.), *Discovering the Nanoscale*, Amsterdam: IOS Press, pp. 35-50.

Schuler, E.: 2004, 'Perception of Risks and Nanotechnology', in: D. Baird, A. Nordmann & J. Schummer (eds.), *Discovering the Nanoscale*, Amsterdam: IOS Press, pp. 279-284.

Schummer, J.: 2001, 'Ethics of Chemical Synthesis', *Hyle: International Journal for Philosophy of Chemistry*, 7, 103-124 [online: www.hyle.org/journal/issues/7/schummer.htm].

Schummer, J.: 2003, 'Aesthetics of Chemical Products: Materials, Molecules, and Molecular Models', *Hyle: International Journal for Philosophy of Chemistry*, 9, 73-104 [online: www.hyle.org/journal/issues/9-1/schummer.htm].

Schummer, J.: 2003, 'The Notion of Nature in Chemistry', *Studies in History and Philosophy of Science*, 34 (2003), 705-736.

Schummer, J.: 2004, 'Interdisciplinary Issues in Nanoscale Research', in: D. Baird, A. Nordmann & J. Schummer (eds.), *Discovering the Nanoscale*, Amsterdam: IOS Press, pp. 9-20.

Schummer, J.: 2004, 'Multidisciplinarity, Interdisciplinarity, and Patterns of Research Collaboration in Nanoscience and Nanotechnology', *Scientometrics*, 59, 425-465.

Schummer, J.: 2004, 'Naturverhältnisse in der modernen Wirkstoff-Forschung', in: K. Kornwachs (ed.), *Technik ' System ' Verantwortung*, Münster: LIT, pp. 629-638.

Schummer, J.: 2004, 'Why do Chemists Perform Experiments?', in: D. Sobczynska, P. Zeidler, E. Zielonacka-Lis (eds.), *Chemistry in the Philosophical Melting Pot*, Peter Lang (Frankfurt/ M.), 2004, pp. 395-410.

Schummer, J.: 2005, 'Reading Nano: The Public Interest in Nanotechnology as Reflected in Book Purchase Patterns", *Public Understanding of Science*, 14 (2) (forthcoming).

Schummer, J.; Baird, D. (eds.): 2004-5, *Nanotech Challenges*, joint special issue of *Hyle: International Journal for Philosophy of Chemistry & Techne: Research in Philosophy and Technology* (forthcoming).

Schwarz, A.E.: 2004, 'Shrinking the 'Ecological Footprint' with NanoTechnoScience?', in: D. Baird, A. Nordmann & J. Schummer (eds.), *Discovering the Nanoscale*, Amsterdam: IOS Press, pp. 203-208.

Schwarz, R.B. 1998. Storage of hydrogen in powders with nanosized crystalline domains. In *R&D status and trends*, ed. Siegel et al.

Siegel, R.W., E. Hu, and M.C. Roco, eds. 1998. *R&D status and trends in nanoparticles, nanostructured materials, and nanodevices in the United States*. Baltimore: Loyola College, International Technology Research Institute. NTIS #PB98-117914.

Stix Gary., "Toward 'Point One'," *Scientific American*, February 1995, pp. 90-95.

Stuckless, J.T, D.E. Starr, D.J. Bald, C.T. Campbell. 1997. *J. Chem. Phys.* 107:5547.

Suchman, M.: 2001, 'Envisioning Life on the Nano-Frontier', in: M.C. Roco and W.S. Bainbridge (eds.), *Societal Implications of Nanoscience and Nanotechnology*, Dordrecht: Kluwer, pp. 211-216.

Suchman, M.: 2002, 'Social Science and Nanotechnologies', M. Roco & R. Tomellini (eds.), *Nanotechnology ' Revolutionary Opportunities and Societal Implications*, Luxembourg: European Communities, pp. 95-99.

Sweeney, A.E.; Seal, S. & Vaidyanathan, P.: 2003, 'The promises and perils of nanoscience and nanotechnology: Exploring emerging social and ethical issues', *Bulletin of Science, Technology & Society*, 23 (4), 236-245.

Swiss Re: 2004, *Nanotechnology: Small Matter, Many Unknowns*, Zurich: Swiss Re [www.swissre.com].

Tanev, P.T., M. Chibwe, and T.J. Pinnavaia. 1994. *Nature* 368:321.

Taylor, K.J., C.L. Pettiette-Hall, O. Cheshnovsky, and R.J. Smalley. 1992. *J. Chem. Phys.* 96:3319.

Tenner, E.: 2001, 'Nanotechnology and Unintended Consequences', in: M.C. Roco and W.S. Bainbridge (eds.), *Societal Implications of Nanoscience and Nanotechnology*, Dordrecht: Kluwer, pp. 241-246.

The Royal Society & The Royal Academy of Engineering: 2004, *Nanoscience and nanotechnologies: opportunities and uncertainties*, London [www.nanotec.org.uk/finalReport.htm].

Toumey, C.: 2004, 'Anticipating Public Reactions to Nanotechnology', *Techne: Research in Philosophy and Technology*, 8(3) (forthcoming).

Trudeau, M.L., and J.Y. Ying. 1996. *Nanostructured Materials* 7:245.

Tschope, A.S., W. Liu, M. Flyzani-Stephanopoulos, and J.Y. Ying 1995. *J. Catal*. 157:42.

Turton, R., *The Quantum Dot*, Oxford University Press, 1995.

Verbeek, A. & Callaert, J.: 2002, *Detailed analysis of the science-technology interaction in the field of Nanotechnology*, Luxemburg: Office for Official Publications of the European Communities (Linking Science to Technology. Bibliographic References in Patents, Vol. 9).

Vermaas, P.E.: 2004, 'Nanoscale Technology: A Two-Sided Challenge for Interpretations of Quantum Mechanics', in: D. Baird, A. Nordmann & J. Schummer (eds.), *Discovering the Nanoscale*, Amsterdam: IOS Press, pp. 77-91.

Wald, C.A.: 2004, 'Working Boundaries on the NANO Exhibition', in: N.K. Hayles (ed.), *Nanoculture: Implications of the New Technoscience*, Bristol, UK: Intellect Books, pp. 83-108.

Weckert, J.: 2001, 'The Control of Scientific Research: The Case of Nanotechnology', *Australian Journal of Professional and Applied Ethics*, 3, 29-44.

Weil, V.: 2001, 'Ethical Issues in Nanotechnology', in: M.C. Roco and W.S. Bainbridge (eds.), *Societal Implications of Nanoscience and Nanotechnology*, Dordrecht: Kluwer, pp. 193-198.

2

Quantum Biology, Nanomedicine and Molecular Nanotechnology

Quantum Biology

Quantum physics and molecular biology are two disciplines that have evolved relatively independently. However, recently a wealth of evidence has demonstrated the importance of quantum mechanics for biological systems and thus a new field of quantum biology is emerging. Living systems have mastered the making and breaking of chemical bonds, which are quantum mechanical phenomena. Absorbance of frequency specific radiation (e.g. photosynthesis and vision), conversion of chemical energy into mechanical motion (e.g. ATP cleavage) and single electron transfers through biological polymers (e.g. DNA or proteins) are all quantum mechanical effects. Hopefully, the merging of disciplines known as nanotechnology will remove the interface between quantum physics and biology.

In a paper titled 'The Importance of Quantum Decoherence in Brain Processes,' Max Tegmark sought to prove that the brain is too warm to maintain the coherence required for quantum computation. From the results of his calculations, Tegmark claims that, "there is nothing fundamentally wrong with the current classical approach to neural network simulations." This statement contradicts the hypothesis that the brain functions as a quantum computer, originally proposed by Roger Penrose. Tegmark's claim was amplified by a recent report in *Science* beginning with the sentence, 'Sir Roger Penrose is incoherent, and Max Tegmark says he can prove it.' However, the computations carried out by Tegmark relied on a value of 310K for the temperature in his model of the neuron. While the average kinetic energy (temperature) of an entire brain cell may be 310K, the most fundamental characteristic of life is that it is not at equilibrium and thus, our statistical method for measuring temperature *breaks down* at small sizes, especially at the nanoscale.

Biological systems are known to have ways of manipulating local temperatures. For instance, *Koichiro Matsuno* has determined by the postulate of black body radiation measurements that actomyosin complexes (abundant in the axons of nerve cells) can reach local temperatures as low as $1.6*10^{-3}$K. Matsuno argues that actomyosin functions as a heat engine (a device that converts heat energy into mechanical energy) that is able to maintain a constant velocity due to quantum mechanical coherence and entanglement.

Nanotechnology: A Historical Perspective

Being made of molecules, and having a human concern for our health, we will apply molecular machines to biomedical technology. Biologists already use antibodies to tag proteins, enzymes to cut and splice DNA, and viral syringes (like the *T4 phage*) to inject edited DNA into bacteria. In the future, they will use assembler-built nanomachines to probe and modify cells.

With tools like *disassemblers*, biologists will be able to study cell structures in ultimate, molecular detail. They then will catalog the hundreds of thousands of kinds of molecules in the body and map the structure of the hundreds of kinds of cells. Much as engineers might compile a parts list and make engineering drawings for an automobile, so biologists will describe the parts and structures of healthy tissue. By that time, they will be *aided by sophisticated technical AI systems*.

Physicians aim to make tissues healthy, but with drugs and surgery they can only encourage tissues to repair themselves. Molecular machines will allow more direct repairs, bringing a new era in medicine.

To repair a car, a mechanic first reaches the faulty assembly, then identifies and removes the bad parts, and finally rebuilds or replaces them. Cell repair will involve the same basic tasks—tasks that living systems already prove possible.

Access. White blood cells leave the bloodstream and move through tissue, and viruses enter cells. Biologists even poke needles into cells without killing them. These examples show that molecular machines can reach and enter cells.

Recognition. Antibodies and the tail fibers of the *T4 phage*—and indeed, all specific biochemical interactions—show that molecular systems can recognize other molecules by touch.

Disassembly. Digestive enzymes (and other, fiercer chemicals) show that molecular systems can disassemble damaged molecules.

Rebuilding. Replicating cells show that molecular systems can build or rebuild every molecule found in a cell.

Reassembly. Nature also shows that *separated molecules can be put back together again*. The machinery of the *T4 phage*, for example, *self-assembles from solution*, apparently aided by a single enzyme. Replicating cells show that molecular systems can assemble every system found in a cell.

Thus, nature demonstrates all the basic operations that are needed to perform molecular-level repairs on cells. What is more, as I described in this encyclopaedia, systems based on nanomachines will generally be more compact and capable than those found in nature. Natural systems show us only lower bounds to the possible, in cell repair as in everything else.

Cell Repair Machines

In short, with molecular technology and technical AI we will compile complete, molecular-level descriptions of healthy tissue, and we will build machines able to enter cells and to sense and modify their structures.

Cell repair machines will be comparable in size to bacteria and viruses, but their more-compact parts will allow them to be more complex. They will travel through tissue as white blood cells do, and enter cells as viruses do—or they could open and close cell membranes with a surgeon's care. Inside a cell, a repair machine will first size up the situation by examining the cell's contents and activity, and then take action. Early cell repair machines will be highly specialized, able to recognize and correct only a single type of molecular disorder, such as an enzyme deficiency or a form of DNA damage. Later machines (but not much later, with advanced technical AI systems doing the design work) will be programmed with more general abilities.

Complex repair machines will need nanocomputers to guide them. A micron-wide mechanical computer like that described in this encyclopaedia will fit in 1/1000 of the volume of a typical cell, yet will hold more information than does the cell's DNA. In a repair system, such computers will direct smaller, simpler computers, which will in turn direct machines to examine, take apart, and rebuild damaged molecular structures.

By working along molecule by molecule and structure by structure, repair machines will be able to repair whole cells. By working along cell by cell and tissue by tissue, they (aided by larger devices, where need be) will be able to repair whole organs. By working through a person organ by organ, they will restore health. Because molecular machines will be able to build molecules and cells from scratch, they will be able to repair even cells damaged to the point of complete inactivity. Thus, cell repair machines will bring a fundamental breakthrough: they will free medicine from reliance on self-repair as the only path to healing.

To visualize an advanced cell repair machine, imagine it—and a cell—enlarged until atoms are the size of small marbles. On this scale, the repair machine's smallest tools have tips about the size of your fingertips; a medium-sized protein, like hemoglobin, is the size of a typewriter; and a *ribosome* is the size of a washing machine. A single repair device contains a simple computer the size of a small truck, along with many sensors of protein size, several manipulators of ribosome size, and provisions for memory and motive power. A total volume ten meters across, the size of a three-story house, holds all these parts and more. With parts the size of marbles packing this volume, the repair machine can do complex things.

But this repair device does not work alone. It, like its many siblings, is connected to a larger computer by means of mechanical data links the diameter of your arm. On this scale, a cubic-micron computer with a large memory fills a volume thirty stories high and as wide as a football field. The repair devices pass it information, and it passes back general instructions. Objects so large and complex are still small enough: on this scale, the cell itself is a kilometer across, holding one thousand times the volume of a cubic-micron computer, or a million times the volume of a single repair device. Cells are spacious.

Will such machines be able to do everything necessary to repair cells? Existing molecular machines demonstrate the ability to travel through tissue, enter cells, recognize molecular structures, and so forth, but other requirements are also important. Will repair machines work fast enough? If they do, will they waste so much power that the patient will roast?

The most extensive repairs cannot require *vastly* more work than building a cell from scratch. Yet molecular machinery working within a cellular volume routinely does just that, building a new cell in tens of minutes (in bacteria) to a few hours (in mammals). This indicates that repair machinery occupying a few percent of a cells volume will be able to complete even extensive repairs in a reasonable time—days or weeks at most. Cells can spare this much room. Even brain cells can still function when an inert waste called *lipofuscin (apparently a product of molecular damage) fills over ten percent of their volume*.

Powering repair devices will be easy: cells naturally contain chemicals that power nanomachinery. Nature also shows that repair machines can be cooled: the cells in your body rework themselves steadily, and young animals grow swiftly without cooking themselves. Handling heat from a similar level of activity by repair machines will be no sweat—or at least not too much sweat, if a week of sweating is the price of health. All these comparisons of repair machines to existing biological mechanisms raise the question of whether repair machines will be able to *improve* on nature. DNA repair provides a clear-cut illustration.

Just as an illiterate "book-repair machine" could recognize and repair a torn page, so a cell's repair enzymes can recognize and repair breaks and cross-links in DNA. Correcting misspellings (or mutations), though, would require an ability to read. Nature lacks such repair machines, but they will be easy to build. Imagine three identical DNA molecules, each with the same sequence of nucleotides. Now imagine each strand mutated to change a few scattered nucleotides. Each strand still seems normal, taken by itself. Nonetheless, a repair machine could compare each strand to the others, one segment at a time, and could note when a *nucleotide* failed to match its mates. Changing the odd nucleotide to match the other two will then repair the damage.

This method will fail if two strands mutate in the same spot. Imagine that the DNA of three human cells has been heavily damaged—after thousands of mutations, each cell has had one in every million nucleotides changed. The chance of our three-strand correction procedure failing at any given spot is then about one in a million million. But compare five strands at once, and the odds become *about one in a million million million*, and so on. A device that compares many strands will make the chance of an uncorrectable error effectively nil.

In practice, repair machines will *compare DNA molecules from several cells, make corrected copies*, and use these as standards for proofreading and repairing DNA throughout a tissue. By comparing several strands, repair machines will dramatically improve on nature's repair enzymes.

Other repairs will require different information about healthy cells and about how a particular damaged cell differs from the norm. Antibodies identify proteins by touch, and properly chosen antibodies can generally distinguish any two proteins by their differing shapes and surface properties. *Repair machines will identify molecules in a similar way.* With a suitable computer and data base, they will be able to identify proteins by reading their amino acid sequences.

Consider *a complex and capable repair system.* A volume of two cubic microns—about 2/1000 of the volume of a typical cell—will be enough to hold a central data base system able to:

Swiftly identify any of the hundred thousand or so different human proteins by examining a short amino acid sequence.

Identify all the other complex molecules normally found in cells.

Record the type and position of every large molecule in the cell.

Each of the smaller repair devices (of perhaps thousands in a cell) will include a less capable computer. Each of these computers will be able to perform over a thousand computational steps in the time that a typical enzyme takes to change a single molecular bond, so the speed of computation possible seems more than adequate. Because each computer *will be in communication with a larger computer* and the central data base, the available memory seems adequate. Cell repair machines will have both the molecular tools they need and "brains" enough to decide how to use them.

Such sophistication will be overkill (overcure?) for many health problems. Devices that merely recognize and destroy a specific kind of cell, for example, will be enough to cure a cancer Placing a computer network in every cell may seem like slicing butter with a chain saw, but having a chain saw available does provide assurance that even hard butter can be sliced. It seems better to show too much than too little, if one aims to describe the limits of the possible in medicine.

Cures

The simplest medical applications of nanomachines will involve not repair but selective destruction. Cancers provide one example; infectious diseases provide another. The goal is simple: one need only recognize and destroy the dangerous replicators, whether they are bacteria, cancer cells, viruses, or worms. Similarly, abnormal growths and deposits on arterial walls cause much heart disease; machines that recognize, break down, and dispose of them will clear arteries for more normal blood flow. Selective destruction will also cure diseases such as herpes in which a virus splices its genes into the DNA of a host cell. A repair device will enter the cell, read its DNA, and remove the addition that spells "herpes."

Repairing damaged, cross-linked molecules will also be fairly straightforward. Faced with a damaged, cross-linked protein, a cell repair machine will first identify it by examining short amino acid sequences, then look up its correct structure in a data base. The machine will then compare the protein to this blueprint, one amino acid at a time. Like a proofreader finding misspellings and strange characters (char#cters), it will find any changed amino acids or improper cross-links. By correcting these flaws, it will leave a normal protein, ready to do the work of the cell.

Repair machines will also aid healing. After a heart attack, scar tissue replaces dead muscle. Repair machines will stimulate the heart to grow fresh muscle by resetting cellular control mechanisms. By removing scar tissue and guiding fresh growth, they will direct the healing of the heart.

This list could continue through problem after problem (Heavy metal poisoning?—Find and remove the metal atoms) but the conclusion is easy to summarize. Physical disorders stem from mis-arranged atoms; repair machines will be able to return them to working order, restoring the body to health. Rather than compiling an endless list of curable diseases (from arthritis, bursitis, cancer, and dengue to yellow fever and zinc chills and back again), it makes sense to look for the limits to what cell repair machines can do. Limits do exist.

Consider stroke, as one example of a problem that damages the brain. Prevention will be straightforward: Is a blood vessel in the brain weakening, bulging, and apt to burst? Then pull it back into shape and guide the growth of reinforcing fibers. Does abnormal clotting threaten to block circulation? Then dissolve the clots and normalize the blood and blood-vessel linings to prevent a recurrence. Moderate neural damage from stroke will also be repairable: if reduced circulation has impaired function but left cell structures intact, then restore circulation and repair the cells, using their structures as a guide in restoring the tissue to its previous state. This will not only restore each cell's function, but will preserve the memories and skills embodied in the neural patterns in that part of the brain.

Repair machines will be able to regenerate fresh brain tissue even where damage has obliterated these patterns. But the patient would lose old memories and skills to the extent that they resided in that part of the brain. If unique neural patterns are truly obliterated, then cell repair machines could no more restore them than art conservators could restore a tapestry from stirred ash. Loss of information through obliteration of structure imposes the most important, fundamental limit to the repair of tissue.

Other tasks are beyond cell repair machines for different reasons—maintaining mental health, for instance. Cell repair machines will be able to correct some problems, of course. Deranged thinking sometimes has biochemical causes, as if the brain were drugging or poisoning itself, and other problems stem from tissue damage. But many problems have little to do with the health of nerve cells and everything to do with the health of the mind.

A mind and the tissue of its brain are like a novel and the paper of its book. Spilled ink or flood damage may harm the book, making the novel difficult to read. Book repair machines could nonetheless restore physical—health" by removing the foreign ink or by drying and repairing the damaged paper fibers. Such treatments would do nothing for the book's *content*, however, which in a real sense is nonphysical. If the book were a cheap romance with a moldy plot and empty characters, repairs would be needed not on the ink and paper, but on the novel. This would call not for physical repairs, but for more work by the author, perhaps with advice.

Similarly, removing poisons from the brain and repairing its nerve fibers will thin some mental fogs, but not revise the content of the mind. This can be changed by the patient, with effort; we are all authors of our minds. But because minds change themselves by changing their brains, having a healthy brain will aid sound thinking more than quality paper aids sound writing.

Readers familiar with computers may prefer to think in terms of hardware and software. A machine could repair a computer's hardware while neither understanding nor changing its software

Such machines might stop the computer's activity but leave the patterns in memory intact and ready to work again. In computers with the right kind of memory (called "nonvolatile"), users do this by simply switching off the power. In the brain the job seems more complex, yet there could be medical advantages to inducing a similar state.

Physicians already stop and restart consciousness by interfering with the chemical activity that underlies the mind. Throughout active life, molecular machines in the brain process molecules. Some disassemble sugars, combine them with oxygen, and capture the energy this releases. Some pump salt *ions* across cell membranes; others build small molecules and release them to signal other cells. Such processes make up the brain's metabolism, the sum total of its chemical activity. Together with its electrical effects, this metabolic activity underlies the changing patterns of thought.

Surgeons cut people with knives. In the mid-1800s, they learned to use chemicals that interfere with brain metabolism, blocking conscious thought and preventing patients from objecting so vigorously to being cut. These chemicals are anesthetics. Their molecules freely enter and leave the brain, allowing anesthetists to interrupt and restart human consciousness.

People have long dreamed of discovering a drug that interferes with the metabolism of the entire body, a drug able to interrupt metabolism completely for hours, days, or years. The result would be a condition of biostasis (from *bio*, meaning life, and *stasis*, meaning a stoppage or a stable state). A method of producing reversible biostasis could help astronauts on long space voyages to save food and avoid boredom, or it could serve as a kind of one-way time travel. In medicine, biostasis would provide a deep anesthesia giving physicians more time to work. When emergencies occur far from medical help, a good biostasis procedure would provide a sort of universal first-aid treatment: it would stabilize a patient's condition and prevent molecular machines from running amok and damaging tissues.

But no one has found a drug able to stop the entire metabolism the way anesthetics stop consciousness—that is, in a way that can be reversed by simply washing the drug out of the patient's tissues. Nonetheless, reversible biostasis will be possible when repair machines become available.

To see how one approach would work, imagine that the blood stream carries simple molecular devices to tissues, where they enter the cells. There they block the molecular machinery of metabolism—in the brain and elsewhere—and tie structures together with stabilizing cross-links. Other molecular devices then move in, displacing water and packing themselves solidly around the molecules of the cell. These steps stop metabolism and preserve cell structures. Because cell repair machines will be used to reverse this process, it can cause moderate molecular damage and yet do no lasting harm. With metabolism stopped and cell structures held firmly in place, the patient will rest quietly, dreamless and unchanging, until repair machines restore active life.

If a patient in this condition were turned over to a present-day physician ignorant of the capabilities of cell repair machines, the consequences would likely be grim. Seeing no signs of life, the physician would likely conclude that the patient was dead, and then would make this judgment a reality by "prescribing" an autopsy, followed by burial or burning.

But our imaginary patient lives in an era when biostasis is known to be only an interruption of life, not an end to it. When the patient's contract says "wake me!" (or the repairs are complete, or the flight to the stars is finished), the attending physician begins resuscitation. Repair machines enter the patient's tissues, removing the packing from around the patient's molecules and replacing it with water. They then remove the cross-links, repair any damaged molecules an structures, and restore normal concentrations of salts, blood sugar, ATP, and so forth. Finally, they unblock the metabolic machinery. The interrupted metabolic processes resume, the patient yawns, stretches, sits up, thanks the doctor, checks the date, and walks out the door.

Function & Structure

The reversibility of biostasis and irreversibility of severe stroke damage help to show how cell repair machines will change medicine. Today, physicians can only help tissues to heal themselves. Accordingly they must try to preserve the *function* of tissue. If tissues cannot function, they cannot heal. Worse, unless they are preserved, deterioration follows, ultimately obliterating structure. It is as if a mechanic's tools were able to work only on a running engine.

Cell repair machines change the central requirement from preserving *function* to preserving *structure*. As I noted in the discussion of stroke, repair machines will be able to restore brain function with memory and skills intact only if the distinctive structure of the neural fabric remains intact. Biostasis involves preserving neural structure while deliberately blocking function.

All this is a direct consequence of the molecular nature of the repairs. Physicians using scalpels and drugs can no more repair cells than someone using only a pickax and a can of oil can repair a fine watch. In contrast, having repair machines and ordinary nutrients will be like having a watchmaker's tools and an unlimited supply of spare parts. Cell repair machines will change medicine at its foundations.

Medical researchers now study diseases, often seeking ways to prevent or reverse them by blocking a key step in the disease process. The resulting knowledge has helped physicians greatly: they now prescribe insulin to compensate for diabetes, anti-hypertensives to prevent stroke, penicillin to cure infections, and so on down an impressive list. Molecular machines will aid the study of diseases, yet they will make understanding disease far less important. Repair machines will make it more important to understand health.

The body can be ill in more ways than it can be healthy. Healthy muscle tissue, for example, varies in relatively few ways: it can be stronger or weaker, faster or slower, have this antigen or that one, and so forth. Damaged muscle tissue can vary in all these ways, yet also suffer from any combination of strains, tears, viral infections, parasitic worms, bruises, punctures, poisons, sarcomas, wasting diseases, and congenital abnormalities. Similarly, though neurons are woven in as many patterns as there are human brains, individual synapses and dendrites come in a modest range of forms—if they are healthy.

Once biologists have described normal molecules, cells, and tissues, properly programmed repair machines will be able to cure even unknown diseases. Once researchers describe the range of structures that (for example) a healthy liver may have, repair machines exploring a malfunctioning liver need only look for differences and correct them. Machines ignorant of a new poison and its effects will still recognize it as foreign and remove it. Instead of fighting a million strange diseases, advanced repair machines will establish a state of health.

Developing and programming cell repair machines will require great effort, knowledge, and skill. Repair machines with broad capabilities seem easier to build than to program. Their programs must contain detailed knowledge of the hundreds of kinds of cells and the hundreds of thousands of kinds of molecules in the human body. They must be able *to map damaged cellular structures* and decide how to correct them. How long will such machines and programs take to be developed? Offhand, the state of biochemistry and its present rate of advance might suggest that the basic knowledge alone will take centuries to collect. But we must beware of the illusion that advances will arrive in isolation.

Repair machines will sweep in with a wave of other technologies. The assemblers that build them will first be used to build instruments for analyzing cell structures. Even a pessimist might agree that human biologists and engineers equipped with these tools could build and program advanced cell repair machines in a hundred years of steady work. A cocksure, far-seeing pessimist might say a thousand years. A really committed nay-sayer might declare that the job would take people a million years. Very well: fast technical AI systems—a millionfold faster than scientists and engineers—will then develop advanced cell repair machines *in a single calendar year*.

Aging is natural, but so were smallpox and our efforts to prevent it. We have conquered smallpox, and it seems that we will conquer aging.

Longevity has increased during the last century, but chiefly because better sanitation and drugs have reduced bacterial illness. The basic human life span has increased little.

Still, researchers have made progress toward understanding and slowing the aging process. They have identified some of its causes, such as uncontrolled cross-linking. They have devised partial treatments, such as *antioxidants* and free-radical inhibitors. They have proposed and studied other mechanisms of aging, such as—clocks" in the cell and changes in the body's hormone balance. In laboratory experiments, special drugs and diets have *extended the life span of mice by 25 to 45 percent*.

Such work will continue; as the baby boom generation ages, expect a boom in aging research. One biotechnology company, Senetek of Denmark, specializes in aging research. In April 1985, Eastman Kodak and ICN Pharmaceuticals were *reported to have joined in a $45 million venture* to produce isoprinosine and other drugs with the potential to extend life span. The results of conventional anti-aging research may substantially lengthen human life spans—and improve the health of the old—during the next ten to twenty years. How greatly will drugs, surgery, exercise, and diet extend life spans? For now, estimates must remain guesswork. Only new scientific knowledge can rescue such predictions from the realm of speculation, because they *rely on new science* and not just new engineering.

With cell repair machines, however, the potential for life extension becomes clear. They will be able to repair cells so long as their distinctive structures remain intact, and will be able to replace cells that have been destroyed. Either way, they will restore health. Aging is fundamentally no different from any other physical disorder; it is no magical effect of calendar dates on a mysterious life-force. Brittle bones, wrinkled skin, low enzyme activities, slow wound healing, poor memory, and the rest all result from damaged molecular machinery, chemical imbalances, and mis-arranged structures. By restoring all the cells and tissues of the body to a youthful structure, repair machines will restore youthful health.

People who survive intact until the time of cell repair machines will have the opportunity to regain youthful health and to keep it almost as long as they please. Nothing can make a person (or anything else) last forever, of course, but barring severe accidents, those wishing to do so will live for a long, long time.

As a technology develops, there comes a time when its principles become clear, and with them many of its consequences. The principles of rocketry were clear in the 1930s, and with them the consequence of spaceflight. Filling in the details involved designing and testing tanks, engines, instruments, and so forth. By the early 1950s, many details were known. The ancient dream of flying to the Moon had became a goal one could plan for.

The principles of molecular machinery are already clear, and with them the consequence of cell repair machines. Filling in the details will involve designing molecular tools, assemblers, computers, and so forth, but many details of existing molecular machines are known today. The ancient dream of achieving health and long life has become a goal one can plan for.

Medical research is leading us, step by step, along a path toward molecular machinery. The global competition to make better materials, electronics, and biochemical tools is pushing us in the same direction. Cell repair machines will take years to develop, but they lie straight ahead.

They will bring many abilities, both for good and for ill. A moment's thought about military replicators with abilities like those of cell repair machines is enough to turn up nauseating possibilities. Later I will describe how we might avoid such horrors, but it first seems wise to consider the alleged benefits of cell repair machines. Is their *apparent* good *really* good? How might long life affect the world?

Long Life

Cell repair machines raise questions involving the value of extending human life. These are not the questions of today's medical ethics, which commonly involve dilemmas posed by scarce, costly, and half-effective treatments. They are instead questions involving the value of long, healthy lives achieved by inexpensive means.

For people who value human life and enjoy living, such questions may need no answer. But after a decade marked by concern about population growth, pollution, and resource depletion, many people may question the desirability of extending life; such concerns have fostered the spread of pro-death memes. These memes must be examined afresh, because many have roots in an obsolete worldview. *Nanotechnology* will change far more than just human lifespan.

We will gain the means not only to heal ourselves, but to heal Earth of the wounds we have inflicted. Since saving lives will increase the number of the living, life extension raises questions about the effect of more people. Our ability to heal the Earth will lessen one cause for controversy.

Still, *cell repair machines* themselves will surely stir controversy. They disturb traditional assumptions about our bodies and our futures: this makes doubt soothing. They will require several major breakthroughs: this makes doubt easy. Since the possibility or impossibility of cell repair machines raises important issues, it makes sense to consider what objections might be raised.

Cell Repair Machines?

What sort of argument could suggest that cell repair machines are impossible? A successful argument must manage some strange contortions. It must somehow hold that molecular machines cannot build and repair *cells*, while granting that the molecular machines in our bodies actually *do* build and repair cells every day. A cruel problem for the committed skeptic! True, artificial machines must do what natural machines fail to do, but they need not do anything *qualitatively* novel. Both natural and artificial repair devices must reach, identify, and rebuild molecular structures. We will be able to improve on existing *DNA* repair *enzymes* simply by comparing several DNA strands at once, so nature obviously hasn't found all the tricks. Since this example explodes any general argument that repair machines cannot improve on nature, a good case against cell repair machines seems difficult to make.

Still, two general questions deserve direct answers. First, why should we expect to achieve long life in the coming decades, when people have tried and failed for millennia? Second, if we can indeed use cell repair machines to extend lives, then why hasn't nature (which has been repairing cells for billions of years) already perfected them?

People have tried and failed: For centuries, people have longed to escape their short life spans. Every so often, a Ponce de Leon or a quack doctor has promised a potion, but it has never worked. These statistics of failure have persuaded some people that, since all attempts have failed, all always will fail. They say "Aging is natural," and to them that seems reason enough. Medical advances may have shaken their views, but advances have chiefly reduced early death, not extended maximum life-span.

But now biochemists have gone to work examining the machines that build, repair, and control cells. They have learned to assemble viruses and reprogram *bacteria*. For the first time in history, people are examining their molecules and unraveling the molecular secrets of life. It seems that molecular engineers will eventually combine improved biochemical knowledge with improved molecular machines, learning to repair damaged tissue structures and so rejuvenate them. This is nothing strange—it would be strange, rather, if such powerful knowledge and abilities did *not* bring dramatic results. The massive statistics of past failure are simply irrelevant, because we have never before tried to build cell repair machines.

Nature has tried and failed: Nature *has* been building cell repair machines. *Evolution* has tinkered with multicelled animals for hundreds of millions of years, yet advanced animals all age and die, because nature's nanomachines repair cells imperfectly. Why should improvements be possible?

Rats mature in months, and then age and die in about two years—yet human beings have evolved to live over thirty times longer. If longer lives were the chief goal of evolution, then rats would live longer too. But *durability has costs*: to repair cells requires an investment in energy, materials, and repair machines. Rat genes direct rat bodies to invest in swift growth and reproduction, not in meticulous self-repair. A rat that dallied in reaching breeding size would run a greater risk of becoming a cat snack first. Rat genes have prospered by treating rat bodies as cheap throwaways. Human genes likewise discard human beings, though after a life a few dozen times longer than a rat's.

But shoddy repairs are not the only cause of aging. Genes turn egg cells into adults through a pattern of development which rolls forward at fairly steady speed. This pattern is fairly consistent because evolution seldom changes a basic design. Just as the basic pattern of the DNA-*RNA*-protein system froze several billions of years ago, so the basic pattern of chemical signals and tissue responses that guides mammalian development jelled many millions of years ago. That process apparently has a clock, set to run at different speeds in different species, and a program that runs out.

Whatever the causes of aging, evolution has had little reason to eliminate them. If genes built individuals able to stay healthy for millennia, they would gain little advantage in their "effort" to replicate. Most individuals would still die young from starvation, predation, accident, or disease. *As Sir Peter Medawar points out*, a gene that helps the young (who are many) but harms the old (who are few) will replicate well and so spread through the population. If enough such genes accumulate, animals become programmed to die.

Experiments by Dr. Leonard Hayflick suggest that cells contain "clocks" that count cell divisions and stop the division process when the count gets too high. *A mechanism of this sort* can help young animals: if cancer-like changes make a cell divide too rapidly, but fail to destroy its clock, then it will grow to a tumor of limited size. The clock would thus prevent the unlimited growth of a true cancer. *Such clocks could harm older animals by stopping the division of normal cells*, ending tissue renewal, The animal thus would benefit from reduced cancer rates when young, yet have cause to complain if it lives to grow old. But its genes won't listen—they will have jumped ship earlier, as copies passed to the next generation. With cell repair machines we will be able to reset such clocks. Nothing suggests that evolution has perfected our bodies even by the brute standard of survival and reproduction. Engineers don't wire computers with slow, nerve-like fibers or build machines out of soft protein, and for good reason. Genetic evolution (unlike memetic evolution) has been unable to leap to new materials or new systems, but has instead refined and extended the old ones.

. The cell's repair machines fall far short of the limits of the possible—they don't even have computers to direct them. The lack of *nano*-computers in cells, of course, shows only that computers couldn't (or simply *didn't*) evolve gradually from other molecular machines. Nature has failed to build the best possible cell repair machines, but there have been ample reasons.

Healing and Protecting the Earth

The failure of Earth's biological systems to adapt to the industrial revolution is also easy to understand. From deforestation to dioxin, we have caused damage faster than evolution can respond. As we have sought more food, goods, and services, our use of *bulk technology* has forced us to continue such damage. With future technology, though, we will be able to do more good for ourselves, yet do less harm to the Earth. In addition, we will be able to build planet-mending machines to correct damage already done. Cells are not all we will want to repair.

Consider the toxic waste problem. Whether in our air, soil, or water, wastes concern us because they can harm living systems. But any materials that come in contact with the molecular machinery of life can themselves be reached by other forms of molecular machinery. This means that we will be able to design *cleaning machines to remove these poisons* wherever they could harm life.

Some wastes, such as dioxin, consist of dangerous molecules made of innocuous *atoms*. Cleaning machines will render them harmless by rearranging their atoms. Other wastes, such as lead and radioactive isotopes, contain dangerous atoms. Cleaning machines will collect these for disposal in any one of several ways. Lead comes from Earth's rocks; *assemblers* could build it into rocks in the mines from which it came. Radioactive isotopes could also be isolated from living things, either by building them into stable rock or by more drastic means. Using cheap, reliable space transportation systems, we could bury them in the dead, dry rock of the Moon. Using nanomachines, we could seal them in self-repairing, self-sealing containers the size of hills and powered by desert sunlight. These would be more secure than any passive rock or cask.

With replicating assemblers, we will even be able to remove the billions of tons of carbon dioxide that our fuel-burning civilization has dumped into the atmosphere. Climatologists project that climbing carbon dioxide levels, by trapping solar energy, will partially melt the polar caps, raising sea levels and flooding coasts sometime in the middle of the next century. Replicating assemblers, though, will make solar power *cheap enough to eliminate the need for fossil fuels*. Like trees, solar-powered nanomachines will be *able to extract carbon dioxide from the air* and split off the oxygen. Unlike trees, they will be able to grow deep storage roots and place carbon back in the coal seams and oil fields from which it came.

Future planet-healing machines will also help us mend torn landscapes and restore damaged ecosystems. Mining has scraped and pitted the Earth; carelessness has littered it. Fighting forest fires has let undergrowth thrive, replacing the cathedral-like openness of ancient forests with scrub growth that feeds more dangerous fires. We will use inexpensive, sophisticated robots to reverse these effects and others. Able to move rock and soil, they will re-contour torn lands. Able to weed and digest, they will simulate the clearing effects of natural forest fires without danger or devastation. Able to lift and move trees, they will thin thick stands and reforest bare hills. We will make squirrel-sized devices with a taste for old trash. We will make treelike devices with roots that spread deep and cleanse the soil of pesticides and excess acid. We will make insect-sized lichen cleaners and spray-paint nibblers. We will make whatever devices we need to clean up the mess left by twentieth-century civilization.

After the cleanup, we will recycle most of these machines, keeping only those we still need to protect the environment from a cleaner civilization based on *molecular technology*. These more lasting devices will supplement natural ecosystems wherever needed, to balance and heal the effects of humanity. To make them effective, harmless, and hidden will be a craft requiring not just *automated engineering*, but knowledge of nature and a sense of art.

With cell repair technology, we will even be able to return some species from apparent extinction. The *African quagga*—a zebra-like animal—became extinct over a century ago, but a salt-preserved quagga pelt survived in a German museum. Alan Wilson of the University of California at Berkeley and his co-workers *have used enzymes to extract DNA fragments* from muscle tissue attached to this pelt. They cloned the fragments in bacteria, compared them to zebra DNA, and found (as expected) that the genes showed a close evolutionary relationship. They have also succeeded in extracting and replicating DNA from a century-old bison pelt and from millennia-old mammoths preserved in the arctic permafrost. This success is a far cry from cloning a whole cell or organism—cloning one gene leaves about 100,000 uncloned, and cloning every gene still doesn't repair a single cell—but it does show that the hereditary material of these species still survives.

As I described in this encyclopaedia, machines that compare several damaged copies of a DNA *molecule* will be able to reconstruct an undamaged original—and the billions of cells in a dried skin contain *billions* of copies. From these, we will be able to reconstruct undamaged DNA, and around the DNA we will be able to construct undamaged cells of whatever type we desire. Some insect species pass through winter as egg cells, to be revived by the warmth of spring. These "extinct" species will pass through the twentieth century as skin and muscle cells, to be converted into fertile eggs and revived by cell repair machines.

Dr. Barbara Durrant, a reproductive physiologist at the San Diego Zoo, is preserving tissue samples from endangered species in a cryogenic freezer. The payoff may be greater than most people now expect. Preserving just tissue samples doesn't preserve the life of an animal or an ecosystem, but it does preserve the genetic heritage of the sampled species. We would be reckless if we failed to take out this insurance policy against the permanent loss of species. The prospect of cell repair machines thus affects our choices today.

Extinction is not a new problem. About 65 million years ago, most then-existing species vanished, including all species of dinosaur. In Earth's book of stone, the story of the dinosaurs ends on a page consisting of a thin layer of clay. The clay is rich in iridium, an element common in asteroids and comets. The best current theory indicates that a blast from the sky smashed Earth's biosphere. With the energy of a hundred million megatons of TNT, it spread dust and an "asteroidal winter" planet-wide.

In the eons since living cells first banded together to form worms, Earth has suffered five great extinctions. Only 34 million years ago—some 30 million years after the dinosaurs died—a layer of glassy beads settled to the sea floor. Above that layer the fossils of many species vanish. These beads froze from the molten splash of an impact. Meteor Crater, in Arizona, bears witness to a smaller, more recent blast equaling that of a four-megaton bomb. As recently as June 30, 1908, a ball of fire split the Siberian sky and blasted the forest flat across an area a hundred kilometers wide.

As people have long suspected, the dinosaurs died because they were stupid. Not that they were too stupid to feed, walk, or guard their eggs—they did survive for 140 million years—they were merely too stupid to build telescopes able to detect asteroids and spacecraft able to deflect them from collision with Earth. Space has more rocks to throw at us, but we are showing signs of adequate intelligence to deal with them. When nanotechnology and automated engineering give us a more capable space technology, we will find it easy to track and deflect asteroids; in fact, we could do it with technology available today. We can both heal Earth and protect it.

Population Pressure

People commonly seek long, healthy lives, yet the prospect of a dramatic success is unsettling. Might greater longevity harm the quality of life? How will the prospect of long life affect our immediate problems? Though most effects cannot be foreseen, some can.

For example, as cell repair machines extend life, they will increase population. If all else were equal, more people would mean greater crowding, pollution, and scarcity—but all else will *not* be equal: the very advances in automated engineering and nanotechnology that will bring cell repair machines will also help us heal the Earth, protect it, and live more lightly upon it. We will be able to produce our necessities and luxuries without polluting our air, land, or water. We will be able to get resources and make things without scarring the landscape with mines or cluttering it with factories. With efficient assemblers making durable products, we will produce things of greater value with less waste. More people will be able to live on Earth, yet do less harm to it—or to one another, if we somehow manage to use our new abilities for good ends.

If one were to see the night sky as a black wall and expect the technology race to screech to a polite halt, then it would be natural to fear that long-lived people would be a burden on the "poor, crowded world of our children." This fear stems from the illusion that life is a zero-sum game, that having more people always means slicing a small pie thinner. But when we become able to repair cells, we will also be able to build replicating assemblers and excellent spacecraft. Our "poor" descendants will share a world the size of the solar system, with matter, energy, and potential living space dwarfing our entire planet.

This will open room enough for an era of growth and prosperity far beyond any precedent. Yet the solar system itself is finite, and the stars are distant. On Earth, even the cleanest assembler-based industries will produce waste heat. Concern about population and resources will remain important because the *exponential growth* of replicators (such as people) can eventually overrun any finite resource base.

But does this mean that we should sacrifice lives to delay the crunch? A few people may volunteer themselves, but they will do little good. In truth, life extension will have little effect on the basic problem: exponential growth will remain exponential whether people die young or live indefinitely. A martyr, by dying early, could delay the crisis by a fraction of a second—but a halfway dedicated person could help more by joining a movement of long-lived people working to solve this long-range problem. After all, many people have ignored the limits to growth on Earth. Who but the long-lived will prepare for the firmer but more distant limits to growth in the world beyond Earth? Those concerned with long-term limits will serve humanity best by staying alive, to keep their concern alive.

Long life also raises the threat of cultural stagnation. If this were an inevitable problem of long life, it is unclear what one could do about it—machine-gun the old for holding firm opinions, perhaps? Fortunately, two factors will reduce the problem somewhat. First, in a world with an open frontier the young will be able to move out, build new worlds, test new ideas, and then either persuade their elders to change or leave them behind. Second, people old in years will be young in body and brain. Aging slows both learning and thought, as it slows other physical processes; rejuvenation will speed them again. Since youthful muscles and sinews make young bodies more flexible, perhaps youthful brain tissues will keep minds somewhat more flexible, even when steeped in long years of wisdom.

Anticipation

Long life will not be the greatest of the future's problems. It might even help solve them.

Consider its effect on people's willingness to start wars. Aging and death have made slaughter in combat more acceptable: *As Homer had Sarpedon, hero of Troy, say*, "O my friend, if we, leaving this war, could escape from age and death, I should not here be fighting in the van; but now, since many are the modes of death impending over us which no man can hope to shun, let us press on and give renown to other men, or win it for ourselves."

Yet if the hope of escaping age and death turns people from battle, will this be good? It might discourage small wars that could grow into a nuclear holocaust. But equally, it might weaken our resolve to defend ourselves from lifelong oppression—if we take no account of how much more life we have to defend. The reluctance of others to die for their ruler's power will help.

Expectations always shape actions. Our institutions and personal plans both reflect our expectation that all adults now living will die in mere decades. Consider how this belief inflames the urge to acquire, to ignore the future in pursuit of a fleeting pleasure. Consider how it blinds us to the future, and obscures the long-term benefits of cooperation. Erich Fromm writes: "If the individual lived five hundred or one thousand years, this clash (between his interests and those of society) might not exist or at least might be considerably reduced. He then might live and harvest with joy what he sowed in sorrow; the suffering of one historical period which will bear fruit in the next one could bear fruit for him too." Whether or not most people will still live for the present is beside the point: the question is, might there be a *significant* change for the better?

The expectation of living a long life in a better future may well make some political diseases less deadly. Human conflicts are far too deep and strong to be uprooted by any simple change, yet the prospect of vast wealth tomorrow may at least *lessen* the urge to fight over crumbs today. The problem of conflict is great, and we need all the help we can get.

The prospect of personal deterioration and death has always made thoughts of the future less pleasant. Visions of pollution, poverty, and nuclear annihilation have recently made thoughts of the future almost too gruesome to bear. Yet with at least a hope of a better future and time to enjoy it, we may look forward more willingly. Looking forward, we will see more. Having a personal stake, we will care more. Greater hope and foresight will benefit both the present and posterity; they will even better our odds of survival.

Lengthened lives will mean more people, but without greatly worsening tomorrow's population problem. The expectation of longer lives in a better world will bring real benefits, by encouraging people to give more thought to the future. Overall, long life and its anticipation seem good for society, just as shortening life spans to thirty would be bad. Many people want long, healthy lives for themselves. What are the prospects for the present generation?

Progress in Life Extension

Four millennia have passed since Sumerian scribes marked clay tablets to record *The Epic of Gilgamesh*, and times have changed. Men no taller than average have now reached the heavens and circled the Earth. We of the Space Age, the Biotechnology Age, the Age of Breakthroughs—need we still despair before the barrier of years? Or will we learn the art of life extension soon enough to save ourselves and those we love from *dissolution*?

The pace of biomedical advance holds tantalizing promise. The major diseases of age—heart disease, stroke, and cancer—have begun to yield to treatment. Studies of aging mechanisms have begun to bear fruit, and researchers have extended animals' life spans. As knowledge builds on knowledge and tools lead to new tools, advances seem sure to accelerate. Even without cell repair machines, we have reason to expect major progress toward slowing and partially reversing aging.

Although people of all ages will benefit from these advances, the young will benefit more. Those surviving long enough will reach a time when aging becomes fully reversible: at the latest, the time of advanced cell repair machines. Then, if not sooner, people will grow healthier as they grow older, improving like wine instead of spoiling like milk. They will, if they choose, regain excellent health and live a long, long time.

In that time, with its replicators and cheap spaceflight, people will have both long lives and room and resources enough to enjoy them. A question that may roll bitterly off the tongue is: "When?... Which will be the last generation to age and die, and which the first to win through?" Many people now share the quiet expectation that aging will someday be conquered. But are those now alive doomed by a fluke of premature birth? The answer will prove both clear and startling.

The obvious path to long life involves living long enough to be rejuvenated by cell repair machines. Advances in biochemistry and molecular technology will extend life, and in the time won they will extend it yet more. At first we will use drugs, diet, and exercise to extend healthy life. Within several decades, advances in nanotechnology will likely bring early cell repair machines—and with the aid of automated engineering, early machines may promptly be followed by advanced machines. Dates must remain mere guesses, but a guess will serve better than a simple question mark.

Imagine someone who is now thirty years old. In another thirty years, biotechnology will have advanced greatly, yet that thirty-year-old will be only sixty. Statistical tables which assume *no* advances in medicine say that a thirty-year-old U.S. citizen can now expect to live almost *fifty* more years—that is, well into the 2030s. Fairly routine advances (of sorts demonstrated in animals) seem likely to add years, perhaps decades, to life by 2030. The mere beginnings of cell repair technology might extend life by several decades. In short, the medicine of 2010, 2020, and 2030 seems likely to extend our thirty-year-old's life into the 2040s and 2050s. By then, if not before, medical advances may permit actual rejuvenation. Thus, those under thirty (and perhaps those substantially older) can look forward—at least tentatively—to medicine's overtaking their aging process and delivering them safely to an era of cell repair, vigor, and indefinite life-span.

If this were the whole story, then the division between the last on the road to early death and the first on the road to long life would be perhaps the ultimate gap between generations. What is more, a gnawing uncertainty about one's own fate would give reason to push the whole matter into the subconscious dungeon of disturbing speculations.

But is this really our situation? There seems to be another way to save lives, one based on cell repair machines, yet applicable today. As described this encyclopaedia, repair machines will be able to heal tissue so long as its essential structure is preserved. A tissues ability to metabolize and to repair itself becomes unimportant; the discussion of *biostasis* illustrated this. Biostasis, as described, will use molecular devices to stop function and preserve structure by *cross-linking* the cell's molecular machines to one another. Nanomachines will reverse biostasis by repairing molecular damage, removing cross-links, and helping cells (and hence tissues, organs, and the whole body) return to normal function.

Reaching an era with advanced cell repair machines seems the key to long life and health, because almost all physical problems will then be curable. One might manage to arrive in that era by remaining alive and active through all the years between now and then—but this is merely the most obvious way, the way that requires a minimum of foresight. Patients today often suffer a collapse of heart function while the brain structures that embody memory and personality remain intact. In such cases, might not *today's* medical technology be able to stop biological processes in a way that *tomorrow's* medical technology will be able to reverse? If so, then most deaths are now prematurely diagnosed, and needless.

Nanotechnology Characterization Lab

Eliminating suffering and death from cancer requires an unprecedented collaborative effort that leverages resources from government, industry, and academia. Working in concert with the National Institute of Standards and Technology (NIST) and the U.S. Food and Drug Administration (FDA), the National Cancer Institute (NCI) established the Nanotechnology Characterization Laboratory to perform preclinical efficacy and toxicity testing of nanoparticles

The NCL will serve as a national resource and knowledge base for all cancer researchers to facilitate the regulatory review of nanotechnologies intended for cancer therapies and diagnostics. By providing the critical infrastructure and characterization services to nanomaterial providers, the NCL can accelerate the transition of basic nanoscale particles and devices into clinical applications, thereby reducing suffering and death from cancer.

Nanomedicine and Molecular Nanotechnology

A mature nanomedicine will require the ability to build structures and devices to atomic precision, hence molecular nanotechnology and molecular manufacturing are key enabling technologies for nanomedicine. The prefix "nano-" (from the Greek root nanos, or dwarf) means one-billionth (10^{-9}) of something. The term "nanotechnology" refers most generally to technology on the scale of a billionth of a meter, or a nanometer (a nanometer is ~6 carbon atoms wide). Similarly, the words "nanomachine," "nanorobot," "nanomotor" and "nanocomputer" may refer to complex engineered objects fabricated by positioning matter with molecular control.

Molecular engineering was discussed as an extension of bulk technologies in the 1960s and 1970s by von Hippel, von Foester, and Zingsheim. The phrase "Nano-Technology" was first used in print in 1974 by N. Taniguchi to refer to the increasingly precise machining and finishing of materials, progressing from larger to smaller scales and ultimately to nanoscale tolerances, following in the path of Feynman's proposed "top-down" approach, a schemata which persisted in Taniguchi's thinking throughout the 1980s and 1990s. In 1981, K.E. Drexler described a new "bottom-up" approach involving molecular manipulation and molecular engineering in the context of building molecular machines and molecular devices with atomic precision, a fundamentally different mind-set. Drexler again described molecular technology in 1982 and molecular mechanical devices in 1983, first using the word "nanotechnology" in 1985 and 1986 as synonymous with molecular technology, finally settling upon "molecular nanotechnology" in 1991 and "molecular machine systems" in 1992 to clarify that his concept involved working devices constructed with atomic precision, as distinguished from nanostructured bulk materials, micromachinery, polymeric self-assembly, pure biotechnology, nanolithography, Langmuir-Blodgett thin films, and the like. Drexler's definition — molecular nanotechnology as the three-dimensional positional control of atomic and molecular structure to create materials and devices with molecular precision — is the usage adopted in this book. The first known use of the term "nanomedicine" was in 1991 by Drexler, Peterson, and Pergamit in their popular book Unbounding the Future.

Is molecular nanotechnology possible? This question is explicitly addressed, but the bottom line is that molecular nanotechnology violates no physical laws and there exist many possible technical paths leading to useful results. Even by 1985, for example, G. Yamamoto had reported on molecular gears, describing "compounds that exist in conformations which are regarded as static meshed gears with two-toothed and three-toothed wheels and some of them behave as dynamic gears." H. Iwamura prepared a system that formed a chain of beveled molecular gears with ~GHz rotation rates. In 1998, it was generally accepted that molecular nanotechnology would be developed, although there was still some disagreement about how long it would take.

It is often noted that molecular biological systems are themselves nanomachines, constituting an existence proof for molecular nanotechnology. Indeed, biotechnology is one possible implementation pathway for molecular nanotechnology that is being pursued reveals the close functional correspondence between the macroscale components of everyday machines and the molecular components of natural biological systems. Such comparisons have a long history. For example, Marcello Malpighi (1628-1694), a professor of medicine in Pisa who discovered the fine structure of the lungs and the capillaries using the microscope, once observed that "our bodies are composed of strings, thread, beams, levers, cloth, flowing fluids, cisterns, ducts, filters, sieves, and other similar mechanisms."

Parallels to living systems as molecular machines were drawn by Changeau, McClaire, Laing, Drexler, and Mitchell, inspiring early thinking in molecular manufacturing. For example, in 1991 Drexler observed: "Technology-as-we-know-it is a product of industry, of manufacturing and chemical engineering. Industry-as-we-know-it takes things from nature — ore from mountains, trees from forests — and coerces them into forms that someone considers useful. Trees become lumber, then houses. Mountains become rubble, then molten iron, then steel, then cars. Sand becomes a purified gas, then silicon, then chips. And so it goes. Each process is crude, based on cutting, stirring, baking, spraying, etching, grinding, and the like." "Trees, though, are not crude. To make wood and leaves, they neither cut, stir, bake, spray, etch, nor grind. Instead, they gather solar energy using molecular electronic devices, the photosynthetic reaction centers of chloroplasts. They use that energy to drive molecular machines — active devices with moving parts of precise, molecular structure — which process carbon dioxide and water into oxygen and molecular building blocks. They use other molecular machines to join these molecular building blocks to form roots, trunks, branches, twigs, solar collectors, and more molecular machinery. Every tree makes leaves, and each leaf is more sophisticated than a spacecraft, more finely patterned than the latest chip from Silicon Valley. They do all this without noise, heat, toxic fumes, or human labor, and they consume pollutants as they go. Viewed this way, trees are high technology. Chips and rockets are not."

Contemplating applications of nanotechnology to medicine, Brian Wowk concluded that 20th century physicians were in a predicament similar to that which would be faced by 18th-century engineers trying to maintain a 20th-century automobile — repairs would be crude at best, and breakdowns inevitable: "Like primitive engineers faced with advanced technology, medicine must 'catch up' with the technology level of the human body before it can become really effective. What is the technology level? Since the human body is basically an extremely complex system of interacting molecules (i.e., a molecular machine), the technology required to truly understand and repair the body is molecular machine technology — nanotechnology. A natural consequence of [our achieving] this level of technology will be the ability to analyze and repair the human body as completely and effectively as we can repair any conventional machine today."

In Engines of Creation, Drexler drew inspiration from the cell's eye view to recognize that nanotechnology could bring a fundamental breakthrough in medicine. Noting that 20th century physicians relied chiefly on surgery and drugs to treat illness, Drexler explained: "Surgeons have advanced from stitching wounds and amputating limbs to repairing hearts and reattaching limbs. Using microscopes and fine tools, they join delicate blood vessels and nerves. Yet even the best microsurgeon cannot cut and stitch finer tissue structures. Modern scalpels and sutures are simply too coarse for repairing capillaries, cells, and molecules. Consider `delicate' surgery from a cell's perspective. A huge blade sweeps down, chopping blindly past and through the molecular machinery of a crowd of cells, slaughtering thousands. Later, a great obelisk plunges through the divided crowd, dragging a cable as wide as a freight train behind it to rope the crowd together again. From a cell's perspective, even the most delicate surgery, performed with exquisite knives and great skill, is still a butcher job. Only the ability of cells to abandon their dead, regroup, and multiply makes healing possible." "Drug therapy, unlike surgery, deals with the finest structures in cells. Drug molecules are simple molecular devices. Many affect specific molecules in cells. Morphine molecules, for example, bind to certain receptor molecules in brain cells, affecting the neural impulses that signal pain. Insulin, beta blockers, and other drugs fit other receptors. But drug molecules work without direction. Once dumped into the body, they tumble and bump around in solution haphazardly until they bump a target molecule, fit, and stick, affecting its function. Drug molecules affect tissues at the molecular level, but they are too simple to sense, plan, and act. Molecular machines directed by nanocomputers will offer physicians another choice. They will combine sensors, programs, and molecular tools to form systems able to examine and repair the ultimate components of individual cells. They will bring surgical control to the molecular domain."

By the end of the 20th century, mainstream military (DoD), NIH, NSF, and other U.S. government and international groups had begun to seriously consider the potential future applications of molecular nanotechnology in medicine. For example, in 1997 a panel of U.S. Department of Defense health science experts known as Military Health Service Systems (MHSS) 2020 concluded in its final report: "If a breakthrough to a [molecular] assembler occurs within ten to fifteen years, an entirely new field of nanomedicine will emerge by 2020. Initial applications will be focused outside the body in areas such as diagnostics and pharmaceutical manufacturing. The most powerful uses would eventually be within the body. Possible applications include programmable immune machines that travel through the bloodstream, supplementing the natural immune system; cell herding machines to stimulate rapid healing and tissue reconstruction; and cell repair machines to perform genetic surgery."

The present book takes as its starting point the assumption that the mass production of nanomachines at modest cost is technically feasible, and then explores the medical implications of this assumption. Proposed systems presented in this trilogy are intended not as final engineering blueprints but merely as points of departure for further analysis and refinement. All designs and projected capabilities are, for the most part, conservatively drawn with generous safety margins, leaving a considerable volume of design space yet to be explored by more intrepid future investigators.

The general idea of biological engineering stretches back at least to the mid-19th century, but the first science fiction story involving actual genetic engineering was "Proteus Island" (1936), written by the chemical engineer Stanley Weinbaum. Artificial engineered organisms subsequently appeared in minor roles in several stories, a notable example being the familiars employed by the fake witches in Fritz Leiber's "Gather, Darkness!" (1943), and A.E. van Vogt used "gene transformation" to create the superman in "Slan" (1940). The first artificially evolved creatures appeared in Theodore Sturgeon's "Microcosmic God" (1941), wherein a biochemist established conditions allowing accelerated artificial evolution, creating the Neoterics, a submillimeter-sized race of intelligent hypermetabolic creatures which could accomplish tasks very rapidly. The first artificially designed microcreatures appeared in James Blish's "Surface Tension" (1952). In this story, crash-landed dying human astronauts create a completely new form of humanity — tiny men and women reduced to protozoan size — who are seeded in the pools and puddles at the surface of the new planet, and who go on to master the biotechnology necessary to travel from one water puddle to another. The scientific tradition of biological nanomachines for medical purposes began in 1964 when Robert Ettinger, an early cryonics pioneer, suggested that cellular-level or even molecular-level repair might be developed for life extension. Ettinger speculated that "...surgeon machines, working 24 hours a day for decades or even centuries, will tenderly restore the frozen brains, cell by cell, or even molecule by molecule in critical areas."

In 1965, the synthesis of artificial life was publicly proposed as a national goal by the president of the American Chemical Society, Professor Charles Price, who pointed out that many new types of life might be made, not "mere imitations" of biology as we know it.

In 1967, Isaac Asimov suggested the future possibility of "factories...where the working machinery consists of submicroscopic nucleic acids" and that a "repertoire of hundreds or thousands of complex enzymes" could be used to "bring about chemical reactions more conveniently than any methods now used" and also for "helping to construct life."

In 1968, G.R. Taylor cited the possibilities for genetic engineering and genetic surgery: "The microsurgery of DNA may possibly be achieved by physical methods: fine beams of radiation (probably laser light or pulsed X-rays) may be used to slice through the DNA molecule at desired points." He also cited predictions that bacteria would soon be programmed.

In 1969, J. White suggested that a modified virus could be used as a cell repair machine: "It has been proposed that appropriate genetic information be introduced by means of artificially constructed virus particles into a congenitally defective cell for remedy; similar means may be used for the more general case of repair. The repair program must use means such as protein synthesis and metabolic pathways to diagnose and repair any damage... [Information] can be preserved by specifying that the repair program incorporate appropriate RNA tapes into itself..."

In 1970, Jeon et al carried out "the reassembly of Amoeba proteus from its major components: namely nucleus, cytoplasm, and cell membrane," taken from three different cells.

In 1972, Ettinger proposed using genetic engineering to make microscopic biorobots: "Genetic engineering's most sensational impact will concern the modification of humans, but it will have other uses as well. Some of the "robots" that will serve us will need to be nanominiaturized." Existing organisms could be modified to make biologically-based programmable biorobots for medical applications: "If we can design sufficiently complex behavior patterns into microscopically small organisms, there are obvious and endless possibilities, some of the most important in the medical area. Perhaps we can carry guardian and scavenger organisms in the blood, superior to the leukocytes and other agents of our human heritage, that will efficiently hunt down and clean out a wide variety of hostile or damaging invaders." Computerized cellular repair machines "must use means such as protein synthesis and metabolic pathways to diagnose and repair any damage...[Information] can be preserved by specifying that the repair program incorporate appropriate RNA tapes into itself...." Also in 1972, Danielli described various possibilities for generating new life forms via "life-synthesis" and genetic engineering, noting that "macromolecular engineering" might enable the development of very powerful and compact macromolecular computer systems.

In 1974, Halacy noted "some rather inglorious ways" to use "the miracle of artificial life," including potential capabilities for growing diverse items ranging from computers to airplanes. Morowitz suggested cooling cells to cryogenic temperatures in order to analyze and determine their structure. Artificial cells could also be assembled at such temperatures and then be set in motion by thawing. Morowitz further reported that microsurgery experiments on amoebas "have been most dramatic. Cell fractions from four different animals can be injected into the eviscerated ghost of a fifth amoeba, and a living functioning organism results."

In 1975, Richard Laing described the theoretical possibility of molecular machine self-replicators using molecular (data) tapes based on the idea of universal Turing machines, examining several ways that such "artificial organisms" might replicate themselves as "a vehicle for the exploration of broad biological possibilities."

In 1976, Donaldson presented the first detailed (and quite ambitious) list of biotechnological techniques that appeared necessary to achieve cell repair and might prove feasible, writing at a time that predated many current capabilities such as automated protein/DNA sequencing and synthesis, and most knowledge of restrictions on cellular developmental pathways and genetic programs and networks. For repair at the level of the cell, Donaldson's techniques would have included:

1. The ability to design enzymes to produce specific repair functions such as renaturing denatured proteins, joining broken lipoprotein complexes, annealing broken strands of DNA or RNA, reading proteins of existing or special types onto RNA and replicating them, and giving a cell the ability to metabolize new substrates, use novel cofactors, or construct essential amino acids;
2. Specially constructed bacteria or macrophages able to replicate themselves, spread throughout a specific target tissue, and carry out specific repairs according to the programs designed into their DNA/RNA; these could be designed to operate at unnatural temperatures or to utilize metabolic pathways not presently found in nature;

3. The abilities to re-introduce lost DNA or lost organelles such as mitochondria into a cell, to introduce entirely new forms of organelles perhaps to perform specific repair functions, and to introduce new metabolic capacities into a cell;
4. The ability to modify at will the developmental program of a cell, as for instance to induce postmitotic cells such as neurons to divide, according to a specific program, forming daughter cells with specific properties; and
5. Several different types of repair bacteria able to work together in an integrated fashion, and linked together by chemical means (e.g. hormones), so as to apply optimal repairs to every body cell in order to (A) diagnose the precise nature of the damage, call other repair bacteria to its location, and report to the attending doctor that new types of repair bacteria other than those already introduced are needed, and (B) identify structures which must be preserved (e.g. memory) and reconstruct them if necessary.

For repairs at the level of the whole organism, Donaldson offered the following rather aggressive biological nanotechnology techniques:

1. Understanding the physiology of aging combined with the ability to reverse it;
2. Control over growth and development, including the ability to program types of growth and development which do not naturally occur, such as growth of new eyes or other organs which have been lost or damaged, growth of an entire and well-formed body from a head alone, and regrowth of injured or lost brain tissue; and
3. Nonpermanent "substitute organs" which will take over from others which have been lost, including (A) the ability to keep a given tissue alive and healthy in vitro for an indefinite time and similar abilities for a body part, and (B) temporary replacements for any body organ which may have been lost. These would be used to support the body while new organs were growing, e.g. as "metabolic crutches." This capacity specifically includes the ability to make temporary replacements for diffuse systems such as the vascular or nervous systems. For instance, a specially created "plant" would grow an entire vascular system into the patient, down to replacement for the capillaries and venules from a single seed, always introducing its fibrils between cells and destroying none or very little of the original structure.

The last of these is a description of a (clearly speculative) "repair net," a concept which Donaldson may have been the first to propose in 1976. A whole-body repair net was later termed a "chrysalis", and twelve years later, in 1988, Donaldson provided an artist's conception and an additional description of these proposed biotechnological instrumentalities: "Severe crushing or mangling injuries require us to provide a new vascular system. The repair device might resemble a fungus, growing mycelia into the injured tissue. A repair net would grow into a [crushed] limb, guided by recognition of the injured cells and a plan for how the limb should look after repair....[In the most severe injuries,] a chrysalis first envelops the patient, then enters in between all his cells. It disassembles the patient, surrounding each cell with its own repair machinery and vascular system. The geometry already preserves information about locations of the patient's cells. If necessary, morphogen chemical gradients could also retain this information. A patient would [locally] swell up to 10 times original diameter. After repair, the chrysalis withdraws the same way it entered."

In 1977, Darwin further developed the theme of tissue repair and cellular repair biomachines, independently proposing a modified white blood cell to perform repair functions.

In 1981, Asimov suggested that we "consider the bacteria. These are tiny living things made up of single cells far smaller than the cells in plants and animals....[We] can, by properly designing these tiniest slaves of ours, use them to reshape the world itself and build it close to our hearts' desire." More importantly, Donaldson elaborated on his earlier discussion of how cryonically suspended human beings might be repaired. He extended the earlier concepts of Ettinger and White, concluding that "with such hybrid technology as micro-miniature biological-mechanical machines the size of viruses" and related technology, "it seems unlikely that (to repair a single cell) there would be any difficulty at all in principle to carrying out any imaginable repair." He estimated that about 10 programmable cell-repair biomachines could be introduced into a cell that was under repair without causing too much mechanical disruption. In 1988, Donaldson described artificial macrophages that "can carry control machinery to recognize target cells, responding only to them, or responding differently depending upon cell type or cell conditions. They can still work even if the target cell isn't functioning (unlike viruses), rebuild target cell machinery other than the genes, and transfer many more genes up to an entire copy of the patient's genome." They can also "communicate with one another [and] release diffusible chemicals to guide one another's behavior."

By the 1990s, bioengineered viruses of various types and certain other vectors were routinely being used in experimental genetic therapies as a means to target and penetrate certain cell populations, with the objective of inserting therapeutic DNA sequences into the nucleus of human target cells in vivo. Retrovirally-altered lymphocytes (T cells) began to be injected into humans for therapeutic purposes. Another example of an engineered cell in therapeutics was the use of genetically modified cerebral endothelial cell vectors to attack glioblastoma, which was being pursued by Neurotech (in Paris) in 1998. Engineered bacteria were being pursued by Vion Pharmaceuticals in collaboration with Yale University. In their "Tumor Amplified Protein Expression Therapy" program, antibiotic-sensitive Salmonella typhimurium (food poisoning) bacteria were attenuated by removing the genes that produce purines vital to bacterial growth. The tamed strain (cell line VNP20009) could not long survive in healthy tissue, but quickly multiplied ~1000-fold inside tumors, which are rich in purines. The next step would be to add genes to the bacterium to produce anticancer proteins that can shrink tumors, or to modify the bacteria to deliver various enzymes, genes, or prodrugs for tumor cell growth regulation. The engineered bacteria were available in multiple serotypes to avoid potential immune response in the host. Phase I human clinical trials were expected to begin in 1999 using clinical dosages produced in 50-liter fermenters, and other possible bacterial vector species were being examined.

Micro-biorobotics was still regarded by many in the biotechnology community as a highly speculative topic in 1998. Glen A. Evans described the possible construction of synthetic genomes and artificial organisms. His proposed strategy involves determining or designing the DNA sequence for the genome, synthesizing and assembling the genome, introducing the synthetic DNA into an enucleated pluripotent host cell, then introducing the host cell into an organism. Evans foresees the high throughput "read-out" of gene products discovered through genomic sequencing, rapid construction of designer genes and genomes, automated designer vector synthesis for gene replacement/ insertional mutagenesis, and finally the construction of synthetic genomes and artificial organisms. Robert Bradbury has considered requirements and costs for genome synthesis and replacement. For example, Bradbury estimates genome synthesis costs by assuming a projected ~$0.03/base for DNA synthesis (compared to ~$0.80-$1.00/base-micromole in 1999), with ~20,000 expressed genes per biorobotic cell with an average gene size of ~3000 bases, giving a ~60 megabase expressed genome per biorobotic cell (1 chromosome) and a raw synthesis cost of ~$3 million per designer cell line (excluding design costs). There has also been discussion of "chemical reaction automata" as precursors for synthetic organisms, synthetic lifeforms and "nanobiology", "biorobotics", "cell rovers", microbial engineering, lymphocyte engineering, and artificial chromosomes, and cell engineering is rapidly gaining popularity.

Biotechnology and Molecular Nanotechnology

Some pragmatic readers may be wondering what is so special about molecular nanotechnology, that existing or anticipated biotechnology could not accomplish just as well? After all, biotechnology is already an established medical capability. It has real applications and real products already on the market. Reflecting upon the future possibility of sophisticated mechanical medical nanorobots equipped with powerful nanocomputers, in 1989, one well-known cryobiologist mused that "what is not clear is just what need we would have for such devices." The simplest answer, suggested, is that each of the three contemporary branches of "nanotechnology" offers something of unique value to the practice of medicine.

Nanoscale materials technology has already found widespread use in medicine, including biocompatible materials and analytical techniques, surgical and dental practice, nerve cell research using intracellular electrodes, biostructures research and biomolecular research using near-field optical microscopy, scanning-probe microscopy and optical tweezers, and vaccine design, and also many 20th century bulk chemical and biochemical manufacturing techniques along with much of classical pharmacology.

As for "biotechnology," the original meaning of this word contemplated "the application of biological systems and organisms to technical and industrial processes". In recent times, the field has expanded to include genetic engineering and now takes as its ultimate goal no less than the engineering of all biological systems, even completely artificial organic living systems, using biological instrumentalities.

The third branch, molecular nanotechnology, takes as its purview the engineering of all complex mechanical systems constructed from the molecular level — potentially offering new tools for medical practice, the principal subject of this book. Observes G.M. Fahy, "the difference between nanotechnologists and biotechnologists is that the former do not restrict themselves to the biological limitations of the latter, and they are much more ambitious about the kinds of accomplishments that they want to achieve."

Doctors can utilize solutions to medical problems from all three approaches. As noted earlier, 80-90% of medical complaints resolve themselves via natural homeostatic processes, or without the necessary involvement of active biotechnological or molecular-nanotechnological agents. But by employing biotechnology, the range and efficacy of treatment options greatly increases. With molecular nanotechnology, the range, efficacy, comfort and speed of possible medical treatments again expands enormously. Molecular nanotechnology is essential when the damage to the human body is extremely subtle, highly selective, or time-critical (as in head traumas, burns, or fast-spreading diseases), or when the damage is very massive, overwhelming the body's natural defenses and repair mechanisms.

An earlier discussion from G.M. Fahy, makes these options more explicit. At every difficulty level, most classes of medical problems may be resolved with varying efficacy within the homeostatic/nanomaterials, biotechnological, or molecular-nanotechnological approaches. As the chosen technology becomes more precise, active, and controllable, the range of options broadens and the quality of the options improves. Thus the question is not whether molecular nanotechnology is required to accomplish a given medical objective. In many cases, it is not — though of course there are some things that only biotechnology and molecular nanotechnology can do, and some other things that only molecular nanotechnology can do. Rather, the important question is which approach offers a superior solution to a given medical problem, using any reasonable metric of treatment efficacy. For virtually every class of medical challenge, a mature molecular nanotechnology offers a wider and more effective range of treatment options than any other approach.

It is quite possible to imagine an advanced biotechnology that uses an engineered white cell, fibroblast, or macrophage chassis, energized by native oxygen and glucose and modified mitochondrial powerplants, driven by pseudopodia, cilia or flagella, communicating and navigating via biochemical signals, and even incorporating onboard digital biocomputers to make microscopic biorobots. Principal arguments favoring the biotechnology approach for medical purposes are: (1) that we are already somewhat familiar with such systems, after half a century of intensive molecular biology research; (2) that we have already "built" precursor systems, such as a whole living amoeba constructed from five distinct parts, bioengineered viruses and bacteria as DNA insertion devices, and natural replication stimulated in genetically engineered starter microbes; (3) that biocompatibility will not be a major issue since fibroblasts (which express no HLA Class II antigens, hence stimulate no rejection response; could be used as the starter material; (4) that both engineered viruses and bacteria are already in wide commercial and research use; and (5) the greater complexity of self-repair in mechanical systems, should it be needed. Many believe that the development pathway to early biorobots may be considerably shorter than for the mechanical nanorobots of the molecular nanotechnology approach, for which in 1998 not a single working prototype yet existed, even in research laboratories.

It is also possible to imagine a molecular nanotechnology that uses mechanical nanorobotic systems. Such systems will have many constitutional differences from biological-based systems. For instance, mechanical systems will transport parts, materials, energy and instructions via fixed channels, whereas most (but not all) biological systems operate by diffusion. Mechanical systems will have structures constrained by specific geometries, whereas biological systems have structures defined by patterns of containment and interconnection — the shape of a membrane compartment in a cell matters less than its continuity and the contents of the volume it defines. Mechanical systems will be deterministically manufactured by operations analogous to manual construction, whereas engineered ribosomes will self-assemble via diffusion and stochastic matching of complementary parts; in other words, biology uses recipes, while mechanical systems use blueprints. Cells grow, with their parts adapting to one another; mechanical nanorobots may be constructed from parts of fixed structure. Biology uses self-repair; mechanical systems generally do not — largely because self-repair (by component exchange) will not be needed in molecular mechanical systems, whose designs may be made more simple by relying upon high component redundancy. These and other differences imply a number of important advantages that mechanical-based medical systems will enjoy over biological-based medical systems, which, taken together, strongly suggest that predominantly mechanical nanosystems may be the approach of choice for a mature medical nanotechnology. The advantages of molecular nanotechnology (e.g. the mechanical tradition) are many:

1. *Speed of Medical Treatment.* Doctors may be surprised by the incredible quickness of nanorobotic action when compared to the speeds available from fibroblasts or leukocytes. Normal homeostatic processes such as dermal wound repair via natural

fibroblasts may require weeks to run to completion. Typical fibroblast movements occur at 0.1-1 microns/sec, but mechanical nanomanipulators can operate at 1-10 cm/sec speeds or faster, a speed advantage of 4-5 orders of magnitude. Even the strongest biological fibers (e.g. intermediate filaments) have a failure strength 3 orders of magnitude below the strongest mechanical fibers. Biological cilia beat at ~30 Hz while mechanical nanocilia may cycle up to ~20 MHz, though practical power restrictions and other considerations may limit them to the ~10 KHz range for most of the time. Flexible mechanical surfaces can complete a morphing motion in ~0.1 millisec, compared to the ~100 millisec snapback time for pinched red cell membrane, again a thousandfold advantage in speed. Thus we expect that mechanical therapeutic systems can reach their targets up to ~10,000 times faster, all else equal, and treatments which require ~10^5 sec for a biological system may need only ~10^2 sec for a mechanical system; tachyiatria improves both patient and physician comfort. In either biological or mechanical systems, large numbers of devices of comparable physical size (e.g. ~microns) may be employed to do the work, so numbers alone cannot offset the mechanical speed advantage.

2. *Power Density and Transduction*. Biological cells typically employ power densities of 10^3-10^4 W/m^3, with maximum densities of ~10^6 W/m^3 in honeybee flight muscle cells and bacterial flagellar motors. By contrast, nanomechanical power systems can produce power densities of 10^9-10^{12} W/m^3, an advantage of 10^3-10^8 for mechanical over biological systems. By 1998, conducting polymer-based actuators generated 20-100 times the force for a given cross-sectional area as mammalian skeletal muscle. Additionally, amoebic locomotion in motile cells requires diffusion-limited cytoskeletal disassembly and reassembly to achieve movement; mechanical motility systems may employ simple cable-pulling, winches, or ratchets, which are faster and more direct. Some biological energy transducers are reversible, but muscle contraction is irreversible — not only cannot muscle actively re-expand, but stretching it doesn't make it produce much useful chemical energy. In contrast, electric motors can be run backwards to generate electricity, forcing pistons makes them pump, and loudspeakers can be used as microphones.

3. *Superior Building Materials*. Typical biological materials have tensile failure strengths in the 10^6-10^7 N/m^2 range, with the strongest biological materials such as wet compact bone having a failure strength of ~10^8 N/m^2, all of which compare poorly to ~10^9 N/m^2 for good steel, ~10^{10} N/m^2 for sapphire, and ~10^{11} N/m^2 for diamond and carbon fullerenes, again showing a 10^3-10^5 advantage for mechanical systems that use nonbiological materials. Nonbiological materials can be much stiffer, permitting the application of higher forces with greater precision of movement, and they also tend to remain stable over a larger range of temperature, pressure, salinity and pH. Proteins are heat sensitive in part because much of the functionality of their structure is due to noncovalent bonds involved in folding, which are broken more easily at higher temperatures; in diamond, sapphire, and many other rigid materials, structural shape

is covalently fixed, hence is far more temperature-stable. Most proteins tend to become dysfunctional at cryogenic temperatures, unlike diamond-based mechanical structures. Biomaterials are not ruled out for all nanomechanical systems, but represent only a small subset of the materials that can be used in nanorobots. Mechanical systems can employ a wider variety of atoms and molecular structures in their design and construction, with novel functional forms that might be difficult to implement in a biological system such as steam engines or nuclear power. As another example, an application requiring the most effective bulk thermal conduction possible should use diamond, the best conductor available, not some biomaterial with inferior thermal performance.

4. *Nondegradation of Treatment Agents*. Diagnostic and therapeutic agents constructed of biomaterials generally are biodegradable in vivo, although there is a major branch of pharmacology devoted to designing drugs that are moderately non-biodegradable — anti-sense DNA analogues with unusual backbone linkages and peptide nucleic acids (PNAs) are difficult to break down. However, suitably designed nanorobotic agents constructed of nonbiological materials are not biodegradable. An engineered fibroblast may not stimulate an immune response when transplanted into a foreign host, but its biomolecules are subject to chemical attack in vivo by free radicals, acids, and enzymes. Even "mirror" biomolecules or "Doppelganger proteins" comprised exclusively of unnatural D-amino acids have a lifetime of only ~5 days inside the human body. Nonbiological materials such as diamond and sapphire are highly resistant to chemical breakdown or leukocytic degradation in vivo, and pathogenic biological entities cannot easily evolve useful attack strategies against these materials.

5. *Control of Nanomedical Treatment*. Present-day biotechnological entities are not programmable and cannot be switched on and off conditionally during task execution. A digital biocomputer, while possible in theory, represents a considerable conceptual departure from the usual biological paradigm. Even assuming that a digital biocomputer could be installed in a fibroblast, and that appropriate effector mechanisms could be attached, such a system would necessarily have slower clock cycles, less capacious memory per unit volume, and longer data access times, implying less diversity of action, poorer control, and less complex executable programs than would be available in nanoscale electromechanical computer systems. The mechanical approach emphasizes precise control of action, including control of physical placement, timing, strength, structure, and interactions with other (especially biological) entities. The biological approach emphasizes the use of poorly controlled natural structures, needlessly sacrificing huge blocks of the available functionality and design space.

6. *Nanodevice Versatility*. Mechanical systems can readily incorporate biological elements if necessary, but artificial biological systems can incorporate nonbiological materials such as carbon nanotubes or diamond/sapphire structural elements only with difficulty, in part because biology has a more limited repertoire of "effector"

mechanisms. Artificial biological systems cannot easily incorporate nonbiological materials where desired because natural biological assembly methods make no provisions for these materials either in the coded instructions in DNA or in the attachment chemistries. Rebuilding or reconstructing the human body with nonbiological components, or augmentation of human body function with unnatural abilities or features, will be very difficult or impossible to achieve using purely biotechnological means.

7. *Avoiding Overspecialization*. M. Krummenacker notes that one of the most glaring shortcomings of bacteria and other naturally occurring molecular machinery — when viewed as systems subject to further engineering — is the rather limited range of molecular substrates they can utilize. For example, bacterial enzymes are highly specialized devices, with very narrow substrate specificities. Thousands of different enzymes are needed in each organism, and the substance classes capable of digestion are limited to some sugars, various amino acids including proteins that can be degraded by excreted proteases, lipids, and a few other smallish oxygen-functionalized carbon molecules such as glycerol and ethanol. Some bacteria can metabolize CO_2 and a handful of aromatic compounds, but there is a vast range of organic chemicals that most bacteria cannot degrade or manufacture. A much smaller set of substantially more general molecular tools can probably be designed using the mechanical approach; mechanosynthesis can fabricate and assemble, or disassemble, a far wider range of molecular structures than are available to the cellular machinery of life.

8. *Faster and More Precise Diagnosis*. The analytic function of medical diagnosis requires rapid communication between the injected devices and the attending physician. If limited to chemical messaging, biotechnology devices will require minutes or hours to complete each diagnostic loop. Nanomachines, with their more diverse set of input-output mechanisms, can outmessage the results of in vivo reconnaissance or testing literally in seconds. Such nanomachines can also run more tests of greater variety in less time. Mechanical nanoinstruments, including molecule-by-molecule disassemblers, will make comprehensive cell mapping and cell interaction analysis possible. Bacterial resistance can be assayed at the molecular level, allowing new treatment agents to more easily be composed, manufactured and immediately deployed.

9. *More Sensitive Response Threshold for High-Speed Action*. Unlike natural systems, an entire population of nanobiotic devices can be triggered globally by just a single local detection of the target antigen or pathogen. The natural immune system takes $>10^5$ sec to become fully engaged after exposure to a systemic pathogen or other antigen-presenting intruder. A biotechnologically enhanced immune system that can employ the fastest natural unit replication time ($\sim 10^3$ sec for some bacteria) will require $\sim 10^4$ sec for full deployment post-exposure. By contrast, a nanobiotic immune system can probably be fully engaged (though not finished) in at most two blood circulation times, or $\sim 10^2$ sec.

10. *More Reliable Operation*. Engineered macrophages would probably individually

operate less reliably than mechanical nanorobots. Many pathogens, such as *Listeria monocytogenes* and *Trypanosoma cruzi*, are known to be able to escape from phagocytic vacuoles into the cytoplasm; while biotech drugs or cell manufactured proteins could be developed to prevent this (e.g. cold therapy drugs are entry-point blockers), nanorobotic trapping mechanisms can be more secure. Proteins assembled by natural ribosomes typically incorporate one error per ~10^4 amino acids placed; current gene and protein synthesizing machines utilizing biotechnological processes have similar error rates. A molecular nanotechnology approach will improve error rates by at least a millionfold. Mechanical systems can also incorporate sensors to determine if and when a particular task needs to be done, or when a task has been completed. Finally, it is unlikely that natural organisms will be able to infiltrate mechanical nanorobots or to co-opt their functions. By contrast, a biological-based robot could be diverted or defeated by microbes that can piggyback on its metabolism, interfere with its normal workings, or even incorporate the device wholesale into their own structures, causing the engineered biomachine to perform some new or different function than was originally intended. There are many examples of such co-option among natural biological systems, including the protozoan mixotrichs found in the termite gut that have assimilated bacteria into their bodies for use as motive engines, and the nudibranch mollusks (marine snails without shells) that steal nematocysts (stinging cells) away from coelenterates such as jellyfish (i.e. a Portuguese man-of-war) and incorporate the stingers as defensive armaments in their own skins, a process which S. Vogel has called "stealing loaded guns from the army."

11. *Verification of Progress and Treatment*. Using a variety of communication modalities, nanorobots can report back to the attending physician, with digital precision, a summary of diagnostically- or therapeutically-relevant data describing exactly what was found, and what was done, and what problems were encountered, in every cell visited. A biological-based approach relying upon chemical messaging is necessarily slow with limited signaling capacity. Also unlike mechanical nanorobots, biotechnological systems generally cannot monitor their own functions while working, and, except for a few highly specialized DNA proofreading systems, cannot directly inspect their work while it is in progress or after it is finished.

12. *Minimum Side Effects*. Almost all drugs have side effects, such as conventional cancer chemotherapy which causes hair loss and vomiting, although computer-designed drugs have high specificity and relatively few side effects. Carefully tailored cancer vaccines under development in the late 1990s were expected unavoidably to affect some healthy cells. Even well-targeted drugs are distributed to unintended tissues and organs in low concentrations, although some bacteria can target certain organs fairly reliably without being able to distinguish individual cells. By contrast, mechanical nanorobots may be targeted with virtually 100% accuracy to specific organs, tissues, or even individual cellular addresses within the human body. Such nanorobots should have few if any side effects, and will remain safe even in large dosages because their

actions can be digitally self-regulated using rigorous control protocols that affirmatively prohibit device activation unless all necessary preconditions have been, and continuously remain, satisfied. G.M. Fahy has noted that these possibilities could transform "drugs" into "programmable machines with a range of sensory, decision-making, and effector capabilities [that] might avoid side effects and allergic reactions...attaining almost complete specificity of action....Designed smart pharmaceuticals might activate themselves only when, where, and if needed." Additionally, nanorobots may be programmed to excuse themselves from the site of action, or even from the body, after a treatment is completed; by contrast, spent biorobotic elements containing ingested foreign materials may have more limited post-treatment mobility, thus lingering at the worksite causing inflammation when naturally degraded or removed R. Bradbury notes that artificial eukaryotic biorobots may possess an apoptotic pathway which could be activated to permit clean and natural self-destruction that avoids inflammation in surrounding tissues. Artificial prokaryotic biorobots could also be designed for human biocompatibility; by replacing bacterial genes with generic human genes, such a device may be undetectable to the immune system.

13. *Reduced Replicator Danger*. Drexler points out that living systems are evolved systems, while nanomechanical replicators would be designed: "The former are shaped to serve the goal of their own survival and replication in a natural environment, whereas the latter will be shaped (whether well or poorly) to serve human goals, perhaps in an artificial environment." Genetic engineering involves not design of replicators from scratch, but tinkering with the molecular machinery of existing bioreplicators. Since bioreplicators were not designed, they are not necessarily structured in ways that lend themselves to complete understanding, and processes based on diffusion and matching allow complex nonlocal interactions that can be hard to trace. Bioreplicators can be crippled, but having evolved in nature, they resemble systems that can survive in nature. Typically, they are able to exchange genetic information with wild organisms, raising the possibility of the introduction of new, unconstrained replicators in the natural environment. Having evolved to evolve, they have a capacity for further evolution — to serve their own survival, not human goals. For example, even when stripped of key pieces of DNA to interfere with its replication powers, a live attenuated AIDS vaccine can slowly recover its virulence and can attack immune cells. R. Bradbury suggests that artificial biorobots could incorporate multiple fail-safe mechanisms including required external essential nutrients or suicide suppressors, self-destruct triggers, countdown timers like telomeres, and engineering to reject foreign DNA (the basis for restriction enzymes), but it remains logically easier for a system with inchoate capacity to evolve to resume doing so, than for a system which has never had this capability to spontaneously develop it.

In contrast, nanomechanical replicators will be designed from scratch and thus will differ fundamentally from biological systems. The parts and structures of designed mechanical systems will be known, and the relationships among their parts will also be designed and fixed. More important, nanoreplicators will be fundamentally alien to the biosphere, unrelated to anything that has evolved to survive in nature. Certainly the capacity to fail can appear by accident, and emergent capabilities cannot be completely ruled out, but engineering experience shows that the ability to perform complex organized activities (such as replication in a natural environment) does not normally appear spontaneously. Also, and purely as a geometrical consideration, adding a new part inside a densely organized geometric structure (as would be found in a nanomachine) typically requires changes in the relative positions of many other parts, and hence corresponding adjustments in design. Adding a part inside a densely organized topological structure (as found in living systems) typically leaves topologies unchanged — room can be made by stretching and shifting other parts, with no change in their essential design, hence permitting easier modification to biological structures by exogenous agencies.

Note that mechanical medical nanodevices need not be capable of replication. There is no requirement for replication in vivo; such replication would be needlessly dangerous, and adding this capability would reduce effectiveness in carrying out the primary medical task. Analogously, viral vectors employed in genetic therapies are modified to be "incapable" of replication.

14. *Assured Patentability*. Microscopic biorobots, unavoidably derived from natural biological material, may someday be deemed unpatentable under a general prohibition on "genetic colonialism" or other emerging legal doctrines. In contrast, mechanical nanorobots, being fully-artificial and designed machines, should always be patentable provided they satisfy the customary legal criteria.

Naturophilia

As already noted, living things in general and the human body in particular are awesome examples of a powerful and intricately woven natural molecular technology to which human engineers, in 1998, still aspire. But embracing Nature is not the same as finding her perfect. Today, the word "natural" has acquired a strong connotation of rightness, even of sanctity. For most of human history, notes biologist Steven Vogel, "the natural and human worlds stood opposed. Nature was something to be tamed and utilized; we had the ordinary attitude of organisms toward other species. Nowadays the natural world intrudes far less but gets venerated far more. And why not? When one's meat is bought in a store, when locusts don't threaten one's corn crop, when central heating and plumbing are the norm, the aesthetics of nature hold greater appeal." And so we embrace the natural rectitude or moral superiority of nature's ways, a kind of pantheism which may be called ethical naturalism, biophilia, or naturophilia.

Many great minds have fallen prey to naturophilia. In the 4th century BC, Aristotle wrote that "if one way be better than another, that you may be sure is Nature's way."

In the 15th century, we have from Leonardo da Vinci: "Human ingenuity may make various inventions, but it will never devise any inventions more beautiful, nor more simple, nor more to the purpose than Nature does; because in her inventions nothing is wanting and nothing is superfluous." However, as Virginia Postrel notes in *The Future and Its Enemies*: "If nature is itself a dynamic process rather than a static end, then there is no single form of `the natural.' An evolving, open-ended nature may impose practical constraints, but it cannot dictate eternal standards. It cannot determine what is good. The distinction between the artificial and the natural must lie not in their source — human or not — but in their characteristics, in the way they relate to the world around them."

According to the dictionary, "artificial" usually means "made by man, rather than occurring in nature." More usefully, Herbert Simon defines the artificial as that which is designed, expresses goals, and possesses external purposes. The artificial is controlled and serves its creators' purposes, subject to the universal laws of physics. Kevin Kelly defines the natural as "out of control." Nature is evolved, not designed, and serves no goal or external purpose save its own survival. Nature, lacking intent, is amoral — it simply is. By building the artificial, observes Postrel, "we do not overthrow nature, but cooperate with it, using nature's own art to create new natural forms. Our artifice alters the path of nature, but it does not end it, for nature has no stopping point, no final shape. It is a process, not an end."

Some naturophilist writers have decried the increasing "medicalization of society" in which formerly natural functions have come to be regarded as medical conditions requiring intervention or treatment. However, history suggests that naturophilia is usually undermined by any new medical technology that offers clear, safe, and immediate benefits to patients. For example, prior to 1842, intense pain was viewed as the natural outcome of being cut with a scalpel during surgery. It had always been so — how could it ever be otherwise? The invention of anesthesia in 1842 suddenly altered this natural outcome and replaced it with a less painful artificial outcome, despite anguished cries from naturophiles within the medical community that eliminating pain might somehow diminish the human character.

Another example of a widely-accepted medicalization of normal function is childbirth, a quite natural activity that can nevertheless be very dangerous to the mother's health. Precise prehistoric death rates are unknown, although archeological evidence shows that Neandertal females tended to die before the age of 30 due to hazards of childbirth. In the worst 19th century maternity hospitals the natural death rate from childbirth was 9-10%, falling to a very artificial 0.4% rate in England by 1930 and to less than 0.01% in the U.S. during the 1990s. As a result, it now seems "natural" for a woman always to survive childbirth, even though the reverse may have been true for most of human history. Warns historian Roy Porter: "We should certainly not hanker after some mythic golden age when women gave birth naturally, painlessly, and safely; the most appalling Western maternal death rate today is among the Faith

Assembly religious sect in Indiana, who reject orthodox medicine and practice home births; their perinatal mortality is 92 times greater than in Indiana as a whole."

A disease seems "natural" to those who suffer from it when no treatment exists. But once a treatment is discovered and is widely employed, the disease becomes rare and its absence now becomes "natural." To those in the past, writes K.E. Drexler, "the idea of cutting people open with knives painlessly would have seemed miraculous, but surgical anesthesia is now routine. Likewise with bacterial infections and antibiotics, with the eradication of smallpox, and the vaccine for polio: each tamed a deadly terror, and each is now half-forgotten history. What amazes one generation seems obvious and even boring to the next. The first baby born after each breakthrough grows up wondering what all the excitement was about." In the next century, says Charles Sheffield, "our descendants will look on angiograms, upper and lower GIs, and biopsies the way we regard the prospect of surgery without anesthetics."

Future generations who take for granted an all-pervasive nanomedicine in their lives may look aghast upon the 20th century, wondering among other things how we managed to retain productive focus given the constant annoyance of our numerous undiagnosed minor disease states. Most of these diseases are not yet recognized as such, and many are still regarded as "natural" and not worthy of treatment. In a few decades, this may change. Some examples:

1. *Addictions* — In 1998, many people laugh off seemingly harmless addictions to chocolate (chocoholics), fats or sugar (sweet tooth), food (gluttony), nicotine (smokers), caffeine (coffee and cola drinkers), work (workaholics), exercise (runner's high), telling falsehoods (pathological liars), gambling (wagerphilia), stealing (kleptomania), medical treatments (hypochondria), marriage (polygamy), power-seeking (domination), skydiving or bungee-jumping (thrillseeking), superstition (astrology), shopping (spendoholics), driving cars that kill 40,000 Americans per year (mobilophilia), unusual sexual preferences (bestiality), sexual activity (nymphomania, satyriasis), or pregnancy (gravidophilia). Without making any value judgements, it is highly likely that most or all of these addictions have genetic or physiological components which, once properly modified, can greatly reduce or eliminate the addiction if so desired. Many on the list are already suspected to have genetic components, much like schizophrenia, drug abuse, bulimia, and alcoholism (dipsomania).

2. *Allergies and Intolerances* — A food allergy is an allergic reaction to a particular food, although true food allergies are much rarer than is generally believed. In the cases of milk, eggs, shellfish, nuts, wheat, soybeans, and chocolate, sufferers may lack an enzyme necessary for digesting the substance. In other cases, dust particles, plant pollens, pet danders, drugs, or foods may be allergens for natural IgE-mediated immunosensitivity. Intolerance, a much more common condition, is any undesirable effect of eating a particular food, including gastrointestinal distress, gas, nausea, diarrhea, or other problems. Urticaria (hives), angioedema and even mild anaphylaxis

are common reactions to various drugs, insect stings or bites, allergy shots, or certain foods, particularly eggs, shellfish, nuts and fruits. Physical allergies to ordinary stimuli such as cold, sunlight, heat, pollen, pet dander, or minor injury can produce itching, skin blotches, pimples, and hives.

3. *Minor Physical Annoyances* — In a world where most major medical maladies are readily treated, numerous minor medical conditions which today escape our notice will rise up from obscurity and present themselves annoyingly to our conscious minds, demanding attention. These conditions may be of several kinds. First is cosmetics, including small moles, freckles and blemishes on the skin; broken fingernails or unevenly-growing cuticles; minor skin reddenings or pimples; old childhood scars, wrinkled skin, birthmarks or stretch marks; unwanted hair growth in unusual places, or differential hair color or texture growing in patches; fingerprint patterns that are aesthetically unappealing; and mismatched leg lengths, hands with different left/right ring sizes, an asymmetrical face, or lopsided breasts. Second is minor aches and pains, which may include headaches; eyebrow hairs trapped in the eyeball conjunctiva; bent-hair pain (folliculalgia); dyspepsia; creaking limb joints and stomach growling; ingrown nails and hairs; earwax plugs and temporary tinnitis; chapped lips, canker sores and heat rashes; stuffy nose or gritty eyes upon rising in the morning; dermal chafing marks from elastic bands in clothing; minor flatulence; PMS (premenstrual syndrome); a leg or arm "falling asleep" in certain postures; nervous tics, itches, and twitches; uncracked knuckles, stiff neck, or backache; blocked middle ear following descent from high altitude; restless leg syndrome (akathisia); rotationally-induced dizziness, as on an amusement park ride; or rock-and-roll neck, wherein active musical performers or listeners bob their heads violently, rupturing small blood vessels in the neck. Third is minor physical or functional flaws, such as poor stream during male urination, female papillary leakage, colorblindness, snoring, unpleasant body odors, nosebleeds, declining visual or aural acuity, handedness (currently ~90% dextromanual, ~ 10% sinistromanual, mild strabismus (eyeball misaiignment), bad moods (neurotransmitter imbalances), or post-intoxication hangover.

4. *Undiscovered Infectious Agents* — Peptic ulcers once were thought to result from a stressful life, a purely natural response to a lifestyle choice. Then it was found that the major cause of ulcers is the presence of *Helicobacter pylori* bacteria in the stomach. Bacteria have been implicated in some cases of atherosclerosis and Alzheimer's disease, and nanobacteria have been proposed as possible nucleation sites for kidney stones. Other seemingly natural but undesirable conditions may also be due to undiscovered microbial agents, especially since bacteria outnumber tissue cells in our highly infested 20th century bodies by more than 10:1.

5. *Unwanted Syndromes* — Syndromes are groups of related symptoms and signs of disordered function that define a disease whose cause remains unknown, that is, idiopathic. A good example is irritable bowel syndrome (IBS), which affects up to 20% of the adult U.S. population and includes symptoms of abdominal distention and pain,

with more frequent and looser stools. Many are unaware they are afflicted. In 1998 there was no known cause or simple complete treatment for this still "natural" disease. Even more mysterious than IBS is our general activity level — some people seem to have high-energy personalities, while others have more phlegmatic low-energy personalities. Either may be regarded as "natural," but nanomedicine can probably bring this ill-defined neurophysiological variable under human control. The need to sleep is another imperfectly understood syndrome. It is experienced by everyone and thus was universally regarded as "natural" in the 20th century. Physiological short sleepers were unusual, insomnia or asomnia was thought of as an abnormal state, and there were a few anecdotal but medically undocumented instances of total nonsomnia, such as the celebrated case of Al Herpin.

6. *Psychological Traits* — Psychological traits which, if identified by a patient as undesirable, might be subject to genetic or physiological modification could include: sexual preference (6-10% of the adult population is homosexual); shyness or boldness; acquisitive or altruistic propensity; misanthropy or philanthropy; theistic or atheistic orientation; loquacity or dourness; childhood imprinting; criminal propensity (up to 1-5% of the population); various recognized personality disorders that affect ~10% of the population such as antisocial, paranoid, schizotypal, histrionic, narcissistic, avoidant, dependent, obsessive-compulsive, and passive-aggressive disorders; panic attacks (experienced at least once by ~33% of all adults each year); and phobic disorders such as social phobias (~13% of the population), specific phobias including fear of large animals (zoophobia), snakes (ophidiophobia), spiders (arachnophobia), needles (belonephobia, ~10%), the dark (noctiphobia or scotophobia), or strangers (xenophobia) (total ~5.7%), the fear of blood or hemophobia (~5%), agoraphobia (~2.8%), and other unusual phobias such as the fears of certain colors (chromophobia), daylight (phengophobia), girls (parthenophobia), men (androphobia), stars in the sky (siderophobia), the number thirteen (triskaidekaphobia), and even the fear of developing a phobia (phobophobia).

The above sampling of minor afflictions, almost all considered "natural" in 1998, may come to be regarded as commonplace correctable medical conditions in the nanomedical era. By the time such petty annoyances are deemed worthy of immediate treatment, biotechnology and nanomedicine already will have defeated the most fearsome illnesses of the late 20th century and will have moved on to other challenges. Naturophiles may dissent, but the emerging trend from medical biotechnology is to characterize health, not as a static standard, but rather as a condition defined by the lives that people want to lead. Affirming the volitional normative model of disease Virginia Postrel concludes: "Different goals will produce different choices about trade-offs and standards. What makes a condition unhealthy is not that it is unnatural but that it interferes with human purposes. Revering nature [would mean] sacrificing the purposes of individuals to preserve the world as given. It [would require] that we force people to live with biological conditions that trouble them, whether diseases such as cystic fibrosis or schizophrenia, disabilities such as myopia or crooked teeth, or simply less beauty, intelligence, happiness, or grace than could be achieved through artifice. In a world where it's no big deal to take hormone therapy, Viagra, or Prozac, to have a face lift, or to know a child's sex before birth, a world in which even such radical interventions as sex-change operations and heart transplants have failed to turn society upside down, it is extremely difficult to argue that medical innovations are dangerous simply because they fool Mother Nature."

The publication of *Nanosystems* in 1992 laid a strong technical foundation for the field of molecular manufacturing. Such technology was already known to have major implications for medicine. Still lacking, however, was an overview and synthesis of the interactions between molecular machine systems and living systems, particularly human living systems.

With *Nanosystems* as the intellectual cornerstone, research on *Nanomedicine*, originally intended to be a single volume, began in 1994. It quickly became clear that the field of nanomedicine would be even more interdisciplinary than molecular manufacturing, in part because of the many essential interfaces between mechanical nanosystems and living systems. This realization prompted ramification of the book into multiple volumes, allowing the field to be defined from diverse technical orientations. Additionally, the need to consider all major aspects of nanomedical device design and operations demanded a substantial increase in the intended length of the work, a regrettable but necessary circumstance for which the author apologizes in advance.

As a result, *Nanomedicine* has become a three-volume technical work with 31 chapters. Its intended audience is technical and professional people who are seriously interested in the future of medical technology. The three Volumes build upon each other cumulatively. The first Volume, now complete, describes basic capabilities common to all medical nanodevices, and the physical, chemical, thermodynamic, mechanical, and biological limits of such devices. Its primary audience is physical scientists, chemists, biochemists, and biomedical engineers engaged in basic research. The second Volume, still in progress, deals with aspects of device control and configuration, biocompatibility and safety issues, and basic nanomedical components and simple systems. Its primary audience is systems and control engineers, research physiologists, clinical laboratory analysts, biotechnologists, and biomedical engineers doing applied research. The third Volume, also in progress, discusses specific treatments for specific conditions and injuries, using nanomedical technology in the context of clinical (e.g. doctor-patient) situations. Its primary audience is clinical specialists and research physicians, and interested general practitioners.

The first molecular assemblers may be able to build only very simple nanomechanical systems, possibly only from very highly ordered substrates with significant rate-limiting intermediate steps, and only in very small numbers, so the earliest functional nanomachines may be laboratory curiosities. As assembler technology slowly improves, progressively more complex and capable nanomachines will be manufactured in vastly larger numbers. This book primarily investigates the rational design and operation of these more complex and capable nanomachines, and assumes that cubic centimeter quantities (e.g. ~10^{12} micron-sized nanorobots) will not be unreasonably expensive to manufacture and thus can be therapeutically deployed routinely by future doctors.

Nanomedicine, Basic Capabilities, describes the set of basic capabilities of molecular machine systems that may be required by many, if not most, medical nanorobotic devices. These include the abilities to recognize, sort and transport important molecules; sense the environment; alter shape or surface texture; generate onboard energy to power effective robotic functions; communicate with doctors, patients, and other nanorobots; navigate throughout the human body, i.e. determining somatographic or cytographic location within vessels, organs, tissues, or cells; manipulate microscopic objects and move about inside a human body; and timekeep, perform computations, disable living cells and viruses, and operate at various pressures and temperatures.

Nanomedicine, Systems and Operations, considers system-level technical requirements in the design and operation of medical nanodevices. Aspects of nanomedical operations and configurations, including scaling factors and general design principles; control issues including teleoperation and haptic controllers, swarm motions, autogenous control systems, and various operational protocols; repair, replacement, and reliability issues; and molecular machine systems design issues such as tradeoffs between special-purpose and general purpose architectures, and deployment configurations such as nano-organs, medical utility fogs, and replicators. Part 2 deals with a multitude of issues involving clinical safety and performance, specifically medical nanorobot biocompatibility including immunoreactivity and thrombogenicity; methods of nanorobotic ingress and egress from the human body; and possible nanodevice failure modes, environmental interactions, side effects of nanomedical treatments, iatrogenic factors, nanodevice software bugs, and other safety issues. Various classes of medical nanosystems, including instruments, tools, and diagnostic systems; specific medical nanorobot devices; rapid mechanical reading and editing of chromatin and protein macromolecules; and various complex nanorobotic systems that will make possible advanced cytopathology and cell repair, tissue and organ manufacturing, and personal defensive systems.

Nanomedicine, Applications, describes the full range of nanomedical applications which employ molecular nanotechnology inside the human body, from the perspective of a future practitioner in an era of widely available nanomedicine. Proof-of-concept designs for whole nanodevices, artificial nano-organs, and nanomedical treatments include rapid cardiovascular repair; treatments for pathogenic disease and cancer, with epidemiological considerations; responses to various physical traumas, burns and radiation exposures, and new methods of first aid, surgery, and emergency or critical care; neurography, spinal restoration and brain repair; improved nutrition and digestion; sex, reproduction, and population issues; cosmetics, recreation, veterinary and space medicine; the control of aging processes, eliminating most causes of death prevalent in the 20th century, and strategies for biostasis; and human augmentation systems. The Volume concludes with a discussion of the sociology of nanomedicine, regulatory issues, nanotechnology implementation timelines, and some speculations on the future of hospitals, pharmaceutical companies, and the medical profession.

Cancer Diagnostics and Treatment

Nanotechnology has the potential to have a revolutionary impact on cancer diagnosis and therapy. It is universally accepted that early detection of cancer is essential even before anatomic anomalies are visible. A major challenge in cancer diagnosis in the 21st century is to be able to determine the exact relationship between cancer biomarkers and the clinical pathology, as well as, to be able to non-invasively detect tumors at an early stage for maximum therapeutic benefit. For breast cancer, for instance, the goal of molecular imaging is to be able to accurately diagnose when the tumor mass has approximately 100-1000 cells, as opposed to the current techniques like mammography, which require more than a million cells for accurate clinical diagnosis.

In cancer therapy, targeting and localized delivery are the key challenges. To wage an effective war against cancer, we have to have the ability to selectively attack the cancer cells, while saving the normal tissue from excessive burdens of drug toxicity. However, because many anticancer drugs are designed to simply kill cancer cells, often in a semi-specific fashion, the distribution of anticancer drugs in healthy organs or tissues is especially undesirable due to the potential for severe side effects. Consequently, systemic application of these drugs often causes severe side effects in other tissues (e.g. bone marrow suppression, cardiomyopathy, neurotoxicity), which greatly limits the maximal allowable dose of the drug. In addition, rapid elimination and widespread distribution into non-targeted organs and tissues requires the administration of a drug in large quantities, which is often not economical and sometimes complicated due to non-specific toxicity. This vicious cycle of large doses and the concurrent toxicity is a major limitation of current cancer therapy. In many instances, it has been observed that the patient succumbs to the ill effects of the drug toxicity far earlier than the tumor burden. This course will address the state of the art in nanotechnologies and nano-medicine, and their ongoing applications focused on addressing the challenges posed by cancer prevention, diagnosis and treatment. Distinguished instructors will summarize the basics of nanotechnology and cancer biology, along with the current technologies, trials and future barriers. This program is designed to inform cancer researchers, clinicians, bio-nano technologists, technology managers, and business developers of the state of the art in bio nano technologies, focusing on applications of these technologies for cancer prevention, diagnosis and treatment.

REFERENCES

Blumenthal, D. 1992. "Academic-Industry Relationships: Extent, Consequences, and Management," *The Journal of the American Medical Association*, 268: 3, December, pp. 3344 -3349.

Brown, J.S., and Paul Duguid. 2001. Don't count society out. *Societal Implications of Nanoscience and Nanotechnology*. National Science Foundation. Arlington, VA. March.

Callon, M.: 1986, 'Some Elements of a Sociology of Translation', in J. Law (ed.), *Power, Action and Belief: A New Sociology of Knowledge?*, Routledge, London, pp. 196-223.

Canguilhem, G.: 1952, *Machine et organisme' in La connaissance de la vie*, Hachette, Paris [quoted from the fourth edition Vrin, Paris, 1971].

Carroll, J.S.: 2001, 'Social Science Research Methods for Assessing Societal Implications of Nanotechnology', in: M.C. Roco, W.S. Bainbridge (eds.), *Societal Implications of Nanoscience and Nanotechnology*, National Science Foundation, pp. 188-192.

Carsley, J.E., R. Shaik, W.W. Milligan, and E.C. Aifantis. 1997. In *Chemistry and physics of nanostructures and related non-equilibrium materials*. ed. E. Ma, B. Fultz, R. Shull, J. Morral, and P. Nash. Warrendale, PA: TMS.

Chang, K. "Smaller Computer Chips Built Using DNA as Template." New York Times, November 21, 2003.

Chopra, N.G., R.J. Luyken, K. Cherrey, H. Crespi, M.L. Cohen, S.G. Louie, and A. Zettl. 1995. *Science* 269: 966.

Christensen, Clayton M. *The Innovator's Dilemma: When New Technologies Cause Great Firms to Fail*. New York: Harperbusiness, 2000.

Colbert, D.T. and R.E. Smalley. 1999. Fullerene nanotubes for molecular electronics. *Trends in Biotechnology*. Vol 17. 46-50.

Collier, C.P., E.W. Wong, M. Belohradsky, F.M. Raymo, J.F. Stoddart, P.J. Kuekes, R.S. Williams, and J.R. Heath. 1999. "Electronically Configurable Molecular-Based Logic Gates." *Science*, 285.

Constant, E.W., II: 1980, *The Origins of the Turbojet Revolution*, Johns Hopkins University Press, Baltimore.

Doering, Robert and Yoshio Nishi. 2000. "Limits of Integrated-Circuit Manufacturing." *Proceedings of the IEEE*. September.

Doi, M. and S.F. Edwards. 1986. *The Theory of Polymer Dynamics*. Oxford University Press (Clarendon), London-New York.

Donald, A.M. and A.H. Windle. 1992. Liquid crystalline polymers. *Cambridge Solid State Science Series*. Cambridge University Press.

Drell, S. D., A. D. Sofaaer, and G. D. Wilson, *The New Terror, Facing the Threat of Biological and Chemical Weapons*, Hoover Institution Press, Stanford University (1999).

Dresselhaus, M.S., G. Dresselhaus, and P. Eklund. 1996. *Science of fullerenes and carbon nanotubes*. San Diego: Academic Press. Gleiter, H. 1989. *Prog. Mater. Sci*. 33: 223.

Drexler K.E.: 1995, 'Introduction to nanotechnology', in: Krummenacker, M. & Lewis, J. (eds.), *Prospects in Nanotechology. Proceedings of the 1st general conference on nanotechnology: developments, applications, and opportunities, November 11-14, 1992, Palo-Alto*, John Wiley & Sons, New York, pp. 1-20.

Drexler, E., Phoenix, C.: 2004, 'Self-Replication in Nanotechnology: Feasible, Potentially Safe, and Unnecessary' (unpublished paper, in preparation).

Drexler, E.; Smalley, R.E.: 2003, 'Point-Counterpoint: Nanotechnology', *Chemical and Engineering News*, 81(48), 37-42.

Drexler, K. Eric. 1986. *Engines of Creation*. Garden City, N.Y.: Anchor Press/Doubleday.

Drexler, K.E.: 1981, 'Molecular Engineering—An Approach to the Development of General Capabilities for Molecular Manipulation', *Proceedings of the National Academy of Sciences of the United States of America-Physical Sciences*, 78, 5275-5278.

Dupuy, J.P. "Common Knowledge, Common Sense." Theory and Decision 27 (1989): 37-62.

Ellul, J. 1964. *The Technological Society*. New York: Alfred Knopf. (First published 1954).

Empedocles, S.; Bawendi, M. 1999. "Spectroscopy of single CdSe nanocrystallites," *Accounts of Chemical Research*, 32, 389-396.

Entingh, D., Dunn, A., Glassman, E., Wilson, J. E., Hogan, E. & Damstra, T. (1975) in *Handbook of Psychobiology*, eds. Gazzinga, M. S. & Blakemore, C. (Academic, New York), pp. 201-238.

Esant. 1999. "Economic and Social Aspects of Nanotechnology." First draft report of working group 6 of the Euroconferences on Nanoscience for Nanotechnology.

Estermann, M., L.B. McCusker, C. Baerlocher, A. Merrouche, and H. Kessler. 1991. *Nature* 331: 698.

Fedor N. Dzegilenko, Deepak Srivastava, Subhash Saini, "Simulations of carbon nanotube tip assisted mechano-chemical reactions on a diamond surface," *Nanotechnology* 9(December 1998): 325-330.

Feynman R. "There's Plenty of Room At the Bottom." Talk given on at the annual meeting of the American Physical Society at the California Institute of Technology, 1959.

Friedlander, S.K. 1998. Synthesis of nanoparticles and their agglomerates: aerosol reactors. In *R&D status and trends*, ed. Siegel et al.

Galison, P.: 1997, *Image and Logic: A Material Culture of Microphysics*, University of Chicago Press, Chicago.

George M. Whitesides, "The Once and Future Nanomachine," *Scientific American* 285(September 2001): 78-83.

Gleiter, H. 1990. *Progress in Materials Science* 33: 4. Guinier, A. 1938. *Nature* 142: 569; Preston, G.D. 1938. *Nature* 142: 570.

Gutte, B., Dannigen, M. & Wittschieber, E. (1979) *Nature (London)* 281, 650-655.

Hadjiconstantinou, N.G. 1999. Combining atomistic and continuum simulations of contact-line motion. *Physical Review E*. Vol. 59(2).2, 2475-2478.

Hadjipanayis, C.G. 1998. Nanostructured magnetic materials. In *R&D Status and Trends*, ed. Siegel et al.

Hamilton, S.: 2003, 'Biotechnology, SF and the Media', *Science Fiction Studies*, 30, 267-282.

Harold J. Morowitz, "A Model of Reproduction," *American Scientist* 47(1959): 261-263.

Haruta, M. 1997. *Catalyst surveys of Japan* 1: 61 and references therein.

Hayes, B., "The Square Root of NOT," *American Scientist*, July-August 1995, pp. 304-308.

Hayles, N.K. (ed.): 2004, *Nanoculture: Implications of the New Technoscience*, Bristol, UK: Intellect Books.

Hayles, N.K.: 2004, 'Connecting the Quantum Dots: Nanotechscience and Culture', in: N.K. Hayles (ed.), *Nanoculture: Implications of the New Technoscience*, Bristol, UK: Intellect Books, pp. 11-26.

Hayward, R.C.; Saville, D. A.; Aksay, I.A. 2000. "Electrophoretic assembly of colloidal crystals with optically tunable micropatterns", *Nature*, 404, 56-59.

Higgins, R.J. 1997. An economical process for manufacturing of nano-sized powders based on microemulsion-mediated synthesis. In *Proc. of the Joint NSF-NIST Conf. on Nanoparticles.*

Hiruma, K., M. Yazawa, T. Katsoyama, K. Ogawa, K. Haraguchi, M. Koguchi, and H. Kakibayashi. 1995. *J. Appl. Phys.* 77(2): 476.

Howland, H.: 1962, Structural, hydraulic and economic aspects of leaf venation and shape ', in: E.E. Bernard & M.R. Kare (eds.), *Biological prototypes and synthetic systems*, Plenum Press, New York, vol. 1, pp. 183-192.

Hoyningen-Huene, P.: 1993, *Reconstructing Scientific Revolutions: Thomas S. Kuhn's Philosophy of Science* (trans. by A. Levin), University of Chicago Press, Chicago.

Huang, J.Y., Y.K. Wu, and H.Q. Ye. 1996. *Acta Mater.* 44: 1211. Inoue, A. 1997. Private communication. Inturi, R.B., and Z. Szklavska-Smialowska. 1992. *Corrosion* 48: 398. Karch, J., R. Birringer, and H. Gleiter. 1987. *Nature* 330: 556.

Hubbell, J.A., and R. Langer. 1995. Tissue engineering. *Chem. Eng. News* (March 13): 42- 54.

Hughes, P. M., *Global Threats and Challenges: The Decades Ahead*, Lt. Gen. Patrick M. Hughes, Director of the Defense Intelligence Agency, in Statement for the Senate Committee on Intelligence, January 28, 1998.

Jameson, F.: 1982, 'Progress Versus Utopia; or, Can We Imagine the Future', *Science Fiction Studies*, 9, 147-158.

Janet Raloff, "Nano Hazards: Exposure to minute particles harms lungs, circulatory system," Science News Online, Week of March 19, 2005; Vol. 167, No. 12.

Jonas, H. The Imperative of Responsibility. In Search of an Ethics for the Technological Age. Chicago: University of Chicago Press, 1985.

Kline, R.: 2000, 'The Paradox of 'Engineering Science': A Cold War Debate about Education in the United States', *IEEE Technology and Society Magazine*, 19, 19-25.

Kline, R.; Pinch, T.: 1996, 'Users as Agents of Technological Change: The Social Construction of the Automobile in the Rural United States', *Technology and Culture*, 37, 763-95.

Kroes, P. & Meijers, A. "The Dual Nature of Technical Artifacts—Presentation of a New Research Programme." Techné 6, 2 (2002): 4-8.

Kuusi, O. and M. Meyer. 2001. "Technology Generalizations and Leitbilds – The Anticipation of Technological Opportunities." Manuscript submitted to *Technological Forecasting and Social Change.*

Kuusi, O., 1999. "Epistemic value analysis – a tool to evaluate technological options." Personal communication.

Kuusi, O.; Meyer, M.: 2002, 'Technological generalizations and leitbilder ' the anticipation of technological opportunities', *Technological Forecasting & Social Change*, 69, 625-639.

Leah Eisenberg, "Medicating Death Row Inmates So They Qualify for Execution," AMA Case in Health Law, Vol. 6, No. 9, September 2004.

Ledley, Fred D. 1995. "Nonviral Gene Therapy: The promise of Genes as Pharmaceutical Products," *Human Gene Therapy*, Vol. 6, Sep, pp 1129-1144.

Lee, M.S. and M. Head-Gordon. 2000. Absolute and relative energies from polarized atomic orbital selfconsistent field calculations and a second order convergence with size and composition of the secondary basis. *Computers and Chemistry*. Vol. 24(3,4). 295-301.

Lee, S.C. 1998. The nanobiological strategy for construction of nanodevices. *Biological molecules in nanotechnology: the convergence of biotechnology, polymer chemistry and materials science*, edited by S.C. Lee and L. Savage (IBC Press, Southborough, MA).

Liang, G., Z. Li, and E. Wang. 1996. *J. Mater. Sci.* Makino, A., A. Inoue, T. Hatanai, and T. Bitoh. 1997. *Materials Science Forum* 235-238: 723.

Lipson, H. and J. B. Pollack, Nature, 406, 974-978, (2000). *Automatic design and manufacture of robotic life forms.* See, also, several additional articles on this subject in contemporary issues of Nature.

M.L. Petrillo, C.J. Newton, R.P. Cunningham, R.-I. Ma, N.R. Kallenbach and N.C. Seeman. The Ligation and Flexibility of 4-Arm DNA Junctions. Biopolymers, 27, 1337-1352 (1988).

Mackenzie, D.: 1996a, 'Economic and Sociological Explanations of Technological Change', in: *Knowing Machines: Essays on Technical Change*, MIT Press, Cambridge, MA, pp. 49-65.

Mackenzie, D.: 1996b, 'Marx and the Machine', in: *Knowing Machines: Essays on Technical Change*, MIT Press, Cambridge, MA, pp. 23-47.

Malinow, R., Z.F. Mainen, and Y. Hayashi. 2000. LTP mechanisms: from silence to four-lane traffic. *Curr. Opin. Neurobiol*. 10(3): p. 352-7.

Mao, C., W. Sun, and N.C. Seeman. 1997. Construction of Borromean rings from DNA. *Nature* 386: 137-138.

Mark, H., Statement of The Honorable Hans Mark, Director, Defense Research and Engineering, before the House Armed Service Committee, October 20, 1999.

Martin, T.P., U. Naher, H. Schaber, U. Zimmerman. 1993. *Phys. Rev. Lett*. 70: 3079.

Marty, A.: 2003, *Testimony before the Committee on Science on HR 766, Nanotechnology Research and Development Act*, US Government Printing Office, Washington, DC.

Matsuki, H., K. Ofuji, N. Chubachi, and S. Nitta. 1996. Signal transmission for implantable medical devices using figure-of-eight coils. *IEEE Trans. on Magnetics.*, vol. 32. No.5, 5121. September.

McConnell, S.A., ed. 1998. *Pharmaceuticals*. Dun and Bradstreet Industry Reference Handbooks, Gale, Detroit, pp. 30-34.

McCulloch, W.S.: 1962, 'The imitation of one form of life by another – Biomimesis', in: E.E. Bernard & M.R. Kare (eds.), *Biological prototypes and synthetic systems*, Plenum Press, New York, vol. 1., p. 393-97.

Meadows, D.L., et al. 1972. *The Limits to Growth*, New York: University.

Mehl, R.F., and R.W. Cahn. 1983. Historical development. In *Physical metallurgy.* North Holland. Milligan, W.W., S.A. Hackney, M. Ke, and E.C. Aifantis. 1993. *Nanostructured Materials* 2: 267.

Mehta, M.D.: 2004, 'From Biotechnology to Nanotechnology: What Can We Learn from Earlier Technologies?, *Bulletin of Science, Technology & Society*, 24, 34-39.

Merz, J.L. 1986. "The Optoelectronics Joint Research Laboratory: Light Shed on Cooperative Research in Japan." *Scientific Bulletin* (ONR Far East Office). 11 (4), 1-30 (Oct.-Dec.).

Meyer, G. and Amer, N.M. (1988) Novel optical approach to atomic force microscopy. Appl. Phys. Lett. 53(12), 1045-1047

Meyer, G. and Amer, N.M. (1990) Simultaneous measurement of lateral and normal forces with an optical-beam-deflection atomic force microscope. Appl. Phys. Lett. 57(20), 2089-2091

Meyer, M. & Kuusi, O.: 2004, 'Nanotechnology: Generalizations in an Interdisciplinary Field of Science and Technology', *Hyle: International Journal for Philosophy of Chemistry*, 10(2), 153-168.

Michael E.J. Holwill, Peter Satir, "II.4 Generation of Propulsive Forces by Cilia and Flagella," in J. Bereiter-Hahn, O.R. Anderson, W.-E. Reif, eds., *Cytomechanics: The Mechanical Basis of Cell Form and Structure*, Springer-Verlag, Berlin, 1987, pp. 120-130.

Michael Page, Donald W. Brenner, "Hydrogen abstraction from a diamond surface: *Ab initio* quantum chemical study using constrained isobutane as a model," *J. Am. Chem. Soc.* 113(1991): 3270-3274.

Mishra, R.S., R.Z. Valiev, and A.K. Mukherjee. 1997. *Nanostructured Materials* 9: 473. Morris, D.G., and M.A. Morris. 1991. *Acta Metall. Mater*. 39: 1763-1779.

Mizrach, S., *Should there be a limit placed on the integration of humans and computers and electronic technology?*

Mnyusiwalla, A.; Daar, A.S.; Singer, P.A.: 2003, '"Mind the Gap": Science and Ethics in Nanotechnology, *Nanotechnology*, 14, R9-R13.

Morris, D.G., and M.A. Morris. 1997. *Materials Science Forum* 235-238: 861. Nagpal, P., and I. Baker. 1990. *Scripta Metall. Mater*. 24: 2381.

Mui, C. and C. Musgrave, "The Hydroxylation and Oxidation of the Ge(100)-2x1 Surface by Hydrogen Peroxide and Water," *Langmuir*, 20, 7604-7609 (2004).

Perkal M., C. Marks, M.I. Lorber, W.H. Marks. 1992. "A three-year experience with serum anodal trypsinogen as a biochemical marker for rejection in pancreatic allografts. False positives, tissue biopsy, comparison with other markers, and diagnostic strategies." *Transplantation*, Feb; 53 (2): pp. 415-419.

Philip Ball, "Synthetic Biology: Starting from Scratch," Nature, 431, pp. 624-626, 7 October 2004.

Philip G. Collins, Michael S. Arnold, Phaedon Avouris, "Engineering carbon nanotubes and nanotube circuits using electrical breakdown," *Science* 292(27 April 2001): 706-709.

Philip G. Collins, Phaedon Avouris, "Nanotubes for electronics," *Scientific American* 283 (December 2000): 62-69.

Philipse, A.P. 1998. "Colloidal Dispersions," section 3.4, A. ten Wolde (ed.), *Nanotechnology: Towards a molecular construction kit*. STT Report #60, The Hague, 171-178.

Pierce, N.A. and M.B. Giles. 2000. Adjoint reconstruction of superconvergent functionals from PDE approximations. *SIAM Rev.* Vol. 42(2). 247-264.

Pinch, T.: 1993, "Testing – One, Two, Three... Testing!': Toward a Sociology of Testing', *Science, Technology, and Human Values*, 18, 25-41.

Pitt, J.C.: 2004, 'The Epistemology of the Very Small', in: D. Baird, A. Nordmann & J. Schummer (eds.), *Discovering the Nanoscale*, Amsterdam: IOS Press, pp. 157-163.

Ramanan, V.R. 1998. Nanocrystalline soft magnetic alloys for applications in electrical and electronic devices. In *R&D Status and Trends*, ed. Siegel et al.

Ramos, L.; Lubensky, T.C.; Dan, N.; Nelson, P.; Weitz, D.A. 1999. "Surfactant-mediated two-dimensional crystallization of colloidal crystals", *Science*, 286, 2325-2328.

Reed, Mark A. and James M. Tour. 2000. "Computing with Molecules." *Scientific American*. June.. SIA (Semiconductor Industry Association). 1999. "1999 International Technology Roadmap for Semiconductors."

Rheinberger, H.-J.: 1997, *Toward a History of Epistemic Things: Synthesizing Proteins in the Test Tube*, Stanford University Press, Stanford.

Rietman, E.A.: 2001, 'Drexler hypothesis of a universal assembler is supported not by theoretical arguments alone but by existence proof in the form of biological life', in: *Molecular Engineering of Nanosystems*, Springer, New York & Berlin.

Rinzler, A.G.; Liu, J.; Dai, H.; Nikolaev, P.; Huffman, C.B.; Rodriguez-Macias, F.J.; Boul, P.J.; Lu, A.H.; . Heymann, D.; Colbert, D.T.; Lee, R.S.; Fischer, J.E.; Rao, A.M.; Eklund. P.C.; Smalley, R.E. 1998.

Roco, M.&Bainbridge, W. (eds.). Converging Technologies for Improving Human Performances, National Science Foundation report, 2002.

Roco, M.C. & Tomellini, R. (eds.): 2002, *Nanotechnology: Revolutionary Opportunities and Societal Implications, Workshop, Lecce (Italy), 31 January—1 February 2002*, Luxemburg: Office for Official Publications of the European Communities, [200 pp.]

Roco, M.C., R.S. Williams, and P. Alivisatos (eds). 1999. *Nanotechnology Research Directions: IWGN Workshop Report, Vision for Nanotechnology R&D in the Next Decade*. (Kluwer Academic Publishers.

Roco, M.C.: 2003a, 'Broader Societal Issues of Nanotechnology', *Journal of Nanoparticle Research*, 5, 181-189.

Roco, M.C.: 2003b, 'Coherence and Divergence in Megatrends in Science and Engineering', in: M.C. Roco, W.S. Bainbridge (eds.), *Converging Technologies for Improving Human Performance*, Kluwer, Dordrecht, pp. 79-96.

Roco, M.C.: 2003c, 'Fundamentally New Manufacturing Processes and Products', in: Roco, M.C. & Bainbridge, W.S., (eds.), *Converging Technologies for Improving Human Performance*, Kluwer, Dordrecht, pp. 300-302.

Roco, M.C.: 2004, 'The Emergence and Policy Implications of Converging Technologies'

Roco, M.C.; Bainbridge, W.S. (eds.): 2002, *Converging Technologies for Improving Human Performance: Nanotechnology, Biotechnology, Information Technology and the Cognitive Science*, Arlington, VA: National Science Foundation.

Rodney Brooks, "The Merger of Flesh and Machines," The Next Fifty Years: Science in the First Half of the Twenty-First Century, ed. by John Brockman, 2002, p. 191.

Rohrer, H.: 1995, 'The Nanometer Age – Challenge and Chance', *Microelectronic Engineering*, 27, 3-15.

Rosenberg, Nathan. 1994. *Exploring the Black Box: Technology, Economics, and History*. New York: Cambridge University Press.

Rothman, B.K.: 1998, *Genetic Maps and Human Imagination*, W.W. Norton, New York & London,

Ruthven, D.M., S. Farooq, K.S. Knaebel. 1994. *Pressure swing adsorption*. New York: VCH Publishers.

Ruthven, D.M., S. Farooq, K.S. Knaebel. 1994. *Pressure swing adsorption*. New York: VCH Publishers. Sayari, A. 1996. *Chem. Mater*. 8: 1840.

Sankey, H.: 1994, *The Incommensurability Thesis*, Avebury, Aldershot.

Santini, John T., Jr., Michael J. Cima and Robert Langer. 1999. "A Controlled-release Microchip" *Nature*, Vol 397, 28 Jan, p 335-338.

Sarewitz, D.; Woodhouse, E.: 2003, 'Small is Powerful', in: A. Lightman, D. Sarewitz & Chr. Desser, (eds.), *Living with the Genie: Essays on Technology and the Quest for Human Mastery*, Washington, DC: Island Press, pp. 63-83.

Sarikaya, M.; Aksay, I. (eds.), 1995, *Biomimetics. Design and Processing of Materials*, AIP Press, Woodbury, New York.

Schmidt, J.C.: 2004, 'Unbounded Technologies: Working Through Technological Reductionism of Nanotechnology', in: D. Baird, A. Nordmann & J. Schummer (eds.), *Discovering the Nanoscale*, Amsterdam: IOS Press, pp. 35-50.

Schuler, E.: 2004, 'Perception of Risks and Nanotechnology', in: D. Baird, A. Nordmann & J. Schummer (eds.), *Discovering the Nanoscale*, Amsterdam: IOS Press, pp. 279-284.

Schummer, J.: 2001, 'Ethics of Chemical Synthesis', *Hyle: International Journal for Philosophy of Chemistry*, 7, 103-124 [online: www.hyle.org/journal/issues/7/schummer.html].

Schummer, J.: 2003, 'Aesthetics of Chemical Products: Materials, Molecules, and Molecular Models', *Hyle: International Journal for Philosophy of Chemistry*, 9, 73-104 [online: www.hyle.org/journal/issues/9-1/schummer.html].

Schummer, J.: 2003, 'The Notion of Nature in Chemistry', *Studies in History and Philosophy of Science*, 34 (2003), 705-736.

Schummer, J.: 2004, 'Interdisciplinary Issues in Nanoscale Research', in: D. Baird, A. Nordmann, J. Schummer (eds.), *Discovering the Nanoscale*, IOS Press, Amsterdam, pp. 9-20.

Schwarz, A.E.: 2004, 'Shrinking the 'Ecological Footprint' with NanoTechnoScience?', in: D. Baird, A. Nordmann & J. Schummer (eds.), *Discovering the Nanoscale*, Amsterdam: IOS Press, pp. 203-208.

Schwarz, R.B. 1998. Storage of hydrogen in powders with nanosized crystalline domains. In *R&D status and trends*, ed. Siegel et al.

Siegel et al. Ogunnaike, B and W. Ray. 1994. *Process dynamics, modeling and control*. Oxford University Press, pp 5-21; 1033-48.

Siegel R.W. and G.E. Fougere. 1994. In *Nanophase materials*, ed. G.C. Hadjipanayis and R.W. Siegel. Netherlands: Kluwer Acad. Publ.

Siegel, H.: 1980, 'Objectivity, Rationality, Incommensurability, and More', *British Journal for the Philosophy of Science*, 31, 359-384.

Siegel, R.W., E. Hu, and M.C. Roco, eds. 1998. *R&D status and trends in nanoparticles, nanostructured materials, and nanodevices in the United States*. Baltimore: Loyola College, International Technology Research Institute. NTIS #PB98-117914.

Siegel, R.W., E. Hu, and M.C. Roco, eds. 1998. *R&D status and trends in nanoparticles, nanostructured materials, and nanodevices in the United States*. Baltimore: Loyola College, International Technology Research Institute.

Smalley, R.E. 2000. Hearing of the Senate Science and Technology Caucus on Nanotechnology. April 5, (unpublished).

Smalley, R.E.: 1999, 'Prepared written statement and supplemental material', Rice University, 22 June.

Smalley, R.E.: 2001, 'Of Chemistry, Love and Nanobots', *Scientific American*, (Sept.), 76-77.

Smith, A., *The Wealth of Nations: An Inquiry into the Nature and Causes*, Random House, New York, 1994.

Smith, D., H. Babcock, and S. Chu. 1999. Single-polymer dynamics in steady shear flow. *Science*. Vol. 283. 1724-1727. March.

Snider, G.L.; Orlov, A.O.; Amlani, I.; Zuo, X.; Bernstein, G.H.; Lent, C.S.; Merz, J.L.; Porod, W. 1999. "Quantum-dot cellular automata: Review and recent experiments (invited)", *Journal of Applied Physics*, 85, 4283-4285.

Snow, E.S., P.M. Campbell, and F.K. Perkins. 1997. Nanofabrication with proximal probes. *Proceedings of the IEEE* 85: 601-11.

Sterman, J. D. (1989b). Modeling Managerial Behavior: Misperceptions of Feedback in a Dynamic Decision Making Experiment. *Management Science*, 35(3), 321-339.

Steve, Jurvetson, "Transcending Moore's Law with Molecular Electronics," Nanotechnology Law & Business Journal, Vol. 1, No. 1, article 9, p. 9.

Stipe, B.C.; Rezaei, M.A.; Ho, W. 1998. "Single-molecule vibrational spectroscopy and microscopy", *Science*, 280, 1732-1735.

Stix, G.: 1996, 'Trends in Nanotechnology: Waiting for Breakthroughs', *Scientific American*, 274(4), 94-99.

Stix, G.: 2001, 'Little Big Science: Nanotechnology Is All the Rage. But Will It Meet Its Ambitious Goals? And What The Heck Is It?', *Scientific American* 285, 32-37.

Stokes, Donald E. 1997. *Pasteur's Quadrant: Basic Science and Technological Innovation*. Washington, D.C.: Brookings Institution Press.

Stranick, S.J.; Atre, S.V.; Parikh, A.N.; Wood, M.C.; Allara, D.L.; Winograd, N.; Weiss, P.S. 1996. "Nanometer-scale phase separation in mixed composition self-assembled monolayers", *Nanotechnology*, 7, 438-442.

Tenner, E.: 2001, 'Nanotechnology and Unintended Consequences', in: M.C. Roco and W.S. Bainbridge (eds.), *Societal Implications of Nanoscience and Nanotechnology*, Dordrecht: Kluwer, pp. 241-246.

Thacker, E.: 2000, 'SF, Technoscience, Net.art: The Politics of Extrapolation (science fiction)', *Art Journal*, 59(3), 64-73.

Thacker, E.: 2004, *Biomedia*, University of Minnesota Press, Minneapolis.

The Royal Society & The Royal Academy of Engineering: 2004, *Nanoscience and nanotechnologies: opportunities and uncertainties*, London [www.nanotec.org.uk/finalReport.htm].

Theis, T.: 2001, 'Information Technology Based on a Mature Nanotechnology', in: M.C. Roco, W.S. Bainbridge (eds.), *Societal Implications of Nanoscience and Nanotechnology*, National Science Foundation, pp. 60-67.

Thomas, E.L., D.M. Anderson, C.S. Henkee, and D. Hoffman. 1988. Periodic area-minimizing surfaces in block copolymers. *Nature*. Vol. 334. 598.

Thompson, P.B.: 2001, 'Social Acceptance of Nanotechnology', in: M.C. Roco, W.S. Bainbridge (eds.), *Societal Implications of Nanoscience and Nanotechnology*, online: National Science Foundation, pp. 198-202.

Tighe, T.S., J.M. Worlock, and M.L. Roukes. 1997. Direct thermal conductance measurements on suspended monocrystalline nanostructures. *Appl. Phys. Lett.* 70: 2687-9.

Tihamer Toth-Fejel, "Modeling kinematic cellular automata: An approach to self-replication," Environmental Research Institute of Michigan, Ann Arbor, MI, 4 October 2000.

Tilton, John E. (ed). 1991. *World Metal Demand*, Washington, DC: Resources for the Future.

Tobio, M., R. Gref, A. Sanchez, R. Langer and M.J. Alonso. 1998. "Stealth PLA-PEG Nanoparticles as Protein Carriers for Nasal Administration", *Pharmaceutical Research*, Vol. 15, No 2, pp 270-275.

Tramontin, A.D. and E.A. Brenowitz. 2000. Seasonal plasticity in the adult brain. *Trends Neurosci.* 23(6): p. 251-8.

3

Nanosafety, Nanolabs & Nanobodies

Nanotechnology's Environmental Risks

"Stakeholders ranging from corporations to start-ups to protest groups are concerned about the environmental, health, and safety (EHS) risks of nanoparticles — the prospect that tiny, engineered particles of matter might harm workers, consumers or the environment. While such EHS risks do exist, they can be appropriately addressed today using well-established risk management techniques, according to a new report from Lux Research entitled 'A Prudent Approach to Nanotech Environmental, Health, and Safety Risks.'"

Approaches to Nanotechnology

Nanotechnology was first proposed by Nobel physicist Richard Feynman in December 1959, in a talk in which he also issued a seemingly "impossible" challenge to build a working electric motor no larger than a 1/64th-inch (400-micron) cube, backed by a $1000 prize to spur interest in the new field. Just 11 months later, engineer William McLellan had constructed a 250-microgram 2000-rpm motor out of 13 separate parts and collected his reward. (McLellan's entire motor is only as big as the period at the end of this sentence.) In 1995, a $250,000 Feynman Grand Prize was made available, this time sponsored by the Foresight Institute, for any engineer who could build the first programmable nanometer-scale robotic arm. How long will it take for history to repeat?

In his famous 1959 talk, Feynman proposed the prototypical top-down strategy for building complex nanomachinery — essentially a completely teleoperated machine shop, including mills, lathes, drills, presses, cutters, and the like, plus master-slave grippers to allow the human operator to move parts and materials around the workshop.

To build a nanomachine using Feynman's scheme, the operator first directs a macroscale machine shop to fabricate an exact copy of itself, but four times smaller in size. After this work is done, and all machines are verified to be working properly and as expected, the reduced-scale machine shop would be used to build a copy of itself, another factor of four smaller but a factor of 16 tinier than the original machine shop. This process of fabricating progressively smaller machine shops proceeds until a machine shop capable of manipulations at the nanoscale is produced. The final result is a nanomachine shop capable of reconstructing itself, or of producing any other useful nanoscale output product stream that is physically possible to manufacture, using molecular feedstock.

As Feynman describes the process: "When I make my first set of slave hands at one-fourth scale, I am going to make ten sets. I make ten sets of hands, and I wire them to my original levers so they each do exactly the same thing at the same time in parallel. Now, when I am making my new devices one-quarter again as small, I let each one manufacture ten copies, so that I would have a hundred hands at the 1/16th size....If I made a billion little lathes, each 1/4000th the scale of a regular lathe, there are plenty of materials and space available because in the billion little ones there is less than two percent of the materials in one big lathe. It doesn't cost anything for materials, you see. So I want to build a billion tiny factories, models of each other, which are manufacturing simultaneously, drilling holes, stamping parts, and so on." "As we go down in size, there are a number of interesting problems that arise. All things do not simply scale down in proportion. There is the problem that materials stick together by the molecular (van der Waals) attractions. There will be several problems of this nature that we will have to be ready to design for....[But] if we go down far enough, all of our devices can be mass produced so that they are absolutely perfect copies of one another. We cannot build two large machines so that the dimensions are exactly the same. But if your machine is only 100 atoms high, you only have to get it correct to one-half of one percent to make sure the other machine is exactly the same size — namely, 100 atoms high!"

Progress is being made on Feynman's top-down approach in a relatively new engineering field known as Micro Electro-Mechanical Systems or MEMS, originally an extension of chip-etch technology rather than micromanipulation. Conventional purely-electronic devices fabricated on silicon chips with ~0.2 micron features must be coupled to external systems to produce mechanical effects, but MEMS devices integrate mechanical components directly with electrical circuitry. Thus while microelectronic chips merely route electrons, microelectromechanical systems give the electronics immediate access to control applications, enabling single microdevices to interact directly with the physical world.

Generic microsystems research has proven to be a rewarding activity for two decades, with each successively smaller motor or machine widely publicized. The field gathered momentum after the first working micromotors were demonstrated in the late 1980s by groups at Berkeley and MIT. By 1990, tiny electrostatic motors with 100-micron rotors displayed operating speeds of ~250 Hz and operating lifetimes of $\sim 10^6$ revolutions. A new *Journal of Micromechanics and Microengineering* appeared in

1991, followed by the *Journal of Microelectromechanical Systems* in 1992 and the first MEMS Application Symposium in Tokyo in June 1996. By the mid-1990s, Sandia's Microelectronics Development Laboratory, one of the most prominent vendors of micromachines, could mass-produce planar micromechanical devices smaller than ~1 mm^2, incorporating motors with gears and cogs each no thicker than a human hair and turning at speeds of >4000 Hz. Wobble micromotors such as those fabricated in silicon by the University of Utah's micromachine lab demonstrated very little friction or abrasion, with measured operating lifetimes in excess of 300 million revolutions. (Diamond coatings could further reduce friction by an order of magnitude over silicon, and the wear rate of polycrystalline diamond is known to be ~10^4 times lower than for silicon.)

MEMS research has produced multi-micron-scale accelerometers, diverse microsensors (e.g., blood pressure microsensors attached to cardiac catheters), microscale cantilevers and jointed crank mechanisms, 5-micron barbs, flow microvalves and pressure microtransducers, micropistons and micropumps, microgear trains, microactuators and piezodriven micromotors, micromirrors and microshutters, Fresnel lens microarrays, microgyroscopes, a 2-mm long combustion chamber suitable for turbine use (making possible a 1 cm^3 gas-turbine generator that would deliver 50 W of power or 0.2 N of thrust), and multidevice microsystems that were available customized or off-the-shelf in mass quantities by 1998. Laser beams and ion beams were used to carve ~300-micron-diameter gears with 50-micron wide teeth in solid diamond. Microelectrodeposition and microcontact printing readily produced topologically complex 3-D microstructures with 1-10 micron feature sizes. Colloidal templating yielded 0.71-micron diameter hollow silica spheres with wall thicknesses from tens to hundreds of nanometers, 10-25 nm trenches, or 4-nm thick graphite sheet layers tiled on spherical surfaces ~0.2 microns in diameter. A CAD/CAM desktop micromanufacturing system invented at Lincoln Laboratory in Lexington, MA, machined three-dimensional structures, including spheres, in silicon, using computer-controlled laser/chemical etching of cubic-micron voxels at the rate of ~20,000 voxels/sec.

By 1998, smaller (~100 nm) accelerometers were being developed by such companies as Analog Devices in the U.S. and Daimler-Benz in Germany, and electronic control systems such as microswitches and microrelays were available commercially from several companies including Integrated Micromachines of California. Micromachining was a well-developed engineering discipline. Westinghouse Science and Technology Center used MEMS sensors to build a miniaturized mass spectrometer. Micro-opto-mechanical systems were under development. Chemists were proposing and building Integrated Chemical Synthesizers consisting of millimeter- to micron-sized chemical reactors, designed for specific applications, along with associated devices for moving reactant and product streams (pumps), mixing reactants (reaction chambers), analyzing streams, and separating products (separation chambers), constructed conventionally using silicon substrates and photolithographic techniques to form the desired structures but also utilizing the high selectivity of biological molecules such as enzymes and antibodies to fabricate sensors. MEMS had also led to the development of microgrippers that could manipulate individual 2.7-micron polystyrene spheres, dried red blood cells of similar size, and various protozoa. Market Intelligence Research Corp. of Mountain View, CA, estimated a total worldwide budget for micromachine research and development of $995 million in 1992 and over $3 billion by 1998. Systems Planning Corporation of Arlington, VA, projected that the microsystems industry could be worth more than $14 billion annually by 2001.

Some MEMS research followed surprising directions. For example, in the mid-1990s MEMS engineers at the University of Tokyo Department of Mechanical Engineering fabricated an 8-hinge, 8-plate wing microstructure capable of independent flight through the air. The structure was fabricated by traditional silicon micromachining techniques and was then released from the wafer surface, with the initially flat sections lifted into flight position using a micromanipulation system with glass probes several microns in diameter. Each wing was ~500 microns square; applying 300 VAC at 10 KHz caused resonant vibration and the flapping amplitude of the wings exceeded 30°. In another 4-wing configuration, the flapping resonance frequency was ~150 Hz and the device freely flew up and down a thin glass pole when placed in an alternating electromagnetic field. Another bi-wing design made of magnetized nickel coated silicon measuring 2000 x 10,000 microns successfully flew without power supply cables and guides in the manner of a butterfly or mosquito, when an alternating magnetic field strength of >400 oersted and 12 Hz was applied; a similar device with 320 x 800 micron wings also flew in a 300 Hz magnetic field. "If we release the silicon mosquito from the silicon chip, it flies off and we cannot find it again," complained researcher H. Miura. "It's very small, like dust." MEMS "gnat robots" have also been described.

Using electron-beam (e-beam) lithography, in 1997 researchers at the Cornell University Nanofabrication Facility built what they believed were the world's smallest mechanical devices, including a 4-micron-wide Fabry-Perot interferometer and, just for fun, the world's smallest guitar, carved out of crystalline silicon and no larger than a single human cell. The "nanoguitar" was 10 microns long and 2 microns wide, with six strings each ~50 nm (~200 atoms) wide that would probably resonate at >10-100 MHz if plucked. Presentation of the nanoguitar enthralled the public, surprising at least one Cornell researcher (in another lab) who remarked that "anyone with a sufficiently advanced e-beam lithography system could make it. I make 50 nanometer silicon objects routinely. Doesn't seem like a big deal to me, but evidently most people have no idea what we're capable of these days."

Direct-write e-beam lithography uses a tightly-focused beam of electrons, steered by electromagnetic deflectors, to trace out patterns on a wafer resist surface. The electrons chemically alter the resist, which is then etched away, leaving the desired pattern. By 1998, e-beam spot sizes had been focused to less than 1 nm, and 5-nm beam widths were routine, although forward scattering of the electrons imposed a limit of resist exposure resolution of ~10 nm for e-beams and ~5 nm for focused ion beams. E-beam writing times were very slow — for example, 80 hours were required to carve a single photolithography mask for a 1G DRAM, though much of this time was required for proximity calculations by the AI (artificial intelligence) software [E.A. Rietman, personal communication, 1999] (a better approach is to use a neural net). By 1998, an electron interferometer containing atomically smooth mirrors spaced a few atomic layers apart had been fabricated, and in 1999, careful electron-energy-loss spectroscopy experiments revealed that silicon dioxide must be at least 4 atoms (~0.7 nm) thick to act as a conventional electrical insulator.

Atom lithography, employing a low-energy neutral atomic beam (usually Na or Cr) cooled and collimated by optical molasses, then concentrated at the nodes of the standing wave formed by the superposition of a laser beam and its reflection, was predicted also eventually to allow surface deposition of 5-10 nm features. The first directional atomic beam (of sodium) with a ~0.002 radian beamspread and ~femtogram pulses (~26 million atoms, or ~$(100 \text{ nm})^3$, per pulse) was demonstrated in early 1999, with one commentator noting that "the longer dream...is atomic holography...[that] could combine beams of atoms to build a 3-D solid object". At the same time, another group achieved a continuous stream of rubidium atoms lasting up to ~0.1 sec in duration, with a beam radius potentially as small as ~1 nm. Current-carrying wires have been used to guide cold lithium atoms "just as optical fibers guide light"; the first curved focusing atom mirror also was demonstrated in early 1999. By 1998, the speculative possibility of quantum-mechanical micromanipulation also had been described.

Other approaches offered comparable results. For instance,engineers at the University of Minnesota NanoStructure Laboratory used nanoimprint lithography (NIL) in 1996 to etch out patterns of lines, grooves and circles as small as 25 nanometers in a polymer, and by 1998 NIL had been demonstrated in nanoscale quantum-wire, quantum-dot, and ring transistors with feature sizes below 10 nm. Nanochannel array glasses containing 33-nm capillaries arranged in a 2-D hexagonal close packing configuration at a number density of ~$300/\text{micron}^2$ had been available since 1992.

Thus by 1998, top-down MEMS techniques could in theory already make nanoscale (though not atomically precise) parts. Might these techniques also produce assemblies of such parts, creating complex machines? Very simple mobile robots of ~1 cm^3 volume were commonplace, so for a more challenging demonstration of MEMS' ability to manufacture complete working microrobots, in 1994 Japanese researchers at Nippondenso Co., Ltd. fabricated a 1/1000th-scale working electric car. As small as a grain of rice, the microcar was a 1/1000-scale replica of the Toyota Motor Corp's first automobile, the 1936 Model AA sedan. The tiny vehicle incorporated 24 parts, including tires, wheels, axles, headlights and taillights, bumpers, a spare tire, and hubcaps carrying the company name inscribed in microscopic letters, all manually assembled using a mechanical micromanipulator of the type generally used for cell handling in biological research. In part because of this handcrafting, each microcar cost more to build than a full-size modern luxury automobile.

The Nippondenso microcar was 4800 microns long, 1800 microns wide, and 1800 microns high, consisting of a chassis, a shell body, and a 5-part electromagnetic step motor measuring 700 microns in diameter with a ~0.07-tesla magnet penetrated by an axle 150 microns thick and 1900 microns long. Power was supplied through thin (18 micron) copper wires, carrying 20 mA at 3 volts. The motor developed a peak torque of 1.3 x 10^6 N-m (mean 7 x 10^7 N-m) at a mean frequency of ~100 Hz (peak frequency ~700 Hz), propelling the car forward across a level surface at a top speed of 10 cm/sec. Some internal wear of the rotating parts was visible after ~2000 sec of continuous operation; the addition of ~0.1 microgram of lubricant to the wheel microbearings caused the mechanism to seize due to lubricant viscosity. The microcar body was a 30-micron thick 20-milligram shell, fabricated with features as small as ~2 microns using modeling and casting, N/C machine cutting, mold etching, submicron diamond-powder polishing, and nickel and gold plating processes. Measured average roughness of machined and final polished surfaces was 130 nm and 26 nm, respectively. The shell captured all features as small as 2 mm on the original full-size automobile body. Each tire was 690 microns in diameter and 170 microns wide. The license plate was 10 microns thick, 380 microns wide and 190 microns high. Nippondenso subsequently used similar manufacturing techniques to build a prototype of a capsule intended to crawl through tiny pipes in a power plant or chemical plant like an inchworm, hunting for cracks. In 1999, three Japanese electronics companies announced the creation of a 0.42-gram, 5-mm long "ant size" robot reportedly able to lift 0.8-gram loads and to move at ~2 mm/sec, as part of the government's ongoing Micro Machine Project.

By 1998, micromanipulators with high placement precision had been demonstrated in various laboratories around the world. For example, the Spider-II micromanipulation robot employed bimorphic piezoactuators within a 260-micron cubic work volume and a three-axis gripper placement accuracy of 8 nm. A microassembly robot using stick-and-slip actuators had 5-nm resolution over a 200-nm scanning range with a maximum speed of 4 mm/sec and 0.4% repeatability at 10 KHz, delivering a maximum driving force of 155 mN and rigidity of 6.3 N/micron.

In terms of system size, micromachined MEMS devices (typically ~10^{-13} m^3 in volume) lie exactly intermediate between the world (~10^{-4} m^3) and the nanoscale world (~10^{-22} m^3). The medical nanorobots described in this book generally will have dimensions, precisions, and component part sizes ~1000-fold smaller than the Nippondenso microcar, which itself was, in turn, ~1000-fold smaller than a macroscale automobile. It is difficult to see how MEMS techniques involving statistical materials deposition and removal could fabricate parts to better than ~10 nm feature sizes or tolerances, or could position parts during assembly to much better than ~10 nm spatial resolution, nor does it appear likely that MEMS techniques, on their own, can manufacture or position any structure to atomic precision. However, this does not rule out the possible use of MEMS techniques to fabricate crude useful parts for subsequent "finishing" to atomic precision by other techniques, or to facilitate the assembly of nanoscale components (of atomic precision or otherwise) that have been fabricated or subassembled by other means.

Carbon nanotubes—tiny straws of pure carbon—have many properties that make them attractive for applications as varied as nanoelectronics and nanofibers. Scientists are recruiting carbon nanotubes in the fight against cancer, too. A report published online this week by the *Proceedings of the National Academy of Sciences* suggests that when paired with a modified laser beam, the rods can act as tiny heaters to selectively destroy tumor cells. When exposed to near-infrared light, carbon nanotubes quickly release excess energy as heat. Nadine Wong Shi Kam and her colleagues at Stanford University exploited this property to attack cancerous cells. "One of the longstanding problems in medicine is how to cure cancer without harming normal body tissue," notes study co-author Hongjie Dai. Cancer cells tend to be coated in folate receptors, whereas normal cells are not. Thus, to ensure that the carbon nanotubes were attracted only to diseased cells, the researchers coated them with folate molecules. The team then shined a flashlight-size near-infrared laser on aqueous solutions of both tumor and normal cells. Although harmless to regular cells, the light heated the nanotubes to 70 degrees Celsius within two minutes, killing the cancer cells they had invaded. The researchers hope to refine the process for future use. "Folate is just an experimental model that we used," Dai says. "In reality there are more interesting ways we can do this. For example, we can attach an antibody to a carbon nanotube to target a particular type of cancer cell." To that end, Dai is currently investigating the possibility of using the technique on mice with lymphoma because lymphoma cells have well-defined surface receptors that can be targeted.

Gene Gun & Human Body Composition

David Pui has big dreams, all generated by an ultrasmall technology, and he's not alone. Leaders from industry, academia, and government, including President Bill Clinton and University president Mark Yudof, agree that nanotechnology could someday revolutionize science, medicine, and industry.

Pui, a Distinguished McKnight University Professor in mechanical engineering, is director of the Particle Technology Laboratory—a hub of nanotechnology research at the University—and a staunch advocate of the emerging field. Pui, who chaired the Minnesota Nanotechnology Summit last March, says that the University has already established itself as a leader in promoting and supporting interdisciplinary nanotechnology research.

Pui should know. He's been involved in one of the University's most promising nanotechnology achievements to date, the development of a "gene gun"—a newly patented electrospraying apparatus for continuous gene transfection. The gun shows great potential as a delivery method for gene therapy.

Nowhere are the hopes for nanotechnology riding higher than among medical researchers. The gene gun project is the result of years of research by Pui and Assistant Professor Da-Ren Chen, also a member of the mechanical engineering faculty. Christine Wendt, an assistant professor in the Medical School who has worked with Pui's team, says she can readily foresee ways in which the gun could have medical applications. The goal of gene therapy is to treat cells with genes that will help the cells produce specific proteins to fight disease. But in order for gene therapy to work, a number of things must happen. The gene first must get to the cell, and then the cell must accept it. Finally, the gene must arrive at the nucleus for its effects to take hold. Traditional methods for this process, called gene transfection, include uptaking plasmid DNA, infecting the cell with viruses that contain the desired gene, fusing cells together, and microinjecting cells. All these methods treat a large number of cells simultaneously, but they tend to have a low efficiency rate. In treating lung cancer, for example, doctors have been frustrated by their inability to successfully deliver genes to diseased cells.

"The traditional ways of treating many diseases have limitations, but gene delivery offers new and exciting possibilities," Wendt says. "When you use chemicals to insert genes into cells, there are efficiency and price issues. Shocking the cell with electricity can cause serious damage, and it's a method that's limited to cells contained in petri dishes. The important thing is to optimize gene transfer carefully, which this gene gun can do. It has the potential to be applied directly to the treatment of human disease."

The physical transfer of genes is not a new concept; however, the technology behind the University's gene transfector is novel and offers distinct advantages over its precursors.

In the late 1980s, horticultural scientist John Sanford and his colleagues at Cornell University designed a gene gun that has been widely used, most commonly in the genetic transformation of plants. To deliver genes to the targeted cells, the Sanford gun uses a bullet-like projectile coated with gold particles that act as gene carriers.

The gun itself consists of a highly pressurized upper chamber and a low-pressure lower chamber with a diaphragm in the middle. When the diaphragm is punctured, pressure from the upper chamber emits a shock wave that hits the projectile, discharging it forward until the projectile hits a porous screen. The force of the impact launches the gold particles through the screen downstream toward the targeted cells. The particles hit the cells at a velocity high enough to puncture the cell membrane, penetrate the cells, and release the genes, which are diffused into the nucleus.

According to the University researchers, the Sanford gun has a number of disadvantages. The individual gold particles that coat the projectile are harmless, but they have a tendency to stick together and crush the cell, an effect called "pit damage."

Sometimes researchers have to shoot the same cell sample repeatedly to obtain the desired result. After each shot, the diaphragm and projectile must be reloaded and the process repeated. If there are thousands of cells to shoot, the preparation of the gene-coated carriers becomes a grueling, time-consuming process.

Although it's similar in concept to the Sanford device, the University's gun uses a very different technology, a patented process called continuous gene transfection. A gene suspension—with or without gold particles—is loaded into a capillary via a syringe. The suspension is sprayed out of the capillary at a controlled pace under a high electric field, which is created by applying voltage. The spray itself consists of highly charged and dispersed gene particles or gene-coated particles. Particles with a similar charge repel each other at a velocity high enough to penetrate the cell membrane and release the genes into the cells.

"It's a simple process," says Chen, who is credited with designing the electrospray process. "But it's just how you engineer it to get it to do what you want it to do."

The University researchers also had to be careful not to infringe upon the Sanford gun's patent. Specifically, they had to avoid a technology that used external forces to discharge the gene carriers into the cells, as the Sanford gun does. Pui and his team had to devise a different technique.

Furthermore, they had to design the gun with a specific application in mind. That issue was resolved after Wendt met Pui and Chen, who discussed their ideas for the gene gun project with her. Wendt, who was working on another project involving the effects of small particles, says she knew instantly that the gene gun had great potential for application in her field.

She worked closely with Pui and Chen in testing the gun. In one experiment, the researchers injected enhanced green fluorescent proteins (EGFP) from jellyfish into different types of cells, including African green monkey cells and human lung-cancer cells. The EGFP proteins glow when exposed to ultraviolet light, so the researchers could see whether or not the genes had entered the cells.

Experiments also proved that the gun's design and technology offered other advantages over existing methods of gene transfection. The charged particles repel each other with such force that there's no risk of pit damage. The gun's capillary-and-syringe setup permits continuous injection of the genes. In fact, Pui says, the process could run continuously for days before the gene suspension had to be reloaded. It's a method that could be adapted to a system of mass production, in which the target cells or samples are placed on a conveyor belt and transported to the spraying assembly.

"We can shoot until it's sufficient," Chen says. "[With the Sanford gun], they have to shoot sometimes five times, sometimes up to 30 times, and that can really be a headache. Plus it takes a while to prepare the gene formula on the projectile. When they have to change, it is a lengthy process."

Because their gene gun worked successfully on different types of cells, Pui and his team believe that it can be used on an even broader range of cell types. The experiments also proved that the gold particles weren't essential to the technology and that the process could be conducted on a mass scale.

The years of design and experimentation have paid off. On July 25, Pui and Chen received U.S. patent No. 6,093,557 for an "electrospraying apparatus and method for introducing material into cells." It's one of three patents Pui and Chen have received for their nanoparticle research over the past two years. The other patents, issued last fall, are for devices used to charge the particles. Another patent request is in the works, but the researchers are still cautious about discussing it.

"We've demonstrated that the devices work," Pui says. "They have been proven, and because of that, outside interest has been high." He says one local company has licensed the nanoparticle charger patents and another large Minnesota company has expressed interest in the gene gun-related devices.

Although test results so far have been very successful, the gun is still in the earliest stages of development. Pui and his team have identified several major goals for the next phase of their research, which will subject the device to more difficult challenges. Besides improving the gun's design and dosage controls, they intend to use it on more-difficult cell types and experiment with different genes.

The team also wants to expand the range of the gun's applications by making it portable. Theoretically, doctors treating lung cancer could insert a portable gun into the patient's lung and spray the diseased cells. Once inside the lung, the gun would release the gene, which the patient then could inhale.

"We need to continue to work on this gene gun project," Pui says. "Furthermore, we need to continue to work with other departments, like we did with Christine (Wendt). This work crosses many fields."

It's work that requires both the visionary's capacity for big dreams and the scientist's meticulous, ceaseless scrutiny. But above all, it's the dreaming that compels researchers like Pui, Chen, and Wendt to probe deeper into the tiny world of nanotechnology. And their research has already given them a taste of what lies ahead.

The human body consists of ~7 x 10^{27} atoms arranged in a highly aperiodic physical structure. Although 41 chemical elements are commonly found in the body's construction, CHON comprises 99% of its atoms. Fully 87% of human body atoms are either hydrogen or oxygen.

Somatic atoms are generally present in combined form as molecules or ions, not individual atoms. The molecules of greatest nanomedical interest are incorporated into cells or circulate freely in blood plasma or the interstitial fluid. The gross molecular contents of the typical human cell, which is 99.5% water and salts, by molecule count, and contains ~5000 different types of molecules. The most common molecular and cellular constituents of human blood, and their normal concentrations in whole blood and plasma. This listing is far from complete. The human body is comprised of ~10^5 different molecular species, mostly proteins — a large but nonetheless finite molecular parts list. By 1998, at least ~10^4 of these proteins had been sequenced, ~10^3 had been spatially mapped, and ~7,000 structures (including proteins, peptides, viruses, protein/nucleic acid complexes, nucleic acids, and carbohydrates) had been registered in the Protein Data Bank which at that time was maintained at Brookhaven National Laboratory. Given the current accelerating pace of improving technology, it is likely that the sequences and 3-D or tertiary structures of all human proteins will have been determined by the second decade of the 21st century.

Transporting and sorting such a broad range of essential molecular species will be an important basic capability of many nanomedical systems. The three principal methods for distinguishing and conveying molecules that are most useful in nanomedicine are diffusion transport, membrane filtration, and receptor-based transport. The Chapter ends with a brief discussion of binding site engineering.

Diffusion Transport Active Intake and Carbon Sortation

Fluidic transfer of material, known as convective-diffusive transport, can occur either by convection due to bulk flow or by diffusion due to Brownian motion. In convective transport, material is carried along fluid streamlines at the mean velocity of the fluid, with a velocity distribution such as that in Poiseuille flow. Bulk flow is customarily regarded as the most important physiological transport mechanism in the human circulation. Only for the smallest molecules, such as water or glucose, does the time required to diffuse across the width of a capillary roughly equal the time taken by a fluid element to flow the same distance (~0.02 sec). Larger molecules such as fibrinogen take ~100 times longer (~2 sec) to diffuse across one capillary width. However, bulk flow in the body is usually laminar. Transported materials travel parallel to (and thus cannot reach) fluid/solid interfaces such as the surfaces of blood vessels or membranes. Wall interactions are made possible by diffusion, a random process in which particles can move transversely to fluid streamlines in response to molecular-scale collisions. Additionally, the movement of micronscale devices within a bulk fluid flow is dominated by viscous, not inertial, forces. Molecular transport to and from such nanodevices is governed by diffusion, not by bulk flow.

Foraging nanodevices operating in aqueous environments may only modestly exceed the maximum rates of passive diffusive intake described by engaging in active physical movements designed to increase access to the desired molecules.

Nanodevices may also use diffusion to sort molecules. One of the remarkable features of diffusive sortation is that an input sample consisting of a complex mixture of many different molecular species can sometimes be completely resolved into pure fractions without having any direct knowledge of the precise shapes or electrochemical characteristics of the molecules being sorted. This can be a tremendous advantage for nanodevices operating in environments containing a large number of unknown substances. Another major advantage is the ability to readily distinguish isomeric (though not chiral) molecules. As one example of many possible, molecules suspended in water will diffuse into an adjacent region of pure water at different speeds, giving rise to dissimilar time-dependent concentration gradients which may be exploited for sortation by interrupting the process before complete diffusive equilibrium is reached. After Dt has elapsed, the gate is closed. (A gate with 10-nm sliding segments moving at 10 cm/sec closes in 0.1 microsec.) The faster-diffusing component D_2 approaches diffusive equilibrium in Chamber B, but the slower-diffusing component does not; it is present only in smaller amounts. This gives a separation factor $c_2/c_1 \sim D_2/D_1$ for each diffusion sortation unit. If n units are connected in series, with each unit receiving as input the output of the previous unit, the net concentration achieved by the entire cascade is $\sim(D_2/D_1)$. Such cascades are commonplace in gaseous diffusion isotope separation and other applications.

shows a 2-dimensional representation of an efficient design for a simple diffusion unit that might be used in a sortation cascade. Each unit consists of 5 chambers of equal volume, 7 dilating gates, 3 flap valves, 3 pistons, and 2 sieves which pass only water (or smaller) molecules. Each chamber is roughly cubical with L ~ 35 nm along the inside edge; including full piston throws and drives, controls, interunit piping and other support structures, each unit measures ~125 nm x 100 nm x 80 nm or ~0.001 micron^3 with a mass of ~10^{-18} kg.

The following is a precise description of one complete cycle of operation for each unit:

1. The cycle begins with fluid to be sorted in Chamber A, Chambers B and W full of pure water with piston W all the way out, Chambers R and D empty with pistons R and D all the way in, and all valves and gates closed.
2. Gate AB is opened for a time Dt, then closed. For a small molecule such as urea (MW = 60 daltons), Dt = 1 microsec; for a large molecule such as the enzyme urease (MW = 482,700 daltons), Dt = 35 microsec.
3. Valves AI- and AI+, and gate AR, are opened. Piston R is drawn fully out, slowly to preserve laminar flow and to prevent mixing. Fluid in Chamber A is drawn into Chamber R. Fluid passing through the DO gate of the previous unit in the cascade enters Chamber A through valve AI-. Fluid passing through the RO valve of the subsequent unit in the cascade enters

Chamber A through valve AI+. All valves and gates are closed. Chamber A is now ready for the next cycle.

4. Gates WB and BD are opened. Piston W is slowly pushed all the way in while piston D is slowly pulled all the way out. Concentrated solution in Chamber B is transferred into Chamber D as pure water in Chamber W is transferred into Chamber B, again preserving laminar flow. Both gates are closed; Chamber B is now ready for the next cycle.
5. Gates RW and DW are opened. Pistons R and D are slowly and simultaneously pushed halfway in while piston W is pulled all the way out. Forced at high pressure (~160 atm) through ~0.3 nm diameter sieve pores, half of the solvent water present in Chambers R and D is pushed into Chamber W, filling Chamber W with water. (This design allows for easy backflushing if sieve pores become clogged.) Both gates are closed; Chamber W is now ready for the next cycle.
6. Valve RO and gate DO are opened. Pistons R and D are slowly and simultaneously pushed the rest of the way in. Concentrated return fluid passes through valve RO and back to the AI+ input port of the previous unit in the cascade for further extraction. Concentrated diffusant fluid passes through gate DO and on to the AI- input port of the subsequent unit in the cascade for further purification. The valve and gate are closed; Chambers R and D are now empty and ready for the next cycle.
7. Return to Step (1). Adjacent units operate in counterphase while previous and subsequent units operate in synchrony, in a twophase system.

Increasingly purified sample passes through a multi-unit sortation cascade as described above. For small molecules, a cascade of n ~1000 units (total device volume ~1 micron^3) completely resolves two mixed molecular species with $D_2/D_1 = 1.01$. As a crude approximation, D ~ 1 / $MW^{1/3}$ for small spherical particles, so this cascade separates small molecules differing by the mass of one hydrogen atom which should be sufficient for most purposes. Structural isomeric forms of the same molecule, such as a-alanine and b-alanine, often have slightly different diffusion coefficients, thus are also easily separable using a diffusion cascade. However, stereoisomeric (chiral) forms cannot be sorted by diffusion through an optically inactive solvent like water.

For large molecules, a 1 million-unit cascade (total device volume ~1000 micron^3) provides D_2/D_1 ~ 1.00001, sufficient to completely separate large molecules differing by the mass of a single carbon atom. The fidelity of such fine resolutions depends strongly upon the ability to hold constant the temperature of the chamber, since D varies directly with temperature. Device temperature stability will be determined by at least three factors: (1) the accuracy of onboard thermal sensors in measuring T (DT/T < 10^{-6}; (2) the rapidity with which the temperature measurement can be taken, and

(3) the time that elapses between the temperature measurement and the end of the diffusive sortation process (which may be of the same order as the gate closing time, $\sim 10^{-6}$ sec).

Most of the waste heat is generated in this device by forced water sieving. To remain within biocompatible thermogenic limits ($\sim 10^9$ watts/m^3), each unit may be cycled once every ~3 millisec, a 0.8% duty cycle of a ~23 microsec sieving stroke. Subject to this restriction, each unit would consume ~1 picowatt in continuous operation. A unit presented with a ~0.1 M concentration of small molecules processes $\sim 10^6$ molecules/sec (e.g., ~1 gm/hour of glucose using 1 cm^3 of n = 1000-unit cascades), or $\sim 10^4$ molecules/sec for a unit presented with large molecules at ~0.001 M, circulating $\sim 10^9$ molecules/sec of water as working fluid while running at 340 cycles/sec. Additional chamber segments on each unit, combined with more complex diffusion circuits among the many units in a cascade, should permit the simultaneous complete fractionation of the input feedstock even if hundreds of distinct molecular species are present. By 1998, diffusion-based separation had been demonstrated in microfluidic devices.

Nano-safety and Nano-lab Accident

Some truisms: Almost any technology is subject to use, misuse, abuse, and accidents. The more powerful a technology is when properly used, the worse it is likely to be when abused. Any powerful technology in human hands can be the subject of accidents. Nanotechnology and molecular manufacturing will be no exception. Indeed, if molecular manufacturing replaces modern industry, and if its nanotechnological products replace most modern technologies, then most future accidents will have to involve nanotechnology. Another truism: In a diverse, competitive world, any reasonably-inexpensive technology with enormous commercial, medical, and military applications will almost surely be developed and used. It is hard to envision a scenario (short of the collapse of civilization) in which nanotechnology will not make its appearance; it seems inevitable. If so, then its problems, however tough, must be dealt with. Like trucks, aircraft, biotechnology, rockets, computers, boots, and warm clothes, nanotechnology has the potential for both peaceful and aggressive uses. In peaceful uses (by definition), harm to people occurs either by accident or as an unintended consequence. In aggressive uses, harm is deliberate. In a peaceful context, the proper question to ask is can fallible people of goodwill, pursuing normal human purposes, use nanotechnology in a way that reduces risk and harm to others? In an aggressive, military context, the proper question to ask is can we somehow keep the peace? Our answer to the first will be a clear yes, and to the second, an apprehensive maybe.

Throughout this discussion, we assume that most people will be alert in matters concerning their personal safety, and that some will be alert in matters concerning world safety. During the 1970s, people awakening to the new large-scale, long-term problems of technology often felt isolated and powerless. They naturally felt that technology was out of their control, in the hands of shortsighted and irresponsible groups. Today, there are still battles to be fought, but the tide has turned. When a concern arises regarding a new, obvious technology, it is now much easier to get a hearing in the media, in the courts, and in the political arena. Improving these mechanisms for social vigilance and the political control of technology is an important challenge. Current mechanisms are imperfect, but they can still give a big push in the right directions.

Though we assume alertness, alertness can be a scarce resource. The total amount of concern and energy available for focusing on long-term problems is so limited that it must be used carefully, not squandered on problems that are trivial or illusory. Part of our aim in this chapter is to help sort out the issues raised by nanotechnology so that attention can be focused on problems that must be solved, but might not be. The next few sections deal with accidents of conventional sorts, where safety benefits are obvious. Later sections discuss more novel problems, some tough enough that we have no good answers.

As countries have grown richer, their people have lived longer despite pollution and automobile accidents. Greater wealth means safer roads, safer cars, safer homes, and safer workplaces. Throughout history, new technologies have brought new risks, including risks of death, injury, and harm to the environment, but prudent people have only accepted new technologies when they offered an improved mix of risks and benefits. Despite occasional dramatic mistakes, the historical record says that people have succeeded in choosing technologies that reduce their personal risks. This must be so, or we wouldn't be living longer. Molecular manufacturing and its products should continue this trend, not as an automatic consequence, but as a result of continued vigilance, of people exercising care in picking and choosing which technologies they allow into their daily lives. Nanotechnology will give better control of production and products, and better control usually means greater safety. Nanotechnology will increase wealth, and safety is a form of wealth that people value. Public debate, product testing, and safety regulations are standard parts of this process.

In common home accidents, a dangerous product is wrongly applied, spilled, or consumed. Homes today are full of corrosive and toxic materials, for cleaning drains, dissolving stains, poisoning insects, and so forth. All too often, children drink them and die. With advanced technology, none of these tasks will require such harsh, crude chemicals. Cleaning could be performed by selective nanomachines instead of corrosive chemicals; insects could be controlled by devices like ecosystem protectors that know the difference between a cockroach and a person or a ladybug. There will doubtless be room for deadly accidents, but with care and hard work, it should be possible to ensure that nanotechnologies for the home are safer than what they replace, saving many lives. It is, of course, possible to imagine safety nightmares: nanotechnology could be used to make products far more destructive than anything we've seen because it could be used to extend almost any ability further than we've seen. Such products presumably won't be commonplace: even today, nerve gas would make a potent pesticide, but it isn't sold for home use. Thinking realistically about hazards requires common sense.

We've already seen how post-breakthrough technologies can eliminate oil spills by eliminating oil consumption. A similar story could be told of almost any class of industrial accident today. But what about accidents—spills and the like—with the new technologies? Rather than trying to paint a picture of a future technology, of how it could fail and what the responses could be, it seems better to try a thought experiment. What could be done to deal with oil spills, if oil were still in use? This will show how nanotechnologies can be used to cope with accidents: If there were a spill and oil on the shore, advanced nanomechanisms could do an excellent job of separating oil from sand, removing oil from rocks, and cleaning crude oil from feathers on birds and the feathery legs of barnacles. Oil contamination is a pollution problem, and nanotechnology will be a great aid in cleaning up pollution. But why should the oil reach the shore? Economical production would make it easy to stockpile cleanup equipment near all the major shipping routes, along with fleets of helicopters to deliver it at the first distress call from a tanker. Oil cleanup equipment built with nanotechnology could surely do an excellent job of scooping oil from the water before it could reach the shore. But why should the oil leave the tanker? Economical production of strong materials could make seamless hulls of fibrous materials far tougher than steel, with double, triple, or quadruple layers. Smart materials could even make punctures self-sealing. Hulls like this could be run into rocks at highway speeds without spilling oil.

But why should anyone be shipping crude oil across the sea? Even if oil were still being pumped (despite inexpensive solar energy and solar-derived fuels), efficient molecule processing systems could refine it into pure, fluid fuels at the wellhead, and inexpensive tunneling machines could provide routes for deeply buried pipelines.

Any one of these advances would shrink or eliminate today's problem with oil spills, and all of them are feasible. This example suggests a general pattern. If nanotechnology can provide so many different ways to avoid or deal with an oil spill—one of the largest and most environmentally destructive accidents caused by today's industry—it can probably do likewise for industrial accidents in general. The most direct approach is the most basic: the elimination of anything resembling today's bulk industrial plants and processes. The shift from messy drilling activities and huge tankers to small-scale distributed systems based on solar cells is characteristic of the style in which nanotechnology can be used. The chemical industry today typically relies on plants full of large, pressurized tanks of chemicals. Not surprisingly, these occasionally spill, explode, or burn. With nanotechnology, chemical plants will be unnecessary because molecules can be transformed in smaller numbers, as needed and where needed, with no need for high temperatures, high pressures, or big tanks. This will not only avoid polluting by-products, but reduce the risk of accidents.

Medicine can be safer too. Drugs often have side effects that can do permanent damage or kill. Nanomedicine will offer alternatives to flooding the body with a possibly toxic chemical. Often, one wants to affect just one target: just the stomach, or perhaps just the ulcer. An antibiotic or antiviral treatment should fight specific bacteria or viruses and not damage anything else. When medicine achieves the sophistication of immune machines and cell-surgery devices, this will become possible. But what about medical accidents and side effects? Molecular manufacturing will make possible superior sensors to tell medical researchers of the effects of a new treatment, thereby improving testing. Better sensors will also help in monitoring any negative effects of a treatment on an individual patient. With care, only a few cells would be damaged and only small concentrations of toxic by-products would be produced before this was noticed and the treatment corrected. The resources of nanotechnology-based medicine would then be available for dealing with the problem. With biostasis techniques available, even the worst medically induced illnesses could be put on hold while a treatment was developed. In short, serious medical mistakes could be made far rarer, and most mistakes could be corrected.

The conclusion that follows from these examples of oil spills, chemical plants, and the effects of medical treatments is straightforward. Today our comparative poverty and our comparative technological incompetence press us in the direction of building and using relatively dangerous and destructive devices, systems, and techniques. With greater wealth and technological competence, we will have the option of accomplishing what we do today (and more) with less risk and less environmental destruction: in short, being able to do more, and do it better. With better-controlled technologies, and with an ample measure of foresight and concern, we will even be able to do a better job of recovering from mistakes. It won't happen automatically, but with normal care we can arrange for our future accidents to be smaller and less frequent than those in our past.

The previous section discussed ordinary accidents that would occur during the use of nanotechnology by generally responsible, yet fallible, human beings. Nanotechnology also raises the specter, however, of what have been termed "extraordinary accidents": accidents involving runaway self-replicating machines. One can imagine building a device about the size of a bacterium but tougher and more nearly omnivorous. Such runaways might blow like pollen and reproduce like bacteria, eating any of a wide range of organic materials: an ecological disaster of unprecedented magnitude—indeed, one that could destroy the biosphere as we know it. This may be worth worrying about, but can this happen by accident?

The so-called "Star Trek scenario" (named after an episode of Star Trek: *The Next Generation* that featured runaway "nanites") is perhaps the most commonly imagined problem. In this scenario, someone first invests considerable engineering effort in designing and building devices almost exactly like the one just described: bacterial-sized, omnivorous, able to survive in a wide range of natural environments, able to build copies of themselves, and made with just a few built-in safeguards—perhaps a clock that shuts them off after a time, perhaps something else. Then, accidentally, the clock fails, or one of these dangerous replicators builds a copy with a defective clock, and away we go with an unprecedented ecological disaster. This would be an extraordinary accident indeed. Note well, though, that this accident scenario starts with someone building a highly capable device that is almost disastrously dangerous, but held in check by a few safeguards. This would be like wiring your house with dynamite and relying on a safety-catch to protect the trigger: a subsequent explosion could be called an accident, but the problem isn't with the safety mechanism, it's with the dynamite installation. Do we need to build nanotechnological dynamite? It's worth considering just how little practical incentive there is for anything even resembling the dangerous replicator just described. (Note that our topic here is accidents; deliberate acts of aggression are another matter.)

With our present technology, which is simpler to build—a car that runs on gasoline, or one that forages for fuels in the forest? A foraging car would be very hard to design, cost more to manufacture, and have more parts to break down. The situation is similar with nanotechnology. Ralph Merkle of Xerox Palo Alto Research Center discussed this issue at the First Foresight Conference on Nanotechnology. He explains, "It's both uneconomical and more difficult to design a self-replicating system that manufactures every part it needs from naturally occurring compounds. Bacteria do this, but in the process they have to synthesize all twenty amino acids and many other compounds, using elaborate enzyme systems tailored specifically for the purpose. For bacteria facing a hostile world, the ability to adapt and respond to a changing environment is worth almost any cost, for lacking this ability they would be wiped out." "But in a factory setting, where adequate supplies of all the needed parts are provided, the ability to synthesize parts from scratch is not only unneeded, it consumes extra time and energy, and produces excess waste. Even if we could design artificial self-replicating systems as flexible as existing natural ones, an inflexible and rigid system is better adapted to the controlled factory setting in which it will find itself than a more complex, more adaptable, less efficient design." What is more, the Desert Rose Industries scenario showed how an expandable factory setup could operate with no self-replicating machines at all: molecular manufacturing doesn't require them. If they are used for some purpose, they will most likely resemble automobiles in their finicky requirements. A self-replicating molecular machine built for industrial purposes (and made as simple as possible) would float in a container of specially selected chemicals. As with the automobile, the best chemicals to use will probably be chemicals not commonly found in nature, and it would be easy to make that a design rule: *never make a replicator that can use an abundant natural compound as fuel.*

If we follow this rule, the idea of a replicator "escaping" and replicating in the wild will be as absurd as the notion of an automobile going feral and refueling itself from tree sap. Whether for replicators or cars, to design a machine that could operate in the wild would not be a matter of a flick of the draftsman's pen, but an intense, sustained research-and-development effort focused on that objective. Crashes and explosions occur in machinery by accident, but complex new capabilities don't. A simple psychological error frequently occurs when someone first hears about nanotechnology, and hears mention of "molecular machines," and "replicators," and "nanocomputers," and "nanomachines that operate in nature." The error is this: the person makes a single new mental pigeonhole for "nanotechnology," throws everything into it, and stirs. After some mental fermentation, the result is the mythical nanomachine that does everything: it's a replicator, it's a supercomputer, it's a Land-Rover, it slices, it dices—and on reflection, this imaginary nanomachine sounds uncontrolled and dangerous. With enough effort, a do-it-all nanomachine could perhaps be built, but it sounds difficult and there's no good reason to try. There are advantages to making systems of molecular machinery that can use inexpensive, abundant chemicals, and devices that can operate in nature, but these machines needn't be replicators. A facility like Desert Rose might be designed to use little but electric power from solar panels and molecules from the air, but a setup like this isn't going to slip away. Nanomachines built for cleaning up pollutants and other outdoor tasks could be manufactured in facilities run like Desert Rose and then spread or installed where they're needed. Extraordinary accidents deserve attention, but with a little care they can be completely avoided. The incentive to build anything resembling a Star Trek-scenario replicator is negligible, even from a military perspective. Any effort toward building such a thing should be seen not as a use of nanotechnology, but as an abuse. Other abuses seem more likely, however, and are quite bad enough.

The chief danger of nanotechnology isn't accidents, but abuse. The safety benefits of nanotechnology, when used with normal care, will free some of our attention to grapple with this far more difficult problem. As Lester Milbrath observes, "Nanotechnologies have such great power that they could be used for evil or environmentally destructive purposes as easily as they could be used for good and environmentally nourishing purposes. This great danger will require a level of political control far beyond that which most nations know how to exercise. We have a prodigious social learning task that we must face." Thus far, we've focused on how increased abilities can serve constructive ends. Not surprisingly, the potential consequences—with the huge exception of social and economic disruption—are overwhelmingly positive. Inherently clean, well-controlled, inexpensive, superior technologies, when applied with care, can yield far better results than inherently dirty, messy, costly, inferior technologies. This should come as no surprise, but it is only half of the story. The other half is the application of those same superior technologies to destructive ends. Readers feeling that all this may be too good to be true can breathe a sigh of relief. This problem looks tough.

Molecular manufacturing will lead to more powerful technologies, but our current, crude technology already has world-smashing potential. We have lived with that potential for decades now. In the coming years, we will need to strengthen institutions for maintaining peaceful security. If most of the political power in the world, and with it most of the police and military power, sees that the course of self interest lies in peace and stability, then solutions seem possible. (The prospect of an arms race in nanotechnology is terrifying and to be avoided at almost any cost. As of this writing, the end of the Cold War offers a better hope of avoiding this nightmare.) James C. Bennett, a high-tech entrepreneur and public policy commentator affiliated with the Center for Constitutional Issues in Technology, explains the goal: "Advanced technologies, particularly as far-ranging a capability as nanotechnology, will create a strong demand for their regulation. The challenge will be to create sufficient controls to prevent the power-hungry from abusing the technologies, without either smothering development or creating an overbearing international regime." In the coming decades, preventing major abuse of nanotechnology will take the form of regulation, arms control, and antiterrorist activities. In the field of arms control, nanotechnology should present strong motivation for international cooperation and for intimate mutual inspection in the form of joint research programs. The sheer productive capabilities of molecular manufacturing will make it possible to move from a working weapons prototype to mass production in a matter of days. In a more exotic vein, dangerous nanomachines could be developed, including programmable "germs" (replicating or nonreplicating) for germ warfare. Either development could bring war. With peace looking so profitable and an arms race looking so dangerous, arms control through cooperative development should look attractive. This does not make it easy, or likely. Terrorism is not an immediate concern. We have lived with nuclear weapons and nerve gas for decades now, and nerve gas, at least, is not difficult to make. As of this writing, no city has been obliterated by terrorists using these means, and no terrorist has even made a credible threat of this sort. The citizens of Hiroshima and Nagasaki, like the Kurds in Iraq, fell victim to nuclear and chemical weapons wielded by governments, not small groups. So long as nanotechnology is technologically more challenging than the simple chemistry of nerve gas, nanoterrorism should not be a primary concern. To keep dangerous nanotechnologies unavailable, however, will require regulation. If anyone were free to build anything using molecular manufacturing, then someday as the technology base improves and designs become available for more and more nanodevices, someone, somewhere—if only out of sheer spite—would figure out how to combine those nanodevices to make a dangerous replicator and turn it loose. There will almost surely be warning signs, however: In the natural course of events, causes attract protesters before stone-throwers, and produce letter bombs before car bombs. Abuse of nanotechnology is likely to be visible long before it is devastating, and this at least gives some time to try to respond.

Abuse of this sort can be delayed, perhaps for a long time, by proper regulation. The goal here isn't to make regulations so tight that people will have to violate them to get anything done. This would encourage holdouts, underground work, and disrespect for the law. Instead, the goal is to draw boundaries loosely enough to cause little difficulty for legitimate work, while making dangerous activities very difficult indeed. This is a delicate balance to strike: those fearful of risk naturally try to apply crude and oppressive regulations, and companies naturally try to loosen and avoid regulation entirely. Nonetheless, the problem must be solved, and this seems the best direction to explore. In one approach, nanomachines could be divided into two classes: *experimental devices* and *approved products*. Approved products could be made widely available through special-purpose molecular manufacturing systems. Thus, once an experimental device had passed regulatory inspection, it could become inexpensive and abundant. In this way, popular demands for a product could be satisfied without anyone needing to break safety rules.

Approved products could include devices like (but superior to) the full range of modern consumer products, ranging from personal supercomputers with 3-D color displays, through smart construction materials, to running shoes with truly amazing features. The main cost of such goods might be the royalty to the designer. In *Engines of Creation* (the first book to examine this topic), this strategy for producing and distributing approved products is called a "limited assembler system." Note that both approved products and the limited assemblers that build them would lack the ability to make copies of themselves, to self-replicate. Ralph Merkle sees this ability as the one to keep an eye on: "Self-replicating systems can and should be appropriately regulated. There seems no need, however, to have any more than normal concerns for devices which cannot replicate. While we might, as with any device, need laws to ensure their appropriate use, they pose no extraordinary problems." For most products, normal medical, commercial, and environmental standards would apply; the regulatory bureaucracies are already in place. There are great advantages to permitting nearly free experimentation in a new technology, allowing creative people to try ideas without seeking prior approval from a cumbersome committee. Surprisingly, this, too, seems compatible with safety. In the world of nanotechnology, one cubic micron is a large space, with room enough for millions of components. For many purposes, a few cubic microns would amount to a large laboratory space. To a device on a micron scale, a centimeter is an enormous distance. Surrounding a micron-scale device with a centimeter-thick wall would be like surrounding a human being with a wall kilometers thick, and just as hard to penetrate. Further, a micron-scale device can be incinerated in an instant by something as small as a spark of static electricity. Based on observations like these, *Engines of Creation* outlined the idea of a sealed assembler lab, in which a researcher could build anything, even something deliberately designed to be dangerous, and yet be unable to get anything out of the microscopic sealed laboratory except for information.

With a good communications network, a researcher or product developer in Texas could equally easily perform experiments in a remote Maine laboratory run with the security and secrecy of a Swiss bank. A lab would have a responsibility to its customers to keep proprietary work confidential, and a responsibility to regulatory authorities to ensure that nothing but information leaves the laboratory. Researchers could then perform any small-scale experiments they wish. Only approved products, of course, would be built outside the sealed laboratories. While this may not be the best pattern of regulation possible, it does show one way in which freedom of experimentation could be combined with strict regulation of use. By providing a clear separation between legitimate and illegitimate activity, it would help with the difficult problem of identifying and preventing research aimed at damaging ends. A sensible policy will have to balance the risk of private abuse of technology against the risk of government abuse of technology and regulation. Low-cost manufacturing can make surveillance equipment less expensive. Increased surveillance can reduce some risks in society, but the watchers themselves often aren't very well watched. Placing bounds on surveillance is a challenge for today's citizens as well as tomorrow's, and lessons learned in the past can be applied in the future. In the long run, it seems wise to assume that someone, somewhere, somehow, will escape the bounds of regulation and arms control and apply molecular-manufacturing capabilities to making novel weapons. If by then we have had several decades of peaceful, responsible, creative development of nanotechnology (or perhaps a few years of help from smart machines), then we may have developed both ecosystem protectors and sophisticated immune machines for medicine. There is good reason to think that distributed technologies of this sort could be adapted and extended to deal with the problem of protecting against novel nanoweaponry. Failure to do so could mean disaster. Nonetheless, building protective systems of this sort will be by far the greatest challenge of any we have discussed. The chief purpose of regulatory tactics like those we have described must be to buy time for those peaceful developments, to maximize the chances that this challenge can be met before time runs out.

Potential accidents richly deserve the attention they will get, and we have confidence that this attention will suffice to make nanotechnology a force for improved human and environmental safety. Abuse is the greater danger, and harder to deal with. When considering a proposed policy, the first question should be, "How will this affect the long-term likelihood of serious abuse?"

Guiding a technology is a complex task. It means grappling with myriad decisions regarding which applications are beneficial and which harmful in such complex areas as medicine, the economy, and the environment. It means making such happy choices as which of several good approaches to apply in cleaning up toxic-waste dumps and reversing the greenhouse effect. It also means making more difficult choices in planning ecosystem restoration and environmental modification. These problems will confront us with a range of choices better than we have today, yet choices that throw values into conflict. Which is a better use for a particular piece of land—the slow restoration of a wilderness ecosystem, or development as a recreational park? Either may be far better than pavement, strip mines, and dumps, but the choices will be controversial. Likewise, in medicine, we will have a choice of developing many different ways to cure cancers, many different ways to cure heart disease, many different ways to cure AIDS. But the technologies that can be used to rebuild damaged heart muscle could be extended to rebuild muscle and connective tissue structures elsewhere in the body, without the harmful side effects of steroid drugs. The range of choices open to people will be enormous, and again will be cause for great debate. When a new medical technology is discussed today, a frequent comment is, "This procedure raises ethical questions." This is often taken as a signal to delay its use, neglecting such ethical questions as "Is withholding this lifesaving treatment while we ponder akin to murder?" When a choice raises ethical questions or throws values into conflict, it is time to make an ethical decision or to step aside and let others choose for themselves. Deciding to avoid whatever raised the question is itself a decision—and often ethically indefensible. New technologies will face us with uncomfortable decisions, but so does life itself. Setting up rules for nanotechnology development will be challenging: finding ways to maximize research freedom while preventing serious abuse and making this stick worldwide is a social challenge of the first rank. Beyond this are decisions regarding rules for its application, and the challenge of maximizing freedom of choice and action while preventing serious abuse, again worldwide. To guide nanotechnology means grappling with a set of decisions that could ultimately remake much of the world—for the better if we are reasonably wise, or for the worse if we are too blundering and incautious. To avoid this responsibility (if we could) would be tempting, yet given the environmental and human stakes, it would, perhaps, be a wrong of historic proportions.

The simplest imaginable approach to "guiding" nanotechnology would be to stop it. The easiest trip to plan is the trip that goes nowhere. This would have a certain appeal, if it were possible. Because of its enormous potential for abuse, nanotechnology has the potential of doing great harm. If we believe that human beings and human institutions are too incompetent to deal with nanotechnology—that they are too likely to turn it to aggressive military use, or too likely to make it freely available to madmen—then the option of stopping the development of nanotechnology may seem attractive indeed. But the ethical question that must guide human actions is not "Would it be better to stop?", but "Would attempts to stop make things better?" One option is to push forward, emphasizing the need for caution but also the potential for good applications. The promise of medical, economic, and environmental applications, joined with the threat posed by a new arms race, provides a powerful motive for international cooperation. With positive goals and an inclusive stance, international cooperation is a promising strategy; it could provide a basis for guiding the development and application of nanotechnology. Another option would be to emphasize the downside, to focus debate on potential abuses in support of a campaign to halt development. In following this strategy, an activist group would want to downplay the civilian applications of nanotechnology and emphasize its military applications. Horror stories of potential abuse (including abuses that regulation could easily prevent) would help to make the technology seem strange and dangerous. This strategy might succeed in suppressing civilian research in many countries, though probably not all. Unfortunately, it would also guarantee funding for classified military research programs in laboratories around the world, even in the most morally honest countries, because of their then-inevitable fear of the consequences if someone else developed nanotechnology first. In a hostile public atmosphere, research would be pushed into secret programs, and in secrecy the prospects for broad international cooperation would disappear. Attempts to stop nanotechnology for fear of a new, unstable arms race become self-fulfilling prophecies. Afterward, the advocates of this view could then say, "We warned you!" as the world slid toward a war they themselves had helped to prepare. Attempting to stop technological development is a simple but dangerous idea. The greater its success, the greater the polarization it would cause between technology advocates and technology critics. A moderate success would push research out of the public universities and into corporate and military research labs. A greater success would push research out of the corporate laboratories and into heavily classified programs. A truly amazing success would end most of these, leaving the only remaining military programs in the hands of those states with thoroughly repressive governments or alien ideologies. This, presumably, is not how one would prefer nanotechnology to be developed. The only genuine success would be a total success, and this would mean banning research not only in the United States, and Germany, and France, and the rest of Western Europe, and Japan, and the Soviet Union, and the People's Republic of China, and Taiwan, but in Korea, South Africa, Iran, Iraq, Israel, Brazil, Argentina, Vietnam, and the part of Colombia controlled by the Medellín Cartel. Later, as computers improve, as chemistry advances, as more and more proximal probe microscopes are built by high school students, total success would require banning kids from tinkering in suburban garages in Pittsburgh.

Competitive pressures are pushing technology toward thorough control of matter, and we have seen that this goal can be reached by many different paths. Preventing one area of research would not prevent the advance, nor would stopping work in one country. When the United States delays drug development through strong regulation by the FDA, drug companies simply switch research overseas, or non-U.S. companies pull ahead. Orbital-launch capability and nuclear weapons capability are other examples. Very seldom has one country given these abilities to another, yet at least eight nations are able to launch satellites to orbit independently, at least seven have detonated nuclear devices, and another two are suspected to be within easy reach of nuclear capability. India and Israel have built bombs and launched satellites, though neither is considered a great power or a leading force in world technology. Where nanotechnology is concerned, many countries are capable of doing the required research, and more will be in the future. South Korea has both the needed educational levels and the ambition; visitors from the People's Republic of China ask about nanotechnology. A decision at the top directing the resources of a nation could get results almost anywhere. The United States is only gradually being shaken from its illusion that it rules the world of technology. This illusion is a poor basis for decisions and action.

For all practical purposes, nanotechnology seems inevitable. With work, it can be made beneficial, but only if we exercise ordinary care in avoiding accidents and extraordinary care in preventing abuse. It's hard to get people to take future technologies seriously. Present-day problems dominate discussions, and ideas about future possibilities take effort to judge. Because of this inertia, broad international regulation of nanotechnology won't be possible until nanotechnology already exists, until people begin to see its results. And then, for regulation to be most effective, researchers and governments in many countries will need to cooperate and be on speaking terms with the technology's critics. What, then, is the socially responsible course of action, the approach most likely to avoid serious abuse of nanotechnology and most likely to deliver some of its potential benefits? It is, we believe, to point out potential dangers and abuses and how they can be avoided, but also to emphasize the civilian applications in medicine, the environment, and the economy. It is these benefits that provide grounds for advocating open civilian development programs, and for international cooperation that can provide a basis for effective international guidance. To guide nanotechnology will not be simple. We will be confronted with a range of choices greater than we have faced before in history. It is only by grappling with those choices that we will be able to affect them for the better.

Dynamic Pore Sizing and Gated Channels

A more efficient nanosieve system can be designed if pore size and shape can be actively modified during device operation, as for example by exchanging filters (from a membrane library stocking various pore sizes) each half cycle. Better, if pores can be reliably dilated or constricted in place during a period of time $Dt << t_p$, then filtration cascades can be more rapidly reconfigured to match changing input feedstock characteristics or to extract varying selections of desired molecules at will. Additionally, fully differentiating sieving cascades can be collapsed into a single unit, providing more compact devices especially useful in chemical sensor systems requiring preconcentration of sample. Control of pore shape should also provide finer discrimination among molecules of similar size but different shape, such as some isomers of nonchiral molecules.

Two or more overlapping surfaces containing regular arrays of perforations of fixed size and shape can conveniently generate a wide variety of pore geometries. Control of pore geometry is achieved by sliding or rotating one surface relative to the other surface by a small increment, as suggested schematically..

Circular dilating apertures can also be constructed using a matched set of overlapping segments, which may be driven either radially or tangentially to enlarge or contract the hole like an irising camera diaphragm, consistent with Akey's model of the nuclear pore complex present on the cell nucleus surface. Diaphragming mechanisms may be vertically staggered to maximize areal hole density in filtration surfaces (at the cost of increased vertical rugosity). Filters constructed of hydrogen-passivated diamondoid can have pores with <0.1 nm feature sizes, although H-free fullerene materials might avoid any possibility of dehydrogenation shearing.

Methods of positioning surfaces to accuracies of ~0.01 atomic diameter (~0.001 nm) are discussed. Assuming pore sizing blades require ~25 nm^2 of diamondoid contact surface per pore and each blade travels 25 nm at 0.01 meter/sec during one cycle, sliding friction dissipates ~0.01 zJ/pore, or ~4 x 10^{-18} watts/pore during each 2.5 microsec resizing cycle. Since fluid friction approaches kT for nanometer-size holes changing size in ~10^{-9} sec, maximum blade speed is ~1 m/sec and the fastest resizing cycle is ~10^{-8} sec.

A single sieving unit with controllable pores can be moderately efficient. Consider a design similar to that described in, except for a separate chamber and piston on either side of the filter block. Suppose that the particles desired to be extracted are of radius r, and the next smallest possible pore size is r - Dr. The device operates in two phases. In the first phase, the sample is placed in the first chamber, pore size is set equal to r, then the first piston forces the fluid through the membrane. Particles larger than r remain behind and are flushed from the first chamber. The pores then contract to r - Dr, and the second piston pushes the remaining filtrate back into the first chamber. After this second phase, particles of radius ~(r ± Dr) remain in the second chamber at significantly higher concentration and may be removed for further use. Analogous double-sieve editing paradigms are commonplace in biological systems.

Other designs might work equally well, such as a 3-chamber flowthrough design using a variable pore membrane with pores of size r between the first and second chamber, a membrane with pores of size r - Dr between the second and third chamber, and a piston at either end (one pushing, one pulling), thus concentrating molecules of size r ± Dr in the central chamber. M. Krummenacker suggests fixed chambers with a moving sieve operated as a dragnet. Filtration processes may be most useful in performing complete separations of complex mixtures. But they are inefficient in the sense that the energy expended to orient molecules passing through pores is wasted if the molecules are allowed to randomize on the other side; a eutactic mill-like molecule handling system might preserve this order and greatly improve energy efficiency.

Besides controlling nanopore size and shape, individual molecular transport channels can be gated either mechanically (e.g., ligand gating) or electrically (e.g., voltage gating). Either method might usefully be employed to control molecular transport through the surfaces of medical nanodevices in a process that could very loosely be described as molecular transistor gating.

A good example of mechanical gating in biology is the nicotinic acetylcholine receptor channel, probably the best understood ligand-gated channel. Nerve impulses are communicated across neuromuscular junctions and autonomic ganglia via neurotransmitters such as acetylcholine. STM images confirm that the receptor itself is cylindrical, a bundle of 5 rod-shaped polypeptide subunits arranged like barrel staves with outside diameter ~6.5 nm. The receptor protrudes 6 nm on the synaptic side of the membrane and 2 nm on the cytoplasmic side. The water-filled channel pore lies along the symmetry axis, lined by 5 a-helices, with a 2.2 nm wide mouth on the synaptic surface, a 0.65 nm waist where the structure dives through the cell membrane, and a 2 nm wide cytosolic exit.

Normally, the channel is closed and no ions may pass. In this closed state, the channel is occluded at the waist by a ridge of large residues forming a tight hydrophobic ring. Each subunit has a bulky leucine at the bend in the a-helix, a critical position. When two acetylcholine molecules bind to the receptor, these helices allosterically tilt, shifting the position of the ridges. The pore becomes open because it is now lined with small polar residues rather than by large hydrophobic ones. This conformational change allows 2.5 x 10^7 Na^+ ions/sec to flow through the channel, about 10% of the diffusion-limited rate. (Anions like Cl^- cannot enter the pore because they are repelled by rings of negatively charged residues positioned at either end of the receptor.)

Acetylcholine binding opens the gate in less than 100 microsec under physiologic conditions. Subsequent rapid destruction of acetylcholine by acetylcholinesterase, an enzyme tethered to the membrane surface by a covalently attached glycolipid group, closes the gate in ~1 millisec. Much faster gating action (~10^{-8} - 10^{-6} sec) could be achieved by nanodevices operating variable-scale nanopores in response to sensor data or other control signals. Such signals could drive the insertion or retraction of diamondoid rods, wedges, or trapdoors across the channel lumen to regulate the transmission of molecules having specific sizes, shapes, and charge distributions.

Transport channels through nanodevice surfaces may also be gated electrically. In contrast to the acetylcholine receptor, which is relatively nondiscriminating and allows both inorganic and organic cations to pass, the voltage-gated calcium channel has a highly discriminating mechanism with a Ca^{++}:Na^+ permeation ratio on the order of 1000:1. (The high specificity of the voltage-gated Ca^{++} channel is a consequence of a single-file pore mechanism involving a pair of specific Ca^{++} ion binding sites. Selectivity is assured if either of the two sites is occupied by Ca^{++}, as monovalent ions do not bind strongly enough to the free site or generate sufficient electrostatic repulsion to push the first Ca^{++} ion through the channel.) Potassium channels are 100 times more permeable to K^+ than to Na^+, and sodium channels favor the passage of Na^+ over K^+ by a factor of 12. All three of these voltage-gated channels are important in the generation and conduction of neural action potentials.

A nerve impulse is an electrical signal produced by the flow of ions across the plasma membrane of a neuron. Neuron interiors have high concentrations of K^+ and low concentrations of Na^+. The resting potential of a neuron is 60 mV. An action potential may be generated when the membrane potential is slightly depolarized to 40 mV. This opens the Na^+ voltage-gated channels, rapidly accelerating depolarization to a peak of +30 mV in ~1 millisec. Then Na^+ channels close and K^+ channels open, allowing K^+ ions to exit the cell, restoring the 60 mV resting potential. Only ~1 ion of every ~10^6 Na^+ and K^+ ions present in the local extracellular medium and the axoplasm participate in each such nerve impulse.

The sodium channel is a single polypeptide chain with four repeating units. Each repeating unit folds into six transmembrane alpha helices, including one that is positively charged called the S4 helix. The S4 helix is the voltage sensor that triggers the opening of the gate. Three positively charged residues on each S4 helix are paired at the resting membrane potential with negative charges on other transmembrane helices in a staircase geometry. The initial small depolarization event produces a spiral motion of each S4 accompanied by the net movement of one or two charges to the extracellular side of the membrane, essentially turning this left-handed hydrogen-bonded "molecular screw" through a ~60° rotation. This outward 0.5-nm translation of the four S4 segments opens the sodium gate by removing a steric barrier to ion flow. The energy cost of moving ~6 electrical charges (~10^{-18} coul) from the cytosolic to the extracellular side of the membrane against a ~100 mV potential (thus opening the gate in ~75 microsec) is ~100 zJ. Quantum tunneling activation of sodium channels, taking 1-1000 microsec, has been analyzed by Chancey.

Artificial ion-gated polymer membranes were reported in 1982, protein engineering of switchable pore-forming proteins is well-known, and "intelligent gels" are being developed that can change size and molecular porosity in response to chemical, electrical or thermal stimuli. In 1998, however, electroporation was a more commonly used method in biological research and a useful technique for "transfecting" cells in genetic studies. Electroporation employs a brief intense pulse of electricity to provide a force that opens cellular pores, enabling the insertion of macromolecules like DNA into cells of interest; laser pulses reduce cell loss to 10% by using a square-wave pulse to effect rapid and reversible pore formation. Artificial pH-gated 200-nm diameter nanopores and natural pH-gating of virion pores in the cowpea chlorotic mottle virus were demonstrated in 1998.

The first true voltage-gated nanomembrane was fabricated by Charles Martin and colleagues in 1995. This membrane consists of cylindrical gold nanotubules with inside diameters as small as 1.6 nm. When the tubules are positively charged, cations are excluded and only negative ions are transported through the membrane. When the membrane receives a negative voltage, only positive ions are transported through the tubules. Nanodevices may combine voltage gating with pore size and electrosteric constraints to achieve precision transport control with moderate molecular specificity at diffusion-limited throughput rates. In 1997, an ion channel switch biosensor with sub-picomolar sensitivity and quantitative detection time of ~600 sec was demonstrated by an Australian research group.

Transporter Pumps & Sorting Rotors

While gated channels enable ions to flow rapidly through membranes in a thermodynamically downhill direction, active pumps use a source of free energy to force an uphill transport of ions or molecules. In biology the energy supply is usually ATP or photons of light; medical nanomachines can make use of vastly more diverse energy sources. Actually, the term "pump" may be something of a misnomer because the action is highly specific — only one or a very small number of molecular species are selectively transported.

Molecular pumps generally operate in a four-phase sequence: (1) recognition (and binding) by the transporter of the target molecule from a variety of molecules presented to the pump in the input substrate; (2) translocation of the target molecule through the membrane, inside the transporter mechanism; (3) release of the molecule by the transporter mechanism; and (4) return of the transporter to its original condition, so that it is ready to accept another target molecule. Such molecular transporters that rely on protein conformational changes are ubiquitous in biological systems, and are illustrated schematically.

Transporter pumps need not be limited to the movement of a single molecular species in a single direction, which biochemists call a uniport transport mechanism. Numerous well-known biological systems are capable of moving two molecules simultaneously in one direction (symport mechanisms), two molecules sequentially in opposite directions (antiport mechanisms), and charged molecules in one direction only, thus building up an electrical charge on one side of the membrane (electrogenic mechanisms). Such pumps exist in nature for numerous ions, amino acids, sugars, and other small biomolecules. Active drug efflux systems and multidrug resistance are made possible by the expression of bacterial genes coding for molecular pumps that are constantly evolving new specificities to increasing numbers of microbicidal drugs.

The action of the Na^+ - K^+ antiporter, a familiar ion pump present in all mammalian cells, is illustrated. Three Na^+ and two K^+ ions are transported per 10 millisec cycle, requiring the hydrolysis of one ATP molecule to ADP to drive the conformational changes. (More than one-third of the ATP consumed by a resting animal is used to pump these two ions.) Hydrolysis of ATP liberates ~80 zJ/molecule of free energy, so the antiporter is transporting Na^+ and K^+ at a cost of 16 zJ/ion at a 0.5 KHz frequency. Pump site density is ~1000/micron^2 of cell membrane in neural C fibers. (The Na^+ - K^+ pump can also be operated in reverse to synthesize ATP from ADP by exposing the mechanism to steep ionic gradients.) By contrast, artificial nanomechanical antiporter and symporter devices will operate at MHz frequencies. They should be able to transport much larger molecules, and may also be fully reversible.

Drexler's molecular sorting rotor is a related class of nanomechanical device capable of selectively binding molecules from solution and then transporting these bound molecules against concentration gradients. The archetypal sorting rotor is a disk with 12 binding site "pockets" along the rim exposed alternately to the external solution and interior chamber by axial rotation of the disk. (Other designs may have more, or fewer, pockets.) Each pocket selectively binds a specific molecule when exposed to the solution. Once the binding site rotates to expose it to the interior chamber, the bound molecules are forcibly ejected by rods thrust outward by the cam surface (or using some other means by which receptor affinity can be adjusted during the inbound transport process). In the case of protein molecules, the debinding geometry must be carefully designed to avoid denaturation during ejection. Also, the rotor implicitly assumes that target molecules remain in the liquid or gaseous state after importation. M. Krummenacker observes that most bloodborne molecular species will precipitate as solids unless they are well-solvated; thus nanomedical sorting systems may require internal solvent or in some cases should be made completely eutactic. The discovery of positionally disordered water molecules resident inside protein hydrophobic cavities suggests that good rotor designs may also need to include solvent drainage channels.

Molecular sorting rotors can be designed from about 10^5 atoms (including housing and pro rata share of the drive system), measuring roughly 7 nm x 14 nm x 14 nm in size with a mass of 2×10^{-21} kg. Rotors turn at ~86,000 rev/sec with a conservative rim speed of 2.7 mm/sec and an almost negligible drag power of $\sim 10^{-16}$ watts against the fluid, sorting small molecules at a rate of 10^6 molecules/sec with laminar flow.The energy cost of small-molecule sortation at 310 K ranges from ~10 zJ/molecule at low pressures ($c_2/c_1 = 10$) up to ~40 zJ/molecule when pumping against the highest head pressures ($c_2/c_1 = 10^4$, ~30,000 atm for natural bloodstream concentrations of salt with osmotic $p_p \sim 3$ atm), consuming 0.01-0.04 pW per device in continuous operation. Rotors are fully reversible, so they can be used to load or unload target molecules depending on the direction of rotor rotation. Cylindrical rotors with many receptor rows are somewhat more energy-efficient, and rotor lifetimes should be $>10^6$ sec.

Typical molecular concentrations in the blood for target molecules of nanomedical interest are $\sim 10^{-11}$ - 10^{-3} molecules/nm^3, which should be sufficient to ensure ~99% occupancy of rotor binding sites. Rotors targeting serum hormones and other low-concentration species at the parts-per-billion level must slow to <1 rev/sec to ensure complete receptor occupancy and to avoid exceeding diffusion limits.

Sorting of ionically charged species can use binding sites that display opposite charge, effectively increasing the affinity of the binding site for the charged species. Many ionic species dissolved in water are actually more complex than their symbolic representation would suggest. As an example, a naked proton (H^+) in water is always highly hydrated, usually as $H_5O_2^+$ or $H_7O_3^+$, or even as $H_9O_4^+$ in strong acid solutions. Small ions from Li^+ to I^- are also found in solvent cages, bound with energies ≥ 330 zJ relative to vacuum; for instance, Li^+ and I^- are coordinated to 46 water molecules. As a result, the design of an appropriate binding site or filtration process for such ions can likewise become more complex. The existence of biological channels that show remarkable selectivity for specific ions provides one design approach for dealing with such species.

Drexler has also proposed cascades of sorting rotors to achieve high fidelity purification and a contaminant fraction of $<10^{-15}$. However, since only $\sim10^{10}$ small molecules can be stored in 1 micron^3 volume (typical for nanomedical devices), 100% process purity requires a contaminant fraction of only $<10^{-10}$ which can be ensured in micron-scale nanosystems using at most 5 stages, starting from a dilute input substrate containing only 1 part per billion of the target molecule with each stage providing a concentration factor of $\sim10^4$. For statistically pure extractions of more common molecules in blood and cytoplasm, a 3- or 4-stage cascade will usually suffice. Note that the optimal receptor structure may differ at different stages in a cascade, and that each 12-arm outbound rotor can contain binding sites for 12 different impurity molecules. The first-stage receptor will likely pass only a relatively small number of different contaminant species, so the number of outbound rotors in the entire system can probably be reduced to a small fraction of the number of inbound rotors.

Internal Transport Streams

After pump mechanisms described above have reliably sorted externally-encountered target molecules into reservoirs filled with species of a single type, a medical nanodevice may require these molecules to be transported to specific internal locations for further processing. Bulk fluid flow or fluidized (solvated or suspended) transport through nanopipes may suffice for some purposes. However, in many cases it will be necessary to present reagent molecules to other subsystems as a well-ordered stream of precisely positioned moieties transported in vacuo, especially for mechanochemical operations.

For this purpose, Drexler proposes molecular mills — eutactic systems of nanoscale belts moving over rollers, with reagent-binding devices mounted on the belt surface. This class of device can be assembled into complex molecular transportation networks using conditional switching, crossed-axis belting, and transit speed/frequency multipliers, and may also be employed to drive mechanosynthetic chemical reactions. The benchmark mechanism uses 10-nm diameter rollers to carry closely packed reagent devices measuring 4 nm x 4 nm x 2 nm, or 32 nm^3. A 20-roller mill mechanism 1 micron long has a ~2 micron long belt with 500 reagent devices and delivers 10^6 molecules/sec at a belt speed of 4 mm/sec. Total power dissipation is $\sim1.4 \times 10^{-18}$ watts, a rate of ~0.001 zJ per moiety (or per reagent device) delivered or $\sim10^{-6}$ zJ/nm traveled per reagent device. Total mill mechanism mass is $\sim6 \times 10^{-20}$ kg. A less energy-efficient, but far more versatile, internal molecular transport device is the 100-nm telescoping manipulator arm described. This flexible $\sim10^{-19}$ kg device employs a binding tip to pick and place small and large molecules alike, moving them at ~1 cm/sec with repeatable placement accuracy of 0.04 nm. Multiple devices can be used to establish an internal ciliary transport system; standardized volume containerization of molecules permits rapid stereotypical handoff motions and efficient parcel routing. Conveyance through a 100-

nm arc takes 10^{-5} sec consuming 0.1 pW while the arm is in motion, or ~10 zJ/nm traveled per reagent molecule or per container transported (vs. ~1000 zJ per typical covalent bond).

In molecular cytobiology, vesicles and organelles are transported throughout the interior of a cell by riding on microtubular cables crisscrossing the cytosol. For example, neural vesicles show transport speeds up to 2-4 microns/sec.

Receptor-Based Transport

The most efficient of all modes of molecular transport involves receptor sites capable of recognizing and selectively binding specific molecular species. Many receptors reliably bind only a single molecular type; others, such as sugar transporters, can recognize and transport several related sugar molecule types. In nanomechanical systems, artificial binding sites of almost arbitrary size, shape, and electronic charge may be created and employed in the construction of a variety of highly efficient molecular sortation and transport devices, described below. A discussion of binding sites and receptor engineering.

Physical Forces in Molecular Recognition

Covalent bonds, which occur when atoms share electrons, are the strongest bonds. Aside from metals and salts, most material objects are made of atoms held together by covalent bonds. The atoms comprising biological molecules like proteins, nucleic acids and lipids are strung together mostly by single, double, or triple covalent bonds, as are receptors and diamondoid nanomechanical structures. Interatomic bond strengths range from 181 zJ/bond for O-F up to 1785 zJ/bond for C-O. Covalent bond lengths range from 0.10-0.16 nm within CHON-atom molecules, giving a typical covalent bond rupture force of ~10 nN/bond.

But as Jean-Marie Lehn points out, "there is a chemistry beyond the molecule" — noncovalent supramolecular chemistry. The bonds employed in molecular recognition are weak noncovalent bonds. Noncovalent bonds are largely responsible for the secondary and higher order structure of macromolecules. On a per-bond basis, noncovalent bonds are 1-3 orders of magnitude weaker than covalent bonds. However, the possibility of combining within a limited area a great number of noncovalent bonds having complementary elements allows the formation of a large specific association whose affinity may be of the same order of magnitude as a covalent bond. The high combinatorial diversity provided by many complementary elements allows numerous orthogonal specific associations, enabling self-assembly of many components; by comparison, covalent chemistry offers a poor diversity of reactivities. An additional advantage is that the formation of noncovalent bonds often is not hindered by high energy barriers. At least five types of noncovalent bonds may be distinguished: electrostatic, hydrogen, van der Waals, p aromatic, and hydrophobic.

1. *Electrostatic Bond* — The electrostatic bond between two charged particles (e.g., the "salt bridge" in proteins) is a dipole interaction whose energy E_e is given by Coulomb's law. Most isolated amino acids in neutral solution are zwitterionic — the molecule has no overall charge but carries both a negatively charged group (carboxyl, CO_2^-) and a positively charged group (amino, NH_3^+). In proteins the individual amino acids are polymerized, giving a peptide backbone which is electrically neutral except for the ends of the chain. Most of the standard amino acids found in proteins have uncharged side chains, although histidine, lysine and arginine each have a positive charge at neutral pH and both glutamic and aspartic acids normally carry a negative charge.

2. *Hydrogen Bond* — A second important noncovalent interaction is the hydrogen bond, a dipole formed when a hydrogen atom covalently bonded to an electronegative atom is shared with a second electronegative atom (typically an oxygen, nitrogen or fluorine atom), such that the proton may be approached very closely by an unshared pair of electrons. Hydrogen bonds are largely responsible for the unusual thermodynamic properties of water and ice, and the DNA double-helical and protein a-helical and b-structure conformations are extensively hydrogen bonded. Highest bonding energies occur when donor and acceptor atoms are 0.26-0.31 nm apart. Typical hydrogen bond strengths in proteins are 7-50 zJ.

3. *Van der Waals Interaction* — A third important noncovalent force is van der Waals interactions (London dispersion forces). There is an attractive component due to the induction of complementary partial charges or dipoles in the electron density of adjacent atoms when the electron orbitals of two atoms approach to a close distance. There is also a strongly repulsive component at shorter distances, when the electron orbitals of the adjacent atoms begin to overlap, commonly called steric hindrance.

4. *Aromatic p Bonds* — A fourth type of interaction (p electron to p electron), called "aromatic" or "pi" bonding, occurs when two aromatic rings (conjugated p systems) approach each other with the plane of their aromatic rings overlapping, with successive p-bonded systems stacked like layers in a cake. This results in a noncovalent attractive force with a bond strength of ~40-50 zJ. (That is, the planes of conjugated p systems attract each other when superimposed.) p bond stacking forces contribute to nucleic acid stability at least as much as the hydrogen bonds between bases.

5. *Hydrophobic Forces* — Finally, there are the strong hydrophobic forces, due entirely to solvent entropy changes. When two nonpolar residues approach each other, the surface area exposed to solvent is reduced, increasing the entropy of all the water present and decreasing the entropy of the residues, adding to the binding energy a hydrophobic free energy of ~17 zJ/nm^2 of contact surface area that was formerly exposed to water.

In designing an artificial binding site, the above forces may be combined to achieve the desired level of affinity and specificity for a given ligand. All forces are not equally useful in this regard, however. For example, hydrophobicity is the major factor in stabilizing protein-protein associations. But hydrophobicity is almost entirely nonspecific, hence contributes little to ligand discrimination. By contrast, the proper formation of hydrogen bonds and van der Waals contacts require complementarity of the surfaces involved. Such surfaces must be able to pack closely together, creating many contact points, and charged atoms must be properly positioned to make electrostatic bonds. Thus van der Waals and polar interactions may contribute little to the dynamic stability of the ligand-receptor complex, but they do determine which molecular structures may recognize each other. Other design elements of binding sites, such as directed channeling of substrates into the receptor, may also prove useful.

In analyzing molecular forces, note that at the nanoscale level, surface/surface, molecule/surface, and molecule/molecule interactions may feature very complicated behaviors. Nanodevices performing work may generate both thermodynamic and mechanical local nonequilibrium conditions, so calculations based on the general forms of interactions and on macroscopic expressions valid at equilibrium conditions should be taken only as basic estimates.

Ligand-Receptor Specificity

While affinity measures the strength of the binding of a ligand to a receptor, specificity defines the degree to which a receptor can distinguish between similar ligands. That is, the affinity of the target molecule for the receptor must be greater than the affinity of any other ligand in the environment that is competing for that same receptor, by some threshold multiple. How much greater is enough? In natural dynamic cellular systems, the threshold multiple appears to be a factor of $\sim 10^2$-10^3. For example, the carrier which expels Ca^{++} from erythrocytes presents a variation in K_d of 10^{-6} to 10^{-3} in going from the interior to the exterior of the cell. The active transport of amino acids by hepatocytes normally involves $K_d \sim 10^{-1}$, but under conditions of deprivation a high affinity carrier with $K_d \sim 10^{-3}$ comes into play.

However, the key to assessing specificity in the nanomedical context would be to tally all the competing molecules in the in vivo environment, determine which are the nearest competitors, and then design to avoid them by imposing appropriate energy barriers. By 1998, only a very few competitive ligand-receptor binding analyses had been performed. Until a complete molecular inventory of the human body becomes available, the following crude estimate of nearest-neighbor differences must suffice.

The human body contains a minimum of $N_{prot} \sim 10^5$ distinguishable proteins. The maximum number of distinguishable proteins in the biosphere was given by Kauffman who estimates a useful biological catalytic task space of $N_{prot} \sim 10^8$ distinct protein forms, a tiny subset of the $\sim 20^{500}$ possible 500-residue protein sequences. If the average protein is constructed from $N_{residue} \sim 500$ amino acids (MW ~50,000 daltons), then the average protein may differ from its most similar neighbor by n_{var} ~ log (N_{prot})/ log ($N_{residue}$) = 1.9-3.0 residues. (The precise magnitudes of N_{prot} and $N_{residue}$ are not crucial to our conclusion.) Most proteins are confined to cells containing N_{prot} ~ 5000 different protein types each; given that evolution has probably optimized local specificity to ensure that closely competing crucial ligands rarely appear in the same cell, it seems reasonable to assume that the average closestneighbor ligand may differ from the average target ligand by at least n_{var} ~ 1 residue.

How much is receptor affinity reduced when binding molecules differing by n_{var} ~1 residue from the target ligand on receptor-accessible surfaces — the minimum threshold required to ensure specificity within a cell? In one experiment the relative affinity of an antibody constructed for succinylglycinamide-linked histamine (histamine-SGA), which was the target molecule, and the same molecule but with one methyl group or one carboxylic group removed and replaced with a hydrogen, was 1.45 x 10^4 or 2.5 x 10^5, respectively, due to steric hindrance. Similar investigations with antibodies for SGA-linked serotonin produced relative affinities of 500-1000, and with antibodies for single alanine substitutions in Human Growth Hormone (HGH), ~1000. The relative affinity of a particular RNA oligomer for theophylline and caffeine, two ligands which differ by only a single methyl group, was measured experimentally as 10,900. Computational receptor experiments in which CH replaces N suggest a decline in relative affinity of ~5x10^4. Indeed, the change of a single hydrogen atom on a ligand is usually sufficient to destroy its specificity for, or activity within, a particular enzyme. The single-residue affinity reduction at a receptor-accessible surface appears to be of order ~10^3 - 10^5. Each increase of ~10 zJ in bonding energy causes the reaction equilibrium constant to decline (hence receptor affinity to rise) by a factor of ~10. If a difference in affinities of ~10^3-10^5 between the target molecule and its nearest competitor likely to be present in the environment provides sufficient receptor specificity for nanomedical purposes, this requirement corresponds to a binding energy differential affinity of ~30-50 zJ at 310 K between target and closest-neighbor ligands.

Ligand-Receptor Dynamics

Diamondoid structures can exhibit a stiffness and rigidity one or two orders of magnitude greater than that available in protein structures. In general, stiffer structures permit greater specificity because they enhance exclusion of non-target ligands based on van der Waals overlap forces (called steric hindrance) and allow narrower tolerances in distinguishing acceptable ligands.

In less-stiff protein-based receptors, each of the atoms is engaged in relatively large, rapid jiggling movements. Experimental and theoretical work has been done on the atomic fluctuations within the basic pancreatic trypsin molecule, a small enzyme with 58 amino acids and 454 heavy (non-hydrogen) atoms. This work established that fluctuations increase with distance from the center of the molecule, with the magnitude of RMS fluctuations ranging from ~0.04 nm for backbone atoms to ~0.15 nm for the ends of long side chains (roughly one atomic diameter), and an average of 0.069-0.076 nm per atom over the entire molecule. A similar experimental analysis of reduced cytochrome c, a common metabolic enzyme, shows that RMS fluctuations of each of the 103 amino acid residues in the molecule averages ~0.11 nm with lattice disorder (~0.05 nm) included, and fluctuations range from 0.09-0.16 nm. Antibody core domain movements display RMS fluctuations of 0.04-0.19 nm. Hence it appears that the average atom within the typical protein receptor oscillates ~0.1 nm every ~10^{-12} sec, although frequently residues with long side chains (e.g., arg, lys) have much higher RMS deviations than average.

By contrast, in stiff diamondoid-based receptors each of the atoms is locked in a rigid crystalline structure and thus is subject to thermal displacements approximately 10 times smaller. The RMS displacement for a quantum mechanical harmonic oscillator.

The ratio of RMS displacements for protein/diamondoid receptors is ~10:1, so the minimum addressable volume (hence inverse maximum specificity) of a diamondoid receptor should be ~10^3 smaller than for protein receptors, a ~30 zJ binding energy advantage for diamondoid receptors.

Diamondoid Receptor Design

Natural enzymes and antibodies are proteins folded into highly organized, preformed shapes that present a ready-made "keyhole" into which a target ligand will fit. The enzyme is folded in such a way as to create a region that has the correct molecular dimensions, the appropriate topology, and the optimal alignment of counterionic groups and hydrophobic regions to bind a specific target molecule. Tolerances in the active sites can be narrow enough to exclude one isomer of a diastereomeric pair. For example, D-amino acid oxidase will bind only D-amino acids, not L-amino acids.

These "keyholes" are extremely floppy, yet still achieve fair specificity. This is a consequence of "induced fit" in protein binding sites. That is, the interaction of the target molecule with an enzyme induces a conformational change in the enzyme, resulting in the formation of a strongly binding site and the repositioning of the appropriate amino acids to form the active site. The receptor flexes, balloons, hinges, or contracts by 0.05-1.0 nm in just the right places to maximize specificity as the selected ligand enters the site. In some cases such as O_2 and CO binding by myoglobin, ligands enter the receptor through a series of temporary voids that appear and disappear in the receptor as ~10 picosec dynamic structural fluctuations. Induced fit can reduce receptivity to undesired proteins that exploit relative geometry by bonding enough to bring portions of their surfaces into alignment with the same receptor sites that bind desired proteins. Folding transitions appear to be the most prevalent and to possess the most possibilities for adaptability or induced fit.

For smaller molecules, it is likely that recognition processes will be relatively inefficient, time-consuming, and more difficult to engineer if they involve a good deal of rearrangement of the receptor's shape. Thus there is considerable interest among chemists in designing artificial receptors that have their cavities already formed into the shape appropriate for the intended substrate. For instance, rigid-cavity "spherand" receptors are exceptionally efficient at binding metal ions. Bowl-shaped molecules such as cryptaspherands, calixarenes, and carcerands can be lined with chemical groups along their walls and with charged groups along their rims to achieve high binding specificity. Self-assembling capsules or "container molecules" made of hydrogen-bonded subunits, capable of limited molecular recognition, have been synthesized. A designed receptor for creatinine was demonstrated in 1995, and in 1998, K. Suslick and colleagues designed metalloporphyrin-dendrimeric artificial receptors that can bind straight, skinny molecules but block out bent or fat molecules. Container molecules with 0.2-0.4 nm portals control entry to their interiors using "French door" and "sliding door" gates and hinges. Active binding sites for small simple molecules such as NO, CO, and C_2H_4 are well-known.

Ultimately, receptors will be designed to nanoscale precision and may be constructed using diamondoid materials. Electrostatic, hydrophobic, and hydrogen-bond forces will add immensely to artificial receptor specificity and are essential for binding small molecules. For example, using a 0.2-nm range in a saline environment, 10-40 charge contacts would be required to bind molecules of various sizes and concentrations using electrostatic force alone, which is ~0.5 charge/nm^2 over the entire surface of a 60,000 dalton globular protein (vs. ~1 charge/nm^2 for the surface of an isolated zwitterionic amino acid). Or, a 7 nm^2 hydrophobic cavity having the exact folded shape of the target ligand generates ~120 zJ binding energy as the molecule stuffs itself into the cavity to exclude its surface from solvent water.

But consider a theoretical receptor that employs van der Waals dispersion forces alone. Atoms comprising the typical protein or CHON target molecule in the human body have an average atomic mass of ~6 amu/atom and an average density of 1500 kg/m^3, giving a mean molecular volume of ~6.7 x 10^{-30} m^3 per atom in the target molecule. Assume for simplicity a spherical receptor surface that forms a negative image of the surface of the target molecule. The receptor surface lies ~1.5 x (minimum van der Waals contact distance) ~0.3 nm from the perimeter atoms of the target molecule, and completely encloses the target molecule (thus requiring at least one moving part). The theoretically available maximum binding energy E_{vdW}, using only dispersion forces, should be sufficient to adequately bind all but the smallest target molecules according to the criteria set forth. (Proteins are actually ellipsoidal with a much larger surface area $A_p = 0.111\ MW^{2/3}$ nm^2, than if they were spherical.

What about specificity? Given the ability to design diamondoid binding sites to at least localized <0.01 nm tolerances, chirality is readily detected and (purely as a design exercise) it may even be possible to distinguish diatomic nitrogen and oxygen on the basis of size alone. The molecular lengths (major axis) of N_2 and O_2 are 0.250

nm and 0.253 nm, but the molecular widths (minor axis) are 0.140 nm and 0.132 nm, respectively. (Diatomic molecules get longer and narrower at higher molecular weight.) Thus N_2 is distinguishable from O_2 on the basis of width (~0.01 nm). With a van der Waals energy well depth of 1.1-1.4 zJ for N_2 and O_2, in a tight receptor these gases are bound at an energy density corresponding to 3000-4000 atm of pressure.

Minimum Feature Size and Positioning Accuracy

Maximum displacement measurement accuracy in nanoscale devices is ~0.01 nm, and RMS thermal displacement in diamondoid bonds is ~0.01 nm at 310 K. Thermal displacements in 10-nm long diamondoid rods are ~0.01 nm at a 1-nm rod width, ~0.02 nm at 0.5-nm width, and ~0.10 nm at 0.3-nm width. However, it is possible to construct components to even narrower tolerances.

For instance, a single C-O bond inserted into a diamondoid rod in a collinear carbon chain extends rod length by 0.1402 nm, the C-O bond length. An adjacent rod into which an N-N bond is similarly inserted is extended by 0.1381 nm, the N-N bond length. By bonding these rods (aligned at one end) it is possible to build diamondoid structures having 0.002-nm features (at the other end), which at 310 K will nonetheless suffer thermal displacements of ~0.01 nm or more. Similarly tiny displacements can be induced in binding cavity surfaces by inserting a foreign atom deep inside the bulk diamondoid structure, causing dislocation strains that decline in magnitude at greater distances from the compositional disturbance. It is also possible to translate diamondoid components through a picometer step size, much smaller than the unavoidable RMS thermal displacements, using any of several methods; for example:

A. *Levers* — Consider a 10-nm lever joined to a fixed bar by a pivot at one end, and driven axially by a ratchet interposed between lever and bar at the other end. Ratchet movements translate to smaller displacements at positions along the lever distant from the ratchet. Thus a follower rod attached to the lever 1 nm from the pivot and driven by a ratchet with 0.01-nm steps moves ~0.001 nm per ratchet step.

B. *Screws* — Consider a 3-nm diameter cylindrical screw with a 1-nm pitch. Rotating the screw through a 0.01-nm circumferential displacement causes the screw to move laterally by ~0.001 nm, which may be transmitted elsewhere in the machine by an attached follower rod. Of course, nanoscale screws or gears are sensitive to the precise cancellation of the potentials and hence cannot be perfectly smooth and circular, producing some unavoidable "knobbiness" under load.

C. *Gear Trains* — Consider a 32-nm diameter worm gear with 1-nm teeth. One rotation of the gear requires 100 rotations of the worm; hence a 0.1 nm displacement applied to the worm produces a 0.001 nm displacement in the gear. More efficient (and coaxial) compound planetary gear trains commonly employed in transmissions achieve displacement ratios up to 10,000:1, a hundred times better than the above example.

D. *Hydraulics* — Consider a sealed, fluid-filled, tapered pipe. A piston 1000 nm^2 in area is mounted at one end; another piston 10 nm^2 in area lies at the other end. A displacement of 0.1 nm applied to the smaller piston produces a 0.001 nm displacement in the larger piston. (Here again the finite size of molecules may produce "knobby" performance as fluid particles slip from one stable configuration to the next.)

E. *Compression* — Consider a rod upon which a compressive force of 100 nN/nm^2 (near the maximum diamondoid strength) has been imposed. Affixed to the rod are two crossbars spaced 4 nm apart. If the force on the rod is increased to 101 nN/nm^2, the gap between the crossbars compresses by ~0.001 nm.

Imprint Model

Molecular imprinting is an existing technique in which a cocktail of functionalized monomers interacts reversibly with a target molecule using only noncovalent forces. The complex is then crosslinked and polymerized in a casting procedure, leaving behind a polymer with recognition sites complementary to the target molecule in both shape and functionality. Each such site constitutes an induced molecular "memory," capable of selectively binding the target species. In one experiment involving an amino acid derivative target, one artificial binding site per $(3.8\ nm)^3$ polymer block was created, only slightly larger than the $(2.7\ nm)^3$ sorting rotor receptors described by Drexler. Chiral separations, enzymatic transition state activity, and high receptor affinities up to $K_d \sim 10^{-7}$ have been demonstrated, with specificity against closely competing ligands up to $DK_d \sim 10^{-2}$ (~20 zJ).

Several difficulties with this approach from a diamondoid engineering perspective include:

1. A sample of the target molecule is required to make each mold.
2. It is currently unknown how to prepare diamondoid castings.
3. Once the imprint has been taken, the site cannot easily be further modified.

Solid Mosaic Model

In the solid mosaic receptor model, the precise shape and charge distribution of the target molecule is already known. Working from this information, a set of diamondoid components could be fabricated which, when fitted together like a Chinese puzzle box, create a solid object having a cavity in the precise shape of the optimum negative image of the target molecule. The mosaic may contain point charges, voids, stressed surfaces, or dislocations to achieve fine positional control. Mosaic components may be as small as individual atoms, so this model is conceptually similar to 3-D printing or raster-scan techniques in which the desired cavity formation is constructed atom by atom inside a nanofactory. This model, like the imprint model, cannot easily be reconfigured once it has been constructed because each of the many unique parts may contribute to the entire structure. The construction of receptors from parts of fixed size and shape is crudely analogous to members of the heterodimer receptor class (a two-component receptor) such as $GABA_B$.

M. Reza Ghadiri has designed a protein mosaic model using cyclic peptides that assemble spontaneously into nanotubes of predefined diameter; incorporation of hydrophobic amino acid side chains on the outside of these tubes leads to spontaneous insertion into bilayers, allowing the tubes to function as transmembrane ion channels. Other examples of mosaic model receptors are mesoporous silica filters with functionalized organic monolayers forming 36 nm sievelike pores, and zeolites and zeolitelike molecular sieves. Zeolites are artificial crystal structures with precise and uniform 0.4-1.5 nm internal void arrays which can also be used as shape-selective catalysts able to favor one product over another that differs in size by as little as 0.03 nm, such as pxylene and oxylene. By 1998, rational de novo computational design of artificial zeolite templates and crystal engineering had begun.

European Research Infrastructures - Advances in Material Science

Advances in material science - which studies the structure, properties and performance of the materials all around us - are crucial to Europe's leadership in the global knowledge economy. Understanding and controlling changes in the structure of materials at the microscopic nano-scale, or even smaller, allows scientists to enhance their physical and chemical properties. This in turn leads to better, stronger and more efficient and sustainable materials for use in industrial systems, home appliances, and medical devices, enhancing the quality of life of European citizens and the competitiveness of our products. The key role played by European research infrastructures - large-scale instruments, installations and other facilities - in this drive for fundamental knowledge and economic performance is the subject of a press briefing, held at the internationally renowned CCLRC Rutherford Appleton laboratory near Oxford. Janez Potocnik, Commissioner for Science and Research, said "Knowledge and innovation must be our absolute priority if we are to increase our economic development and create jobs. Without knowledge, there is no real future for Europe as a global economic leader. Research infrastructures directly support technological innovation, by offering the essential conditions and the necessary critical mass to carry out cutting edge research."

Research Infrastructures in support of Material Sciences

What is Materials Science?: Materials science looks at the nano-meter and micro-meter scales, where atoms organise themselves into continuous three-dimensional structures that grow into the ceramics, plastics and metals of the visible world. The properties of these structures – and how they can be modified and made suitable for technological applications – are the domain of the materials scientist.

Materials science has always fascinated physicists and scientists. Throughout history, mankind has tried to master this science to unlock the mysteries of life itself. Society as we know it today owes a huge debt to the early efforts of scientists working to develop better, stronger, more efficient materials for use in industrial systems, home appliances, leisure equipment, medical devices and just about everything we touch in our daily lives.

By understanding and controlling changes in the structure of a material – at the microscopic scale, nano-scale and potentially smaller – scientists can greatly enhance its physical and chemical properties on the large scale, making it possible to build better, more sustainable materials and improve our quality of life.

The role of Research infrastructures in materials science: But do we fully understand all the details behind the properties of materials? In recent years science has gone to great lengths to study and develop finer and finer materials to the nano-level. The prefix 'nano' indicates extreme smallness, so small that the naked eye could only appreciate the fine detail of a nano-structure if it was magnified 1 000 million times its original size.

To understand the nano-world, there are tools used to measure and manipulate ultra-small structures, such as nano-scale resolution microscopes, lasers and spectroscopes. Research Infrastructures are those major instruments, installations, or facilities that provide top-class research services to support the work of materials scientists.

Research Infrastructures are clearly vital in helping Europe develop top-class research – fundamental and applied – in materials science. The EU recognises the strategic role that such science plays in stimulating new knowledge which drives growth and industrial competitiveness but the challenge is to develop the right approaches and the tools to support all the scientific developments needed in the future more effectively.

Staying with the example of nano-science, it becomes clear that this is a truly multidisciplinary field involving materials scientists, mechanical and electronic engineers, who team up with the likes of physicists, chemists and biologists to study and produce new materials, medical devices, food and packaging, chemical coatings, sensors, electronic devices and much more. Research infrastructures play an important role in bringing scientists from all these different fields together, and providing them with common tools.

The Success of European Materials Science: The impact of the Research Infrastructures action can be seen in a number of critical scientific endeavours spanning almost two decades. In the materials science field, in particular, there are a number of very positive examples, which are presented today.

Lasers for materials science: the LaserLab project: Another project, Integrated European Laser Laboratories (Laserlab-Europe), is a network of 17 laser infrastructures from nine European countries which has been set-up to increase the contribution of lasers and their applications in all faculties of the sciences, in particular in materials-related technologies such as material modification (e.g cutting and welding), and non-invasive analytical techniques as holography (three dimensional reconstruction of objects) and laser spectroscopy (penetrating matter with light to collect information on its chemical make-up).

Synchrotron radiation to illuminate materials: the IA-SFS project: A third case is a project called Integrating Activity on Synchrotron and Free Electron Laser Science (IA-SFS), whose two main goals are: to help external users gain access to major synchrotron and free electron laser (FEL) facilities, and to foster joint research activities which boost the performance of these facilities. Synchrotrons are large circular accelerators of particles that can be used to produce light/radiation between infrared and x-rays which are used to analyse materials. Synchrotron radiation offers new potential for materials characterisation, in the optimisation of technical processes and in in-situ investigations of microstructure developments.

EU Materials Research in Action: the FAME Network of Excellence: This network focuses on smart nano-materials, which are at the heart of a whole range of new devices and processes. Fame is a practical example on how research infrastructures are key elements for the excellence of research in the field of materials.

Research Infrastructures and Nano-Materials: the Gennesys, Dynasync, NanoQuanta Cases: Three other test cases illustrate the value of European research co-operation. In particular, the Gennesys project illustrates the need to join forces and to integrate actions for the future of research in nanomaterials and nanotechnology using synchrotron and neutron installations.

Other Basic Capabilities

This study describes a miscellany of important technical capabilities that may prove useful in some or all medical nanodevices, in various scenarios or theaters of operation. Any one of these subjects deserves an entire chapter to itself, but unfortunately there is only space in this introductory text for a brief survey of each area. The most important of these topics is computation, including nanomechanical, nanoelectronic, and biological computing, as well as nanoscale data storage technologies. However, the fields of organic and fullerene nanoelectronics, biocomputing, and quantum computing are advancing so fast that whatever is written here will quickly become obsolete. Thus our coverage in this Volume is limited to a broad overview. Interested readers are strongly advised to consult the current literature for the latest results.

Other important basic capabilities covered in this Chapter include timers and nanoclocks with long-term nanosecond stability ; high-pressure materials storage ; defensive cytocidal and antiviral weaponry to allow the efficient destruction of pathogenic intruders and unrepairable tumor cells ; and the effects of very hot or very cold temperatures on nanorobot materials and operations.

Human Chronobiology

The human body incorporates numerous biological clocks. The best-known and most-studied example is the daily (24-hour) endogenous circadian oscillator. This internal clock is normally reset by natural sunlight, which is much brighter than indoor lighting. The clock, in turn, sets the cadence for most of the other 24-hour body rhythms — for example, sleep/wakefulness cycles (e.g., melatonin), urine production, body temperature, blood cortisol and ACTH cycles, and the diurnal rhythm of mitosis in epidermal epithelium (e.g., greatest during sleep or inactivity, least during wakefulness or activity). The circadian clock stability (variation in free-running period) is ~1% in *Drosophila* and ~0.2% in humans.

Mammalian circadian rhythms are regulated by a master pacemaker within the suprachiasmatic nuclei (SCN) of the hypothalamus (just above the point at which the optic nerves from each eye cross in mid-brain). In humans, the SCN is comprised of ~10,000 special cells that send out electrochemical signals in a 24-hour timing pattern. Entrainment of the clock to light-dark cycles is mediated by photoreceptors in the retina, with light information conveyed directly from ganglion cells of the retina to the SCN via the retinohypothalamic tract. Mammals and other vertebrates have another independent circadian clock in each retina, possibly driven by the retinal cryptochromes CRY1 (also made in the SCN) and CRY2, producing rhythms in local physiology such as the diurnal renewal cycles of rods and cones. CRY1 is also abundant in skin tissues. Clock-resetting photoreceptors have been found in the human popliteal region (behind the knees), and such photoreceptors may exist in many different tissues throughout the human body.

The current model for a core circadian oscillator comprises, in part, a transcription/translation-based negative feedback loop in which clock genes are rhythmically expressed, giving rise to cycling levels of clock RNAs and proteins (negative elements). The proteins then feed back, after a lag, to depress the level of their own transcripts, perhaps by interfering with positive elements that increase transcription of the clock genes. In 1998 this genetic network was still in the discovery process. One ~100,000-base-pair gene, aptly named "clock," was known to produce an 855-residue protein (mCLOCK in mice, 115.7 kD) that serves as a major regulator of the ~10 other, still unidentified, genes thought to affect circadian rhythm. (One portion of mCLOCK is the same as in a related 1023-residue clock protein, dCLOCK, found in *Drosophila* fruit flies). Circadian rhythms will almost certainly be made up of many interconnected feedback loops that interact with the CLOCK-related core pathway. It is believed that this interconnected ensemble will ultimately determine all of the classical circadian properties — period length, temperature compensation, and resetting by light or temperature. In 1998, most chronobiologists believed that many of these outer loops would be organism specific and that only the core loop would be more universal.

In addition to the 24-hour circadian oscillator, many other biological clocks have been identified in humans. For example, the 28-day menstrual cycle is timed by two ovarian clocks, each delimiting a period of 14 days. One of these two clocks is the Graafian follicle and its production of estradiol, while the other clock is the corpus luteum and its secretion of progesterone. Both are obligatorily dependent upon the proper functioning of a third clock located in the arcuate region of the hypothalamus, known as the GnRH pulse generator, which ensures the rhythmic production and release of the gonadotropic hormones into the peripheral circulation with a cyclical period of ~1 hour (±25% stability), coincident with a 1-hour cycle of hypothalamic electrical activity (±12% stability). A 90-minute clock paces the development of the somites, blocks of tissue that form in regular arrays along the spinal cord of vertebrate embryos. One regulatory gene, named "chairy," undergoes repeated 90-minute activity cycles and is known not to require protein synthesis; gene expression follows the repeating pattern even when protein synthesis is biochemically blocked. A much longer half-weekly (circasemiseptan) pattern has been observed in mitotic activity in cancer patients.

Other biological clocks abound. The stomach produces rhythmic contractions of ~3/minute during digestion. Rhythmic segmentation contractions of the bowel at ~0.2 Hz are driven by a thin layer of specialized pacemaker cells called the intestinal cells of Cajal, which demonstrate electrical oscillations at about the same rate as the contractions. Cardiac pacemaker cells in the sinoatrial node emit a cyclical pulse that triggers a heartbeat at a ~1 Hz basal rate. The respiratory centers in the medulla oblongata produce an autonomic contraction cycle of the diaphragm at ~0.3 Hz. The human eye produces many spontaneous rhythmic dilations and constrictions, including high-frequency ocular microtremors present in all people (mean frequency ~84-88 Hz), accommodation microfluctuations (~1-2 Hz) with the pupil varying <300 microns in diameter, and low-frequency pupillary oscillations (e.g., ~0.2 Hz "hippus"). Electrical rhythms generated by the brain that are detectable by an electroencephalograph include alpha waves predominantly in the occipital region at 8-13 Hz, beta waves in the frontal and central areas at 18-30 Hz, delta waves at <4 Hz during deep sleep or abnormal function, and theta waves in the temporal and parietal areas at 4-7 Hz.

Individual cells display a number of oscillatory biochemical pathways as well, including calcium ion oscillations, cyclic AMP signalling, and various other cell cycles. For instance, the glycolytic enzyme oscillator, controlled primarily by phosphofructokinase, may mediate various rhythmic physiological behaviors such as slow waves of contraction in smooth muscle, electrical activity in neurons, and insulin release from b islet cells of the pancreas. In one experiment, this oscillator mediated a 55 mV voltage change in guinea pig cardiomyocytes at 0.010 Hz (±10% stability) in an almost perfectly sinusoidal pattern, although a few cells displayed oscillations with irregular phase, amplitude, or both. In another experiment, fluorescently-labeled puffs of tumor-cell cytoplasm were observed being released through plasma membrane ruptures, regularly at ~0.05 Hz during a neutrophil-mediated cytolytic attack, matching the nicotinamide-adenine dinucleotide phosphate and superoxide release periods for neutrophils. By 1998 there was a growing belief that many, if not most, of the cells in an animal may possess individual biological clocks; fruit flies were already known to have clocks distributed throughout their wings, legs, and abdomen.

The human brain also possesses an interval timer that allows a person to gauge the passage of seconds or minutes to a mean accuracy of ±15%. The interval timer functions like a stopwatch and resides in the basal ganglia, a region of the brain that coordinates voluntary muscle movements. A population of neurons in the substantia nigra releases regular pulses of dopamine into an accumulator, called the caudate-putamen, a major part of the basal ganglia. These pulses are then read via neural pathways to the cerebral cortex which are known as striato-cortical loops. Other internal "alarm clocks" that reliably allow waking from sleep at desired times have been described.

In vivo nanorobots should be able to eavesdrop on most of these active chronobiological channels, gaining complete knowledge of the oscillator frequencies and phase settings of virtually all of the human body's biological clock systems. For example, blood plasma cortisol typically peaks in the morning (e.g., 6-23 x 10^{-8} gm/cm^3 at 8 AM), then declines throughout the day (e.g., 3-15 x 10^{-8} gm/cm^3 at 4 PM, then ~50% of 8 AM value at 10 PM). Melatonin (also detectable in saliva) cycles in the opposite direction, normally peaking at 6-7 x 10^{-11} gm/cm^3 at night and then falling to 1.4 x 10^{-11} gm/cm^3 during the day. Longer cycles are easily tracked too. For instance, blood viscosity in healthy young women shows the rhythmic variation of the menstrual cycle, probably due to changes in serum fibrinogen and globulin levels which can be measured more directly. These nanorobot-accessible readings may be used either to measure "body time" or to control it, and in some cases can recalibrate relevant onboard nanochronometers.

Environmental time cues may also be detected by in vivo nanorobots (e.g., temporal macrosensing), such as the level of illumination penetrating the epidermis (sometimes permitting inference of day or night) or the all-pervasive 60/50 Hz (U.S./Europe) electromagnetic hum generated by alternating current electrical supply systems in common use throughout the world, a crude frequency standard whose voltage can sometimes display time-of-day load-related fluctuations.

Nanobodies

Antibodies, often described as magic bullets, are actually more like tanks: big, complicated and expensive. Tinier "nanobodies," derived from camels and llamas, may be able to infiltrate a wider range of diseases at lower cost. That is the hope, at least, of one small start-up in Belgium.

Like many biotech companies, Ablynx emerged from the confluence of a serendipitous discovery, an open window of opportunity and an unreasonable ambition. Housed on two floors in a nondescript gray laboratory on a technology campus outside the university town of Ghent, Belgium, the three-year-old company employs just 45 people, 33 of them scientists and bioengineers. It is a minimal staff with a simply stated mission: find the tiniest sliver of protein that will do the job of a full-size antibody, then turn it into a billion-dollar medicine—or better yet, into the first of a whole new class of "nanobody" drugs against cancer, rheumatoid arthritis, inflammatory bowel disease, perhaps even Alzheimer's disease.

Despite being backed by $40 million of venture capital and partnerships with Genencor, Procter & Gamble and the National Research Council of Canada, Ablynx faces long odds. Its ambitious goal might seem altogether futile were it not for the recent surge in antibody therapies, the problems that still nag these sophisticated drugs, and the insights that Ablynx scientists have into the peculiar biology of the camel family.

Aside from the brain, the most complicated part of the human body is undoubtedly the immune system—and thank goodness. It's a bacteria-eat-man world out there, filled with a nearly endless variety of germs that see us as spawning grounds. Defending against this onslaught are antibodies, which are manufactured by B cells in an equally impressive panoply of models. Antibodies are huge Y-shaped proteins that float about in the blood and the fluid between cells, arms extended, using a chemical sense of touch to interrogate other molecules they encounter. Each model of antibody has its own mission; it patrols for a distinct chemical signature of a certain microbe, allergen or toxin.

Yet despite the sophistication of our immune defenses, we still get sick. No police force is perfect. The immune system is sometimes too slow or complacent in its reaction—for example, to cancers or to infection by respiratory syncytial virus. Other times it overreacts, as happens in organ-transplant rejection and asthma. And when it mistakenly attacks the body's own cells, the immune response can itself cause a degenerative disease such as rheumatoid arthritis.

For years, drugmakers sought to create artificial antibodies that can correct—or at least moderate—these immunological failures. But most early attempts ended in failure and financial disaster. In the two decades following the 1975 invention of a way to produce large batches of antibodies that are identical, or "monoclonal," just two such therapies survived review by the U.S. Food and Drug Administration.

The logjam finally broke in 1997, and by the end of 2004 the fda had approved 17 therapeutic antibodies, including promising treatments for all the ailments mentioned above [see "Magic Bullets Fly Again," by Carol Ezzell; *Scientific American*, October 2001]. Pharmaceutical firms reaped $11.2 billion in sales of these medicines in 2004, the consultancy AS Insights reports.

And the market for monoclonal antibodies (usually abbreviated, idiosyncratically, as MAbs) is still in a formative stage of rapid growth. Dozens more MAbs are now in development or clinical trials, and last year Janice M. Reich-ert of the Center for the Study of Drug Development at Tufts University projected that 16 of them will gain FDA licenses within the next three years. In 2008, she forecast, MAbs will command roughly $17 billion in worldwide sales.

As Ablynx lines up for its first clinical trials in late 2006, it is aiming for a small slice of that large pie, says Mark Vaeck, the company's chief executive. Nanobodies—relatively simple proteins about a tenth the size of antibodies and just a few nanometers in length—may one day yield new medicines for Alzheimer's and other diseases beyond the reach of current antibodies, but that is not the opening strategy Vaeck chose. Instead he directed his scientists to create nanobodies that do what some of the best-selling antibodies do, only better.

Nanoelectronic Computers

In 1998, the metal-oxide-semiconductor field-effect transistor (MOSFET) was still the most common type of transistor in general use. These solid-state devices are characterized by an electrical source, an electrical drain, and a gate that controls the flow. Since its inception, the FET has scaled well to lower sizes. Individual working transistors with 40-nm gate lengths have been demonstrated in silicon, and 25-nm gate-length transistors have been fabricated in gallium arsenide. However, as feature sizes shrink below <~100 nm, a number of obstacles to stable operation begin to appear and the cost-effective downscaling of dense circuitry may not persist. A number of solid-state replacements for the bulk-effect semiconductor transistor have been suggested that could overcome these obstacles by taking advantage of nanometer-scale quantum mechanical effects. Such replacements include quantum dots (QDs), resonant tunneling devices (RTDs), electron turnstiles, and single-electron transistors (SETs). However, while hybrid RTD-FET circuits and SETs have been successfully switched at room temperature, many of these devices must be operated at cryogenic temperatures and other practical challenges remain;such top-down approaches to nanoelectronics are not considered further here.

In contrast, molecular nanoelectronics uses primarily covalently bonded molecular structures, electrically isolated fròm a bulk substrate, producing wires, switches and devices composed of individual molecules and nanometer-scale supramolecular structures. While it is relatively difficult and expensive to sculpt trillions of identical nanoscale structures in bulk materials (e.g., using 2-nm resolution electron-beam lithographic techniques), individual molecules are natural nanometer-scale structures that can be manufactured to be exactly the same in near-mole quantities. The great versatility of organic chemistry offers more options for designing and fabricating nanoelectronic devices than are available in silicon. Investigators are designing, modeling, fabricating, and testing individual molecules and nanometer-scale supramolecular structures that act as electrical switches and even exhibit some of the same properties as small solid-state transistors. For example, three and four terminal devices with ~0.3-nm feature sizes have been examined computationally.

Molecular electronics is a rapidly growing area in the literature. J.M. Tour expects experimental demonstration of molecular-sized transistor devices by 2000-2001, and commercial high-level computers using molecular electronics by 2008-2013. A current list of major research groups is maintained by the International Society for Molecular Electronics and Biocomputing. Some of the following discussion is appreciatively drawn from an extensive 1997 review by MITRE Corporation.

Molecular Wires

Before one can seriously discuss molecular electronic devices embedded in single molecules, it must be asked whether a small single molecule can be made to conduct electrical current at all. A series of difficult and sensitive experiments and theoretical investigations during the 1990s answered this question affirmatively. Single molecule conduction was demonstrated by Tour using a conducting molecule characterized by repeating structures — in this case, a series of benzene-like rings connected by acetylene linkages — each part of which is linked to the next by bonds including many p-electrons above and below the plane of the structure. These electrons in the p orbitals conjugate with each other, or interact, to form a single large orbital throughout the length of the wire to permit mobile electrons to flow. Thiol (-SH) functional groups at either end adsorb well to gold surfaces and act as "alligator clips" for attaching molecular electronic units to metal substrates. Tour's polyphenylene wires can carry ~nanoampere currents. Such molecular wires also have the desirable property that they can be made quite long, if necessary, because they can be lengthened systematically using chemosynthetic methods. Another experiment using a single benzene ring attached between two gold electrodes through thiol groups across a 0.846-nm gap found a singlering threshold resistance of ~20 megohm and a capacitance on the order of $\sim 0.1\ e^2 / kT \sim 10^{-19}$ farad. Fullerene carbon nanotubes or "buckytubes" are hollow cylindrical tubes, essentially rolled-up sheets of graphite, measuring >~1 nm in diameter that (in many chiral and compositional forms;) can conduct electricity. For instance, both individual 10-20 nm diameter carbon nanotubes and atomic wires pulled from an unraveling carbon nanotube can conduct currents of ~1-10 microamps (vs. nanoamps via tunneling). Maximum nanotube current density is $\sim 10^{10}$-10^{11} amp/m^2 over a 0.2-6.0 volt range, superior to the $\sim 10^9$ amp/m^2 typically achieved in superconductors. Nanotube resistivity measures ~8-20 ohm-m for straight tubes, ~38-49 ohm-m for slightly curved (5-30°) tubes, and >100 ohm-m for highly curved (65-80°) tubes. Interestingly, electronic transport in carbon nanotubes is ballistic (e.g., like electron waveguides, permitting only a few propagating modes), with all waste heat being dissipated in the leads to the nanotube element and none in the nanotube itself. Buckytubes are stiff enough to serve as supports for molecular circuit elements, and if filled with conducting metal can create one of the structurally strongest nanowires that is chemically possible. Other 0.6-3 nm molecular wires, 3-20 nm semiconductor wires, and nanoscale metallic wires have been studied as well.

Electromechanical Molecular Switching Devices

In 1998, the best-studied category of molecular electronics device is probably the electromechanical molecular switch. This class of switch employs electrically or mechanically applied forces to turn a current on and off, either by changing the conformation of the molecule or by moving a switching molecule or group of atoms in the manner of a gate. These switches are promising because they could be laid down in a dense network on a solid substrate, or in some cases in three-dimensional arrays; unfortunately, no integrated CPU-scale designs have yet been attempted, although plans for a half-adder and a molecular shift register have been published. Some examples:

A. *Electromechanical Amplifier* — Joachim and Gimzewski measured conductance through a single buckyball held between an STM tip and a conducting substrate. By pressing down harder on the STM tip, the buckyball deforms, tuning the conduction off-resonance and reducing current ~50%. Although the STM tip could be replaced with a small in-situ piezoelectric gate, the ideal actuator would be a molecular-scale electromechanical actuator akin to a logic rod; the major fundamental limit to the speed of this switch is the natural mechanical vibrational resonance frequencies of a buckyball, ~10^{13} Hz.

B. *Linear Atomic Relay* — Researchers at Hitachi Corporation simulated a two-state electronic switch wherein a mobile atom that is not firmly attached to a substrate moves back and forth between two terminals. If the switching atom is in place, the device conducts electricity; if the switching atom is displaced from the two wires, the resulting gap greatly reduces the current that can flow through the atom wire. A small negative charge placed on the third atom wire "gate" near the switching atom moves the switching atom out of its place in the wire; the switching atom is pulled back into place by a second "reset" gate after each use of the switch. Actual experiments approximating this design created a bistable atom switch using a xenon switching atom moving back and forth between an STM tip and a substrate, demonstrating that the movement of a single atom can be the basis of a nanometer-scale switch. In a more mature device, relays could be ~10 nm^2 in size with a speed limited only by the intrinsic vibrational frequency of atoms, typically ~100 THz, but it seems likely that atom relays could only operate at very low temperatures because not much energy would be required to evaporate a switching atom off the substrate and out of the plane of the atom wires, destroying the switch.

C. *Rotational Molecular Relay* — A more reliable two-state device based on atom movement uses the rotation of a molecular group to affect an electric current. The switching atom is part of a rotating group, or "rotamer," which itself is part of a larger molecule, perhaps affixed to the same surface as the atom wires. A conceptual diagram illustrating this approach, wherein the electric field of a nearby gate forces the switching atom to rotate in or out of the atom wire. When the switching atom is "in," wire conductance is high and the switch is "on"; when the switching atom is "out," a second group takes its place, hindering current flow and turning the switch "off." A third large group may provide resistance to thermal free rotation. Alternatively, hydrogen bonding might provide enough rotational resistance to stop the rotamer in the conducting position, but not so much that reversing the gate voltage would be insufficient to turn the rotamer. Controlled rotational state switching of a Pt-adsorbed oxygen molecule by STM using 0.15-volt 20-microsec pulses was demonstrated in 1998.

While a rotating switch with a methyl-like rotamer group has three distinct switch positions, a hinging switch that rotates back and forth between two distinct states might be preferable. Cyclohexane, a simple example of this kind of molecule, can bend into two different forms, commonly known at the "boat" and "chair" conformations. As shown schematically, a voltage on a nearby gate might force the cyclohexane switch into one of its two conformations, affecting the conductivity of a nearby atom wire. The cyclohexane-type molecule attaches to a molecular framework while the remaining ring carbons are replaced by groups tailored to use steric repulsions or van der Waals attractions to reduce undesired switching caused by thermal energy. Switch speed is limited by molecular rotational and torsional frequencies, typically ~1-1000 GHz, which is slower than the atom relay but more reliable. Also in contrast to the atom relay, molecular relays could be packed in three dimensions, possibly up to ~1 nm^3 densities.

D. *Molecular Shuttle Switch* — A rotaxane "shuttle switch" has been synthesized by F. Stoddart's group that consists of one ring-shaped molecule that encircles and slides along a shaft-like chain molecule. Two large terminal groups at either end of the shaft prevent the shuttle ring from slipping off the shaft. The shaft contains two other functional groups, a biphenol group and a benzidine group, that serve as natural stations between which the shuttle moves. The shuttle molecule contains four positively charged functional groups that cause it to be attracted to shaft sites with extra negative change, so the shuttle spends 84% of its time at the benzidine station, which is a better electron donor than the biphenol station. Removing an electron from the benzidine station forces the shuttle to switch to the other station; switching may be controlled chemically or electrochemically. A charged gate could be added at one or both ends to force the shuttle to move between stations, with switch state probed by arranging for the ring to complete an electric circuit in one of its two positions. Switching speed is slow, limited by the speed of electron transfer to and from the benzidine station and the sluggish motion of the ring which is ~10^6 times heavier than an electron, but three-dimensional packing densities of ~0.01 nm^3 appear plausible. Other rotaxane shuttle systems have been studied.In 1999, the first chemically-assembled rotaxane-based logic gate was demonstrated, although the gates could be opened only irreversibly.

REFERENCES

Attard, G.S., et al. 1997. *Science* 278:838. Averback, R.S., J. Bernholc, and D.L. Nelson. 1991. *MRS Symposium Proceedings* Vol. 206.

Axelbaum, R.L. 1997. Developments in sodium/halide flame technology for the synthesis of unagglomerated non-oxide nanoparticles. In *Proc. of the Joint NSF-NIST Conference on Nanoparticles: Synthesis, Processing into Functional Nanostructures and Characterization* (May 12-13, Arlington, VA).

B. W. Matthews, H. Nicholson, and W. J. Becktel, "Enhanced protein thermostability from site-directed mutations that decrease the entropy of unfolding." *Proc. Nat. Acad. Sci., 84:* 6663-6667 (1987), and included references.

Baibich, M..N., J.M. Broto, A. Fert, F. Nguyen Van Dau, F. Petroff, P. Etienne, G. Creuzet, A. Friederich, and J. Chazelas. 1998. *Phys. Rev. Lett.* 61:2472.

Baird, D.: 2004, *Thing Knowledge: A Philosophy of Scientific Instruments*, University of California Press, Berkeley.

Baker, R.T.K. Synthesis, properties and applications of graphite nanofibers. In *R&D status and trends*, ed. Siegel et al.

Beck, J.S., J.C. Vartuli, W.J. Roth, M.E. Leonowicz, C.T. Kresge, K.D. Schmitt, C.T.-W. Chu, D.H. Olsen, E.W. Shepard, S.B. McCullen, J.B. Higgins, and J.L. Schlenker. 1992. *J. Am. Chem. Soc.* 114:10834.

Becker, M.F., J.R. Brock, H. Cai, N. Chaudhary, D. Henneke, L. Hilsz, J.W. Keto, J. Lee, W.T. Nichols, and H.D. Glicksman. 1997. Nanoparticles generated by laser ablation. In *Proc. of the Joint NSF-NIST Conf. on Nanoparticles.*

Berger, R., C. Gerber, H.P. Lang, and J.K. Gimzewski. 1996. Micromechanics: a toolbox for femtoscale science: "Towards a laboratory on a tip." *Microelectronic Engineering*, 35:373-9. (International Conference on Micro- and Nanofabrication, Glasgow, U.K., 22- 25 Sept. 1996).

Berkowitz, A.E., J.R. Mitchell, M.J. Carey, A.P. Young, S. Zhang, F.E. Spada, F.T. Parker, A. Hutten, and G. Thomas. 1992. *Phys. Rev. Lett.* 68:3745.

Berndt, C.C., J. Karthikeyan, T. Chraska, and A.H. King. 1997. Plasma spray synthesis of nanozirconia powder. In *Proc. of the Joint NSF-NIST Conf. on Nanoparticles.*

Bimberg, D., N. Kirstaedter, N.N. Ledentsov, Zh.I. Alferov, P.S. Kopev, and V.M. Ustinov. 1998. InGaAs-GaAs quantum-dot lasers. *IEEE J. of Selected Topics in Quant. Electronics* 3:196-205.

Bowes, C.L., A. Malek, and G.A. Ozin. 1996. *Chem. Vap. Deposition* 2:97.

Braun, P.V., P. Osenar, and S.I. Stupp. 1996. *Nature*. 368: 2. Brinker, C.J. 1996. *Curr. Opin. Solid State Mater. Sci.* 1:798. Chen, C-Y., S.L. Burkett, H.-X. Li, and M.E. Davis. 1993. *Microporous Mater*. 2:27.

Brotzman, R. 1998. Nanoparticle Dispersions. In *R&D status and trends*, ed. Siegel et al. Bu, X., P. Feng, and G.D. Stucky. 1998. Large-cage zeolite structures with multidimensional 12-ring channels. *Science* 278:2080-2085.

Brotzman, R. 1998. Nanoparticle dispersions. In *R&D status and trends*, ed. Siegel et al. Brus, L.E. 1996. Theoretical metastability of semiconductor crystallites in high pressure phases with applications to beta tin structures of silicon. *J. Am. Chem. Soc*. 118:4834-38.

Brus, L. 1996. Semiconductor colloids: Individual nanocrystals, opals and porous silicon. *Current Opinion in Colloid & Interface Science* 1:197-201. *Chemical Engineering News*. 1997. Particulate matter health studies to be reanalyzed (August 18):33.

Calcote, H.F., and D.G. Keil. 1997. Combustion synthesis of silicon carbide powder. In *Proc. of the Joint NSF-NIST Conf. on Nanoparticles*.

Canguilhem, G.: 1952, *Machine et organisme' in La connaissance de la vie*, Hachette, Paris [quoted from the fourth edition Vrin, Paris, 1971].

Carsley, J.E., R. Shaik, W.W. Milligan, and E.C. Aifantis. 1997. In *Chemistry and physics of nanostructures and related non-equilibrium materials*. ed. E. Ma, B. Fultz, R. Shull, J. Morral, and P. Nash. Warrendale, PA: TMS.

Chance, B., Mueller, P., DeVault, D. & Powers, L. (1980) *Phys. Today* 33 (10), 32-38.

Chang, K. "Smaller Computer Chips Built Using DNA as Template." New York Times, November 21, 2003.

Chantrell. 1994. *J. Appl. Phys.* 76:6811. Erb, U., G. Palumbo, R. Zugic, and K.T. Aust. 1996. In *Processing and properties of nanocrystalline materials*, ed. C. Suryanarayana, J. Singh, and F.H. Froes. Warrendale, PA: TMS. Gertsman, V.Y., M. Hoffman, H. Gleiter and R. Birringer. 1994. *Acta Metall. Mater*. 42:3539-3544.

Chopra, N.G., R.J. Luyken, K. Cherrey, H. Crespi, M.L. Cohen, S.G. Louie, and A. Zettl. 1995. *Science* 269:966.

Clasen, R. 1990. *Int. Journal of Glass and Science Technology* 63: 291. Crandall, B.C. and J. Lewis, (eds.). 1992. *Nanotechnology: Research and perspectives*. Cambridge: MIT Press.

Czekai, D., et al. 1994. *Use of smaller milling media to prepare nanoparticulate dispersions*. U.S. Patent application. Docket 69802(02) Filed 2/25/94.

Dai, H., E.W. Wong, Y.Z. Lu, S. Fan, and C.M. Leiber. 1995. *Nature* 375:769. deHeer, W.A., W.S. Bacsa, A. Chatelain, T. Gerfin, R. Humphrey-Baker, L. Forro, and D. Ugarte. 1995. *Science* 268:845.

Dresselhaus, M.S., G. Dresselhaus, and P. Eklund. 1996. *Science of fullerenes and carbon nanotubes*. San Diego: Academic Press. Gleiter, H. 1989. *Prog. Mater. Sci*. 33:223.

Drexler K.E.: 1995, 'Introduction to nanotechnology', in: Krummenacker, M. & Lewis, J. (eds.), *Prospects in Nanotechology. Proceedings of the 1st general conference on nanotechnology: developments, applications, and opportunities, November 11-14, 1992, Palo-Alto*, John Wiley & Sons, New York, pp. 1-20.

Drexler, E., Phoenix, C.: 2004, 'Self-Replication in Nanotechnology: Feasible, Potentially Safe, and Unnecessary' (unpublished paper, in preparation).

Drexler, E.: 1986, *Engines of Creation: The Coming Era of Nanotechnology*, Anchor Books, New York (expanded edition with a new afterword, 1990).

Drexler, E.: 1992, *Nanosystems: Molecular Machinery, Manufacturing, and Computation*, John Wiley & Sons, New York.

Drexler, K.E.: 1981, 'Molecular engineering: An approach to the development of general capabilities for molecular manipulation', *Proceedings of the National Academy of Sciences*, 78, no. 9, chemistry section, 5275-78.

Drexler, K.E.: 1992, *Nanosystems. Molecular machinery, manufacturing and computation*, John Wiley & Sons, New York.

Drexler, K.E.: 2001, 'Machine-Phase nanotechnology', *Scientific American*, (Sept.), 66-67.

Drexler, K.E.: 2003a, 'Open Letter', *Chemical & Engineering News*, 81, no. 48, 37-42.

Drexler, K.E.: 2003b, 'An open letter to Richard Smalley', April 16, [published on KurzweilAI.net].

Dupuy, J.P. "Common Knowledge, Common Sense." Theory and Decision 27 (1989): 37-62.

Dupuy, J.P. "Two Temporalities, Two Rationalities: A New Look at Newcomb's Paradox." in Bourgine, P.&Walliser, B. (eds.), Economics and Cognitive Science. New York: Pergamon. 1992, 191-220.

Dupuy, J.P.: 2000, *The Mechanization of the Mind*, Princeton University Press, Princeton N.J.

Ellsberg, D. "Risk, Ambiguity and the Savage Axioms." Quarterly Journal of Economics 75 (1961): 643-669.

Entingh, D., Dunn, A., Glassman, E., Wilson, J. E., Hogan, E. & Damstra, T. (1975) in *Handbook of Psychobiology*, eds. Gazzinga, M. S. & Blakemore, C. (Academic, New York), pp. 201-238.

Estermann, M., L.B. McCusker, C. Baerlocher, A. Merrouche, and H. Kessler. 1991. *Nature* 331:698.

Esumi, A., A. Suzuki, N. Aihara, K. Uswi, and K. Torigoe. 1998. *Langmuir*, 14:3157.

Evans, D.F.; Wennerstrom, H.: 1999, *The Colloidal Domain: Where Physics, Chemistry, Biology and Technology Meet*, John Wiley & Sons, New York.

Faraday, M.: 1857, 'Experimental relations of gold (and other metals) to light', *Philosophical Transactions of the Royal Society London*, 147, 145-181.

Fennema, O. R. (1973) in *Low-Temperature Preservation of Foods and Living Matter*, eds. Fennema, O. R., Powrie, W. D. & Marth, E. H. (Dekker, New York), pp. 436-475.

Feyerabend, P.: 1981, *Realism, Rationalism and Scientific Method* (Philosophical Papers, volume 1), Cambridge University Press, Cambridge.

Feynman R. "There's Plenty of Room At the Bottom." Talk given on at the annual meeting of the American Physical Society at the California Institute of Technology, 1959.

Feynman, R.: 1960, 'There's Plenty of Room at the Bottom', *Engineering and Science*, 23, 22-36.

Fleming, D. "The Economics of Taking Care: An Evaluation of the Precautionary Principle." in Freestone, D.&Hey, E. (eds.), The Precautionary Principle and International Law. La Haye: Kluwer Law International, 1996.

Foerster, H.v.: 1962, 'Bio-Logic', in: E.E. Bernard & M.R. Kare (eds.), *Biological prototypes and synthetic systems*, Plenum Press, New York, vol. 1., pp. 1-12

Friedlander, S.K. 1993. Controlled synthesis of nanosized particles by aerosol processes. *Aerosol Sci. Technol.* 19:527.

Friedlander, S.K. 1998. Synthesis of nanoparticles and their agglomerates: Aerosol reactors. In *R&D status and trends*, ed. Siegel et al.

Friedlander, S.K., H.D. Jang, and K.H. Ryu. 1998. *Appl. Phy. Lett.* 72(2):173.

Gleiter, H. 1989. *Prog. Mater. Sci.* 33:223. Goddard, W.A. 1998. Nanoscale theory and simulation. In *R&D status and trends*, ed. Siegel et al.

Gleiter, H. 1990. *Progress in Materials Science* 33:4. Guinier, A. 1938. *Nature* 142:569; Preston, G.D. 1938. *Nature* 142:570.

Goddard, W.A. 1998. Nanoscale theory and simulation. In *R&D status and trends*, ed. R. Siegel et al.

Godsell, D.: 2003, *Living Machinery. Bionanotechnology: Lessons from Nature*, Wiley-Liss, New-York.

Guarnieri, F., M. Fliss, and C. Bancroft. 1996. Making DNA Add. *Science* 273:220-223. Hanes, J., J.L. Cleland, and R. Langer. 1997.

Günther, B., A. Baalmann, and H. Weiss. 1990. *Mater. Res. Soc. Symp. Proc.* 195:611-615.

Gutte, B., Dannigen, M. & Wittschieber, E. (1979) *Nature (London)* 281, 650-655.

Hadjipanayis, C.G. 1998. Nanostructured magnetic materials. In *R&D Status and Trends*, ed. Siegel et al.

Han, W., S. Fam, O. Li, and Y. Hu. 1998. Synthesis of gallium nitride nanorods through a carbon nanotube-confined reaction. *Science* 277:1287-1289.

Hansson, S.O. "The False Promises of Risk Analysis." Ratio 6 (1993): 16-26.

Held, R., T. Heinzel, A.P. Studerus, K. Ensslin, and M. Holland. 1997. Semiconductor quantum point contact fabricated by lithography with an atomic force microscope. *Appl. Phys. Lett.* 71:2689-91.

Hellemans, A. 1998. X-rays find new ways to shine. *Science* 277:1214-15.

Hietpas, P.B., S.D. Gilman, R.A. Lee, M.R. Wood, N. Winograd, and A.G. Ewing. 1996. Development of votammetric methods, capillary electrophoresis and tof sims imaging for constituent analysis of single cells. In *Nanofabrication and biosystems*, ed.

Higgins, R.J. 1997. An economical process for manufacturing of nano-sized powders based on microemulsion-mediated synthesis. In *Proc. of the Joint NSF-NIST Conf. on Nanoparticles.*

Hiruma, K., M. Yazawa, T. Katsoyama, K. Ogawa, K. Haraguchi, M. Koguchi, and H. Kakibayashi. 1995. *J. Appl. Phys.* 77(2):476.

Hoch et al. Ho, S.V., P.W. Sheridan, and E. Krupetsky. 1996. Supported polymeric liquid membranes for removing organics from aqueous solutions.

Hoch, H.C., L.W. Jelinski, and H.G. Craighead, eds. 1996. *Nanofabrication and biosystems.* New York: Cambridge University Press.

Howland, H.: 1962, Structural, hydraulic and economic aspects of leaf venation and shape', in: E.E. Bernard & M.R. Kare (eds.), *Biological prototypes and synthetic systems*, Plenum Press, New York, vol. 1, pp. 183-192 .

Hoyningen-Huene, P.: 1993, *Reconstructing Scientific Revolutions: Thomas S. Kuhn's Philosophy of Science* (trans. by A. Levin), University of Chicago Press, Chicago.

Huang, J.Y., Y.K. Wu, and H.Q. Ye. 1996. *Acta Mater*. 44:1211. Inoue, A. 1997. Private communication. Inturi, R.B., and Z. Szklavska-Smialowska. 1992. *Corrosion* 48:398. Karch, J., R. Birringer, and H. Gleiter. 1987. *Nature* 330:556.

Hubbell, J.A., and R. Langer. 1995. Tissue engineering. *Chem. Eng. News* (March 13): 42- 54.

Huo, Q., D.I. Margolese, U. Ciesla, P. Feng, T.E. Gier, P. Sieger, R. Leon, P.M. Petroff, F. Schuth, and G.D. Stucky. 1994. *Nature* 368:317.

Imae, Y., and T. Atsumi. 1989. T. Na+-driven bacterial flagellar motors: A mini-review. *J. Bioenergetics and Biomembranes* 21:705-716.

Itakura, K. & Riggs, A. D. (1980) *Science* 209, 1401-1405.

Jaworek, T., D. Deher, G. Wegner, R.H. Wieringa, and A.J. Schouten. 1998. Electromechanical properties of an ultrathin layer of directionally aligned helical polypeptides. *Science* 279:57-60.

Jena, P., S.N. Khanna, and B.K. Rao. 1996. In *Science and technology of atomically engineered materials*, ed. P. Jena. River Edge, NJ: World Scientific.

Jonas, H. The Imperative of Responsibility. In Search of an Ethics for the Technological Age. Chicago: University of Chicago Press, 1985.

Jounet, C., W.K. Maser, P. Bernier, A. Loiseau, M. Lamy de la Chapelle, S. Lefrant, P. Deniard, R. Lee, and J.E. Fischer. 1997. *Nature* 388:756.

Junno, T., S.-B. Carlsson, H. Xu, L. Montelius, and L. Samuelson. 1998. Fabrication of quantum devices by angstrom-level manipulation of nanoparticles with an atomic force microscope. *Appl. Phys. Lett.* 72:548-550.

K. E. Drexler, "Molecular engineering: An approach to the development of general capabilities for molecular manipulation." *Proc. Nat. Acad. Sci.*, 78: 5275-5258 (1981).

K. E. Drexler, *Engines of Creation*, Anchor/Doubleday (New York, 1986).

Karak, N., and S. Maiti. 1997. Dendritic polymers: A class of novel material. *J. Polym. Mater*.14:105.

Karger, J., and D.M. Ruthven. 1992. *Diffusion in zeolites*. New York: J. Wiley. Krätschmer, W., L.D. Lamb, K. Fostiropoulos, and D.R. Huffman. 1990. *Nature* 347:354.

Karplus, M. & Weaver, D. L. (1976) *Nature (London)* 260, 404-406.

Kasuga, T., M. Hiramatsu, A. Hoson, T. Sekino, and K. Niihara. 1998. *Langmuir* 14:3160.

Katari, J.E.B., V.L. Colvin, and A.P. Alivisatos. 1994. *J. Phys. Chem*. 98:4109.

Ke, M., S.A. Hackney, W.W. Milligan, and E.C. Aifantis. 1995. *Nanostructured Mater*. 5:689.

Kear, B., and G. Skandan. 1998. Nanostructural bulk materials: Synthesis, processing, properties and performance. *In R&D status and trends*, ed.

Kear, B.H., R.K. Sadangi, and S.C. Liao. 1997. Synthesis of WC/Co/diamond nanocomposites. In *Proc. of the Joint NSF-NIST Conf. on Nanoparticles*.

Kirschvink, J.L., A. Koyayashi-Kirschvink, and B.J. Woodford. 1992. Magnetite biomineralization in the human brain. *Proc. Nat'l. Acad. Sci.* USA 89:7683-7687.

Kishida, M., T. Fujita, K. Umakoshi, J. Ishiyama, H. Nagata, and K. Wakabayashi. 1995. *Chem. Commun*. 763.

Klein, J.D., et al. 1993. *Chem. Mater*. 5:902. Koch, C.C. 1989. Materials synthesis by mechanical alloying. *Annual Review of Mater. Sci.* 19:121-143.

Knight, F.H. Risk, Uncertainty and Profit. London School of Economics and Political Science, London, 1933.

Koch, C.C. 1998. Bulk behavior. In *R&D status and trends*, ed. Siegel et al. 32 *Evelyn L. Hu and David T. Shaw.*

Kortan, A.R., R. Hull, R.L. Opila, M.G. Bawendi, M.L. Steigerwald, P.J. Carroll, and L.E. Brus. 1990. *J. Am. Chem. Soc.* 112:1327.

Kourilsky, Ph.&Viney, G. Le Principe de précaution. Report tc the Prime Minister, Paris, Éditions Odile Jacob, 2000.

Krejchi, M.T., E.D.T. Atkins, A.J. Waddon, M.J. Fournier, T.L. Mason, and D.A. Tirrell. 1994. Chemical sequence control of beta-sheet assembly in macromolecular crystals of periodic proteins. *Science* 265:1427-1432.

Krejchi, M.T., S.J. Cooper, Y. Deguchi, E.D.T. Atkins, M.J. Fournier, T.L. Mason, and D.A. Tirrell. 1997. Crystal structures of chain-folded antiparallel beta-sheet assemblies from sequence-designed periodic polypeptides. *Macromolecules* 30:5012-5024.

Kresge, C.T., M.E. Leonowicz, W.J. Roth, J.C. Vartuli, and J.S. Beck. 1992. *Nature* 359:710. Kupperman, A., S. Nadimi, S. Oliver, G. Ozin, J. Garcés, and M. Olken. 1993. *Nature* 365:239.

Kroes, P. & Meijers, A. "The Dual Nature of Technical Artifacts—Presentation of a New Research Programme." Techné 6, 2 (2002): 4-8.

Kuhn, T.: 1970, *The Structure of Scientific Revolutions* (second edition), University of Chicago Press, Chicago.

Kumar, A., and G.M. Whitesides. 1993. Features of gold having micrometer to centimeter dimensions can be formed through a combination of staming with an elastomeric stamp and an alkanethiol ink followed by chemical etching. App. Phys. Lett. 63:2002-2004.

Kurzweil, R.: 1998, *The Age of Spiritual Machines, How We Will Live, Work and Think in the New Age of Intelligent Machines*, Phoenix, New York.

Kyprianidou-Leodidou, T., W. Caseri, and V. Suter. 1994. *J. Phys. Chem.* 98:8992. Kung, H.H., and E.I. Ko, 1996. *Chem. Eng. J.* 64:203.

L. J. Perry and R. Wetzel, "Disulfide bond engineered into T4 lysozyme: stabilization of the protein toward thermal inactivation." *Science,* 226: 555-557 (1984).

L. Regan and W. F. DeGrado, "Characterization of a helical protein designed from first principles." *Science,* 241: 976-978 (1988).

Laudan, L.: 1984, *Science and Values: The Aims of Science and their Role in Scientific Debate*, University of California Press, Berkeley.

Lehn, J.M., 2004: 'Une chimie supramoléculaire foisonnante', *La lettre de l'Académie des sciences*, 10, 12-13.

Leonard, D., M. Krishnamurthy, C.M. Reaves, S.P. Denbaars, and P.M. Petroff. 1993. Direct formation of quantum-sized dots from uniform coherent islands of InGaAs on GaAs surfaces. *Appl. Phys. Lett.* 63:3203-5.

Liang, G., Z. Li, and E. Wang. 1996. *J. Mater. Sci.* Makino, A., A. Inoue, T. Hatanai, and T. Bitoh. 1997. *Materials Science Forum* 235-238: 723.

Lide, D.R., ed. 1993-1994. *CRC Handbook of Chemistry and Physics*, 74th ed. Lin, H.-P., and C.-Y. Mou. 1996. *Science* 273:765.

Lukens, R.: 2004, 'Historic Nanotechnology', *Chemical Heritage,* 22, 29.

Majetich, S.A. and A.C. Canter. 1993. *J. Phys. Chem.* 97:8727.

Mao, C., W. Sun, and N.C. Seeman. 1997. Construction of Borromean rings from DNA. *Nature* 386:137-138.

Martin, T.P., N. Malinowski, U. Zimmerman, U. Naher, and H. Schaber. 1993. *J. Chem. Phys.* 99:4210.

Martin, T.P., U. Naher, H. Schaber, U. Zimmerman. 1993. *Phys. Rev. Lett.* 70:3079. Prigogine, I., and S. Rice. 1988. *Advances in chemical physics*, Vol. 70, Parts 1 & 2. New York: J. Wiley.

McCammon, J. A., Gelin, B. R. & Karplus, M. (1977) *Nature (London)* 267, 585-590.

McConnell, H.M. 1996. Light-addressable potentiometric sensor: Applications to drug discovery. In *Nanofabrication and biosystems*, ed. Hoch et al.

McCulloch, W.S.: 1962, 'The imitation of one form of life by another – Biomimesis', in: E.E. Bernard & M.R. Kare (eds.), *Biological prototypes and synthetic systems*, Plenum Press, New York, vol. 1., p. 393-97.

Mehl, R.F., and R.W. Cahn. 1983. Historical development. In *Physical metallurgy.* North Holland. Milligan, W.W., S.A. Hackney, M. Ke, and E.C. Aifantis. 1993. *Nanostructured Materials* 2:267.

Minsky, M.: 1995, 'Virtual Molecular Reality', in: Krummenacker, M. & Lewis, J. (eds.), *Prospects in Nanotechology. Proceedings of the 1st general conference on nanotechnology: developments, applications, and opportunities, November 11-14, 1992, Palo-Alto*, John Wiley & Sons, New York, pp. 187-205.

Mishra, R.S., and A.K. Mukherjee. 1997. Oral presentation at TMS meeting, Indianapolis, Indiana, 16-18 September 1997, to be published in proceedings of Symp. "Mechanical Behavior of Bulk Nano-Materials."

Mishra, R.S., R.Z. Valiev, and A.K. Mukherjee. 1997. *Nanostructured Materials* 9:473. Morris, D.G., and M.A. Morris. 1991. *Acta Metall. Mater.* 39:1763-1779.

Moore, J.C., H.M. Jin, O. Kuchner, and F.H. Arnold. 1997. Strategies for the in vitro evolution of protein function: Enzyme evolution by random recombination of improved sequences. *J. Mol. Biol.* 272:336-347.

Morris, D.G., and M.A. Morris. 1997. *Materials Science Forum* 235-238:861. Nagpal, P., and I. Baker. 1990. *Scripta Metall. Mater.* 24:2381.

Nieman, G.W., J.R. Weertman, and R.W. Siegel. 1991a. *Mater. Res. Soc. Symp. Proc.* 206:581-586.

Nomura, M. & Held, W. (1974) in *Ribosomes,* eds. Nomura, M., Tissiers, A. & Lengyel, P. (Cold Spring Harbor Laboratory, Cold Spring Harbor, NY), pp. 193-203.

Nordmann, A. (rapp.) Converging Technologies—Shaping the Future of European Societies, European Commission report, 2004.

Parker, J.C., et al. 1995. U.S. Patent 5,460,701. POST (Parliamentary Office of Science and Technology). 1995.

Phoenix, C.; Drexler, E.: 2004, 'Safe Exponential Manufacturing', *Nanotechnology*, 15, 869-72.

Rademann, K., B. Kaiser, U. Even, F. Hensel. 1987. *Phys. Rev. Lett.* 59:2319. Rao, C.N.R., B.C. Satishkumar, and A. Govindaraj. 1997. *Chem. Commun.* 1581.

Ramanan, V.R. 1998. Nanocrystalline soft magnetic alloys for applications in electrical and electronic devices. In *R&D Status and Trends*, ed. Siegel et al.

Rao, M.B., and S. Sircar. 1993. *Gas Separation and Purification* 7:279. Reetz, M.T., et al. 1995. *Science* 267:367.

Rechard, R.P. "Historical Relationship Between Performance Assessment for Radioactive Waste Disposal and Other Types of Risk Assessment." Risk Analysis 19, 5 (1999): 763-807.

Rietman, E.A.: 2001, 'Drexler hypothesis of a universal assembler is supported not by theoretical arguments alone but by existence proof in the form of biological life', in: *Molecular Engineering of Nanosystems*, Springer, New York & Berlin.

Rip, A., Misa, Th. J.,&Schot, J. W. (eds.). Managing Technology in Society. The Approach of Constructive Technology Assessment. London: Pinter Publishers, 1995.

Roco, M.&Bainbridge, W. (eds.). Converging Technologies for Improving Human Performances, National Science Foundation report, 2002.

Rofagha, R., R. Langer, A.M. El-Sherik, U. Erb, G. Palumbo, and K.T. Aust. 1991. *Scr. Metall. Mater*. 25:2867.

Roher, H. 1993. *Jpn. J. Appl. Phys*. 32:1335. Rohlfing, E.A., D.M. Cox, and A. Kaldor. 1984. *J. Chem. Phys*. 81:3846.

Romanov, A.E., V.I. Vladimirov. 1992. In *Dislocations in solids*, ed. F.R.N. Nabarro, Vol. 9. Amsterdam: North-Holland. Salishekev, G.A., O.R. Valiakhmetov, V.A. Valitov, and S.K. Mukhtarov. 1994. *Materials Science Forum*. 170-172:121.

Rosenberg, N. "Why Technology Forecasts Often Fail." Futurist July/August 1995, 16- 21.

Ruthven, D.M., S. Farooq, K.S. Knaebel. 1994. *Pressure swing adsorption*. New York: VCH Publishers. Sayari, A. 1996. *Chem. Mater*. 8:1840.

S. P. Ho and W. F. DeGrado, "Design of a 4-Helix bundle protein: Synthesis of peptides which self-associate into a helical protein." *J. Am. Chem. Soc.,* 109: 6751-6758 (1987).

Sanders, P.G., J.A. Eastman, and J.R. Weertman. 1996. In *Processing and properties of nanocrystalline materials*, ed. Suryanarayana et al.

Sanders, P.G., M. Rittner, E. Kiedaisch, J.R. Weertman, H. Kung, and Y.C. Lu. 1997. *Nanostructured Mater*. 9:433.

Sankey, H.: 1994, *The Incommensurability Thesis*, Avebury, Aldershot.

Sarikaya, M.; Aksay, I. (eds.), 1995, *Biomimetics. Design and Processing of Materials*, AIP Press, Woodbury, New York.

Scanlan, R.M., W.A. Fietz, and E.F. Koch. 1975. *J. Appl. Phys*. 46:2244.

Scheraga, H. A. (1978) in *Versatilty of Proteins*, ed. Li, C. H. (Academic, New York), pp. 119-132.

Schiefsky, M.J.: forthcoming, 'Art and Nature in Ancient Mechanics', in: W.R. Newman & B. Bensaude-Vincent (eds.), *The Artificial and the Natural: An Ancient Debate and its Modern Descendants*, MIT Press, Cambridge, MA.

Schultz, L., J. Wecker, and E. Hellstern. 1987. *J. Appl. Phys*. 61:3583.

Schultz, L., K. Schnitzke, and J. Wecker. 1989. *J. Magn. Mater*. 80:115.

Schwarz, R.B. 1998. Storage of hydrogen powders with nanosized crystalline domains. In *R&D Status and Trends*, ed. Siegel et al.

Shen, T.D., C.C. Koch, T.Y. Tsui, and G.M. Pharr. 1995. *J. Mater. Res.* 10:2892.

Shull, R.D. 1998. NIST activities in nanotechnology. In *R&D Status and Trends*, ed.

Shull, R.D., R.D. McMichael, and J.J. Ritter. 1993. *Nanostructured Mater.* 2:205.

Siegel et al. Ogunnaike, B and W. Ray. 1994. *Process dynamics, modeling and control*. Oxford University Press, pp 5-21; 1033-48.

Siegel et al. Siegel, R.W. 1997. *Materials Science Forum* 235-238:851.

Siegel R.W. and G.E. Fougere. 1994. In *Nanophase materials*, ed. G.C. Hadjipanayis and R.W. Siegel. Netherlands: Kluwer Acad. Publ.

Siegel, H.: 1980, 'Objectivity, Rationality, Incommensurability, and More', *British Journal for the Philosophy of Science*, 31, 359-384.

Siegel, R.W., E. Hu, and M.C. Roco, eds. 1998. *R&D status and trends in nanoparticles, nanostructured materials, and nanodevices in the United States*. Baltimore: Loyola College, International Technology Research Institute.

Smalley, R.: 2001, 'Of Chemistry, Love and Nanobots', *Scientific American*, 285, 76-77.

Smalley, R.: 2003a, 'Smalley Responds', *Chemical & Engineering News*, 81, 39-40.

Smalley, R.: 2003b, 'Smalley Concludes', *Chemical & Engineering News*, 81, 41-42.

Smalley, R.E.: 1999, 'Prepared written statement and supplemental material', Rice University, 22 June [http//www.house.gov/science/smalley_062299.htm].

Smalley, R.E.: 2001, 'Of Chemistry, Love and Nanobots', *Scientific American*, (Sept.), 76-77.

Smith, C.S. 1981. *A search for structure*. Cambridge, Mass.: MIT Press.

Snow, E.S., P.M. Campbell, and F.K. Perkins. 1997. Nanofabrication with proximal probes. *Proceedings of the IEEE* 85:601-11.

Staden, H.v.: forthcoming, '*Physis* and *Technê* in Greek Medicine', in: W.R. Newman & B. Bensaude-Vincent (eds.), *The Artificial and the Natural: An Ancient Debate and its Modern Descendants*, MIT Press, Cambridge, MA.

Stupp, S., and P.V. Braun. 1997. Molecular manipulation of microstructures: biomaterials, ceramics, and semiconductors. *Science* 277:1242-1248.

Sunstrom, J.E., IV, W.R. Moser, and B. Marshik-Guerts. 1996. *Chem. Mater.* 8:2061.

Surowiecki, J.: 2004, 'Bring on the Nanobubble', *The New Yorker*, March 15, p. 68.

4

Nanobiology & Nanomedicine: Tools & Techniques

Nanomedicine

Nanomedicine research is being funded by government sources, such as the National Institutes of Health (NIH), and by companies in various sectors, including pharmaceutical, biotechnology and medical devices. Because of the focused and shorterterm development horizon of commercial applications, nanomedicine research is more easily characterized in these areas. The nanomedicine taxonomy summarized below and further described in this document, classifies some of the leading areas that nanotechnology tools, materials, devices, and intelligent materials and machines are currently applied in medical research. Specific company and university research activities should be interpreted as examples and are only a subset of the vast scope of nanomedicine research currently underway.

Nanotechnology provides a wide range of new technologies for developing customized solutions that optimize the delivery of pharmaceutical products. To be therapeutically effective, drugs need to be protected during their transit to the target action site in the body while maintaining their biological and chemicals properties. Some drugs are highly toxic and can cause harsh side effects and reduced therapeutic effect if they decompose during their delivery. Depending on where the drugs will be absorbed (i.e. colon, small intestine, etc), and whether certain natural defense mechanisms need to be passed through such as the blood-brain barrier, the transit time and delivery challenges can be greatly different.

Once a drug arrives at its destination, it needs to be released at an appropriate rate for it to be effective. If the drug is released too rapidly it might not be completely absorbed, or it might cause gastro-intestinal irritation and other side effects. The drug delivery system must positively impact the rate of absorption, distribution, metabolism, and excretion of the drug or other substances in the body. In addition, the drug delivery system must allow the drug to bind to its target receptor and influence that receptor's signalling and action, as well as other drugs, which might also be active in the body. Drug delivery systems also have severe restrictions on the materials and production processes that can be used. The drug delivery material must be compatible and bind easily with the drug, and be bioresorbable (i.e. degrade into fragments after use which are either metabolized or eliminated via normal excretory routes). The production process must respect stringent conditions on processing and chemistry that won't degrade the drug, and still provide a cost effective product. Nanotechnology can offer new drug delivery solutions in the following areas.

One major class of drug delivery systems is materials that encapsulate drugs to protect them during transit in the body. Drug encapsulation materials include liposomes and polymers (i.e. Polylactide (PLA) and Lactide-co-Glycolide (PLGA)) which are used as microscale particles. The materials form capsules around the drugs and permit timed drug release to occur as the drug diffuses through the encapsulation material. The drugs can also be released as the encapsulation material degrades or erodes in the body. When encapsulation materials are produced from nanoparticles in the 1 to 100nm size range instead of bigger microparticles, they have a larger surface area for the same volume, smaller pore size, improved solubility, and different structural properties. This can improve both the diffusion and degradation characteristics of the encapsulation material.

In addition to liposomes and polymers, other types of nanoparticles are available for encapsulation. Materials such as silica and calcium phosphate (hydroxyapatite) have demonstrated superior properties at the nanoscale than the microscale, and can potentially be better suited for certain drug delivery challenges mentioned above. Nanoparticle encapsulation is also being investigated for the treatment of neurological disorders to deliver therapeutic molecules directly to the central nervous system beyond the blood-brain barrier, and to the eye beyond the blood-retina barrier. Applications could include Parkinson's, Huntington's, Alzheimer's, ALS and diseases of the eye. In particular, ophthalmic disorders represent a rapidly growing area that is threatened by various diseases such as age related macular degeneration (AMD), diabetic retinopathy, glaucoma, and retinitis pigmentosa (RP).

Neurotech is developing an encapsulated cell therapy to treat eye diseases. It uses a semipermeable membrane to encapsulate cells, which also permits therapeutic agents produced by the cells to diffuse through the membrane. The membrane isolates the cells from the local environment and minimizes immune rejection. The encapsulated cells are administered by a device implanted in the eye to permit the continuous release of therapeutic molecules from living cells. This avoids direct injections into the eye, which may not be practical for regular administrations. Advectus Life Sciences is developing a nanoparticle-based drug delivery system for the treatment of brain tumors. The anti-tumor drug doxorubicin is adhered to a nanopolymer Poly ButylCyano Acrylate (PBCA) particle and coated with polysorbate 80. The drug is injected intravenously and circulates through the blood stream. The polysorbate 80 attracts plasma apolipoproteins and is used by the blood stream to carry lipids. This is intended to create a camouflage effect similar to LDL cholesterol, which allows the drug to pass the blood-brain barrier.

Another class of drug delivery systems where nanotechnology offers interesting solutions is in the area of nanomaterials that carry drugs to their destination sites and also have functional properties. Certain nanostructures can be controlled to link with a drug, a targeting molecule, and an imaging agent, then attract specific cells and release their payload when required. Because of their nanoscale size, nanostructures have the ability to enter the cells, as cells will typically internalize materials below 100nm. Some of the leading nanostructures being used for this purpose include fullerenes, dendrimers, and nanoshells. Fullerenes (or Buckyballs) are natural hollow spheres, one nanometer in diameter, made up of 60 carbon atoms. Fullerenes create a unique drug delivery platform that allows active pharmacopheres to be grafted to its surface in three-dimensional orientations for precise control in matching fullerene compounds to biological targets, in entrapping atoms within the fullerene cage, and for attaching fullerene derivatives to targeting agents. C Sixty is developing fullerene-based drug delivery platforms which link fullerenes with antibodies and other targeting agents. Some of C Sixty's drug delivery systems include fullerene-decorated chemotherapeutic constructs, fullerene-radiopharmaceuticals, and fullerene-based liposome systems, called Buckysomes, for the delivery of single drug loads or multiple drug cocktails. Employing rational drug design, C Sixty has produced several drug candidates using its fullerene platform technology in the areas of HIV/AIDS, neuro-degenerative disorders and cancer. Another nanomaterial used as a drug delivery scaffold is the dendrimer, a polymer molecule discovered by Don Tomalia of Dendritic Nanotechnologies. Researchers such as James Baker of the University of Michigan are using dendrimers to get genetic material or tumor-destroying therapies into a cell without triggering an immune response. This is due to the dendrimer's small size and branched structure. Dendrimers can be designed to release attached compounds in response to a specific molecule or chemical reaction. A layered sphere called a nanoshell is being developed by Nanospectra for drug delivery. The nanoshell has a gold exterior layer which covers interior layers of silica and drugs. Nanoshells can be made to absorb light energy and then convert it to heat. As a result, when nanoshells are placed next to a target area such as tumor cell, it can release tumorspecific antibodies when infrared light is administered.

Nano and micro technologies are part of the latest advanced solutions and new paradigms for decreasing the discovery and development times for new drugs, and potentially reducing the development costs. Traditional trial-and-error methods have contributed to a discovery process lasting 10 years or more for new drugs to reach the market. In recent years, a number of new and complementary technologies have been developed which considerably impact the drug discovery process. High-throughput arrays and ultra-sensitive labeling and detection technologies are being used to increase the speed and accuracy of identifying genes and genetic materials for drug discovery and development. These micro and nano technologies along with information technology solutions such as combinatorial chemistry, computational biology, computer-aided drug design, data mining, and data processing tools are addressing the challenges related to eliminating critical bottlenecks in drug discovery. In addition to the capacity for processing information, the number of drug candidates that have been screened in the last ten years has increased by three orders of magnitude from 500,000 drug compounds to approximately 1.5 billion according to BCC.

Nanotechnology provides a new generation of biocompatible nanomaterials for repairing and replacing human tissues. Human tissue that is diseased or traumatically compromised may require synthetic materials for its repair or replacement. While most types of tissues repair the interaction of stem cells with chemical modulators, there are differences in the ways that various tissues heal. "Hard" tissues such as bone and teeth heal by reproducing tissues indistinguishable from the original. However in cases where a dental or artificial bone implant is required, the structural material used in the implant may trigger immune rejection, corrode in the body fluids, or no longer bond to the host bone. This can require additional surgery or result in the loss of the implant's function. In many cases, the failure occurs at the tissue-implant interface, which may be due to the implant material weakening its bond with the natural material. To overcome this, implants are often coated with a biocompatible material to increase their adherence properties and produce a greater surface area to volume ratio for the highest possible contact area between the implant and natural tissue. "Soft" tissues such as skin, muscle, nerves, blood vessels and ligaments repair damaged areas with fibrous tissue. Damaged tissue from various sources such as burns and ulcers can be self-repaired by the body, but can also result in scar formation. Graft material using artificial sheets can replace skin and other tissue with reasonable graft stability and cosmetic outcome. In other types of tissue, notably "Ultrasoft" tissue such as cell membrane and organelles that exhibit metabolic function, tissue replacement can best occur when living cells are transplanted in a mesh-structured synthetic scaffold. The scaffold incorporates signalling ligands or DNA fragments to elicit specific cellular responses, and molecular sensors to accept feedback from the in vivo environment. The scaffold is typically a temporary structure that is bioresorbable when the tissue is regenerated. The scaffold material needs to be fabricated into a desired shape or threedimensional structure, with specific surface properties to support the site where the organized growth of multiple cell types will take place. Nanotechnology can new offer new solutions for tissue repair and replacement in the following areas.

Nanotechnology brings a variety of new high surface area biocompatible nanomaterials and coatings to increase the adhesion, durability and lifespan of implants. Ceramic materials such as calcium phosphate (hydroxyapatite or HAP) are made into implant coatings using nanosized particles instead of microsized particles. In addition to the higher surface areas and improved adhesion properties of the nanoparticle coatings, improved coating techniques are also being developed. While high temperature processes such as plasma spray can melt ceramic particles and reduce their surface area and adhesion properties, new low temperature processes with electromagnetic fields can maintain the nanomaterial properties. This provides the maximum possible contact area between the implants and bone surface to improve the potential for in-growth in the host bone. New types of nanomaterials are being evaluated as implant coatings to improve interface properties. For example, nanopolymers such as polyvinyl alcohol (PVA) can be used to coat implantable devices that are in contact with blood (e.g. artificial hearts, vascular grafts, catheters) for dispersing clots or preventing their formation.

Nanostructures are being researched for the preparation and improvement of tissue regeneration scaffolds. Research areas include the ability to develop molecularly sensitive polymers using the optical properties of nanoparticles as control systems, manipulating the stiffness and strength of scaffolds using hybrid nanostructures, and the use of nanotechnology to prepare molecular imprints to maximize long-term viability and function of cells on scaffold surfaces. With the ultimate objective of growing large complex organs, a variety of nanomaterials and nanotechnology fabrication techniques are being investigated as tissue regeneration scaffolds that provide improved structural requirements and guide the activity of seeded cells. Some examples are as follows:

- Nanoscale polymers such as Polyvinyl alcohol (PVA) are being molded into heart valves and seeded with fibroblasts and endothelial cells.
- PVA is also being investigated for the cornea by having corneal epithelia cells seeded in a PVA hydrogel structure. This polymer material can absorb more than 20 per cent its weight in water while maintaining a distinct three-dimensional structure.
- A polyglycolic ball is being experimented with muscle cells and bladder endothelial cells.
- Polymer nanocomposites are being researched for bone scaffolds.

Commercially viable solutions are thought to be 5 to 10 years away. Scientific challenges related to a better understanding of molecular/cell biology and fabrication methods for producing large three-dimensional scaffolds are among the many obstacles yet to be overcome.

Nanostructures are also being used to study the fundamental properties of implanted tissues. In areas of in vivo analysis, nanostructures are used as tracers for implanted cells, and to study the response of host to implanted tissues.

Nanotechnology provides a new generation of biocompatible materials that can be used as implants or temporary biosorbable structures. Bone is a high strength material that is used as both weight bearing and non-weight bearing structures. Bones are more than just structural materials as they also contain interconnected pores that allow body fluids to carry nutrients and permit interfacial reactions between hard and soft tissues. In the case of bone fractures, grafts, disorders, dental applications and other types of surgery, bones may require repair or replacement. A variety of natural materials are used as bone substitutes. These include autograft from the patient's pelvis, allograft from another human, bovine material or coral blocks. Natural materials tend to be brittle and can lose mechanical strength during sterilization. They can also cause inflammation, pain at the pelvis graft site, and potentially transmit disease. Bone cavities can also be filled with synthetic bone cement. Current bone cements containing polymethylmethacryate (PMMA) act as a filler or grout, which is injected as a flowable paste and then hardens in vivo. While PMMA cement can offer adequate mechanical properties and bonding, it is typically recommended only for non-weight bearing bones. PMMA has also been linked to tissue damage, nerve root pain and other side effects.

Nanotechnology brings a variety of new high surface area biocompatible nanomaterials that can be used for bone repair and cavity fillers. High strength nanoceramic materials, such as calcium phosphate apatite (CPA) and hydroxyapatite (HAP), can be made into a flowable, moldable nanoparticle paste that can conform to and interdigitate with bone. As natural bone is approximately 70 per cent by weight. CPA including hydroxyapatite (HAP), biocompatibility is thought to be extremely high with minimal side effects. As its dense surface and tight three dimensional crystalline structure will allow for a superior compressive strength to PMMA, nanoceramics may be suitable for both weight bearing and non-weight bearing bones. Nanoceramics such as CPA is nanocrystalline and can have grain size of under 50nm. Its nanocrystals can be connected to each other to form connective hard tissue required for bones. Bone cement material being developed at the University of South Carolina and Competitive Technologies, can harden in vivo and in the presence of serum, form a solid bone-like structure capable of stabilizing a fractured bone or used as a bone substitute.

Nanotechnology also brings advances in bioresorbable materials. Bioresorbable polymers are currently being used in degradable medical applications such as sutures and orthopaedic fixation devices. With new production methods, nanostructures are being fabricated which could be used as temporary implants. Bioresorbable implants will biodegrade and do not have to be removed in a subsequent operation. In one application, nanostructured implants are being designed to degrade at a rate that will slowly transfer load to a healing bone that it is supporting. Research is also being done on a flexible nanofiber membrane mesh that can be applied to heart tissue in open-heart surgery. The mesh can be infused with antibiotics, painkillers and medicines in small quantities and directly applied to internal tissues. The nanomaterial will degrade over time and not stick to surgeons' wet gloves, which can complicate the use of some materials currently in use.

Smart materials are a class of nanomaterials that respond to changes in the environment such as a drop in temperature or pH. An environmental change could trigger a physical or chemical effect that mimics a natural mechanism in the body. For example, applications could include a smart polymer that flexes with mechanical strength as an artificial muscle, or a hydrogel that dissolves according to body chemistry to more efficiently deliver drugs. When attached to or embedded in a structure, smart materials can potentially sense disturbances, or trigger physical reactions with force in actuators, piezo sensors, and shape memory metals. It is possible to have smart materials used in the body to exceed current human performance on multiple dimensions, and detect environmental conditions beyond current human limits.

Nanotechnology offers sensing technologies that provide more accurate and timely medical information for diagnosing disease, and miniature devices that can administer treatment automatically if required. Health assessment can require medical professionals, invasive procedures and extensive laboratory testing to collect data and diagnose disease. This process can take hours, days or weeks for scheduling and obtaining results. Some medical information is extremely time sensitive such as finding out if there is sufficient blood flow to an organ or tissue after transplant or reconstructive surgery, before irreversible damage occurs. Certain medical tests such as biopsies are subjective and can provide inconclusive or incorrect results.

In a false negative result where a needle misses the tumor and then samples a normal tissue, the cancer may go untreated and can impact a patient's chances for long-term survival. Some tests such as diabetes blood sugar levels require patients to administer the test themselves to avoid the risk of their blood glucose falling to dangerous levels. Certain users such as children and the elderly may not be able to perform the test properly, timely or without considerable pain. People who are exposed to radiation or hazardous chemicals in their work environment are at a higher risk of illness. Occasional testing is typically done but may not detect a disease in its early stage. Early detection could initiate timely treatment with a higher chance of success, and have a worker removed from the hazardous environment to prevent further damage.

Nanotechnology can new offer new implantable and/or wearable sensing technologies that provide continuous and extremely accurate medical information. Complementary microprocessors and miniature devices can be incorporated with sensors to diagnose disease, transmit information and administer treatment automatically if required. Example applications are as follows.

Micro and nanosized sensors can make use of a wide range of technologies that most effectively detect a targeted chemical or physical property. Researchers at Texas A&M and Penn State use polyethylene glycol beads coated with fluorescent molecules to monitor diabetes blood sugar levels. The beads are injected under the skin and stay in the interstitial fluid. When glucose in the interstitial fluid drops to dangerous levels, glucose displaces the fluorescent molecules and creates a glow. This glow is seen on a tattoo placed on the arm. Another approach by researchers at the University of Michigan is using dendrimers attached with fluorescent tags to sense pre-malignant and cancerous changes inside living cells. The dendrimers are administered transdermally and because of their small size, pass through membranes into white blood cells to detect early signs of biochemical changes from radiation or infection. Radiation changes the flow of calcium ions within the white blood cells and eventually triggers apoptosis, or programed cell death (PCD) due to the radiation or infection. The fluorescent tags attached to the dendrimers will glow in the presence of the death cells when passed with a rentinal scanning device using a laser capable of detecting the fluorescence. A similar application is being researched with NASA for detecting radiation levels in astronauts. Sensor microchips are also being developed to continuously monitor key body parameters including pulse, temperature and blood glucose. A chip would be implanted under the skin and transmit a signal that could be monitored continuously. Another application uses optical microsensors implanted into subdermal or deep tissue to monitor tissue circulation after surgery. Instrumentation is linked to transmit data to a nearby sensor to provide very early indication of inadequate circulation if the surgery is unsuccessful. Another type of implantable sensor uses MEMS devices and accelerometers for monitoring and treating paralysed limbs. Implantable MEMS sensors can measure strain, acceleration, angular rate and related parameters to determine normal and problem data.

Implantable sensors can also work with a series of medical devices that administer treatment automatically if required. Tiny implantable fluid injection systems can dispense drugs electrically on demand making use of microfluidic systems, miniature pumps, and reservoirs. Initial applications may include chemotherapy that directly targets tumors in the colon and are programed to dispense precise amounts of medication at convenient times, such as after a patient has fallen asleep. Lupus, diabetes and HIV/AIDS applications are also being investigated. Implantable sensors that monitor the heart's activity level can also work with an implantable defribulator to regulate heartbeats. These devices are used with microprocessors to deliver electricity that keeps the heart in rhythm when levels go above or below the person's normal heart range. Implantable MEMS devices are also fitted on prostheses to mimic the stability and strain of natural limbs. An artificial leg being developed uses sensors to measure load on the foot, knee angles and motion over 50 times per second. The sensors work with an electronically controlled hydraulic knee to improve its stability. Implantable medical devices that are greater than 1mm in diameter might unfavorably alter the functions of surrounding tissue. Smaller implantable devices with non-intrusive or minimally-intrusive systems will likely contain nanoscale materials and smaller systems approaching the nanoscale. Functional Electrical Stimulation (FES) is a method for treating people to regain the use of their paralyzed limbs by electrically stimulating paralyzed muscles with implanted electrodes. Researchers at Aalborg University in Denmark are applying nanostructures to the electrode surfaces to improve biocompatibility and acceptance in the neural/muscle tissue. When placed within a cell membrane, the nanostructures form a bioelectric interface with the neuron or muscle cell that enables the intracellular potential of the cell to be observed and manipulated. Researchers at Aalborg are also using nanostructures to activate denervated muscles caused by injuries to the lower motor neurons located in the spinal cord. This can result from traumatic spinal cord injury, strokes in the spinal cord, repeated vertebral subluxation, brachial plexus injuries and peripheral myopathies such as polio which destroys the nerve cells controlling muscle (i.e. denervated muscle). The muscle fibre membrane is incorporated with potential-generating nanostructures that change the transmembrane potential of the muscle fibre and improves the extracellular electrical stimulation. Muscle fibre activation is achieved by illuminating the incorporation site on the muscle fibre to optically activate denervated muscles.

Nano and related micro technologies are being used to develop a new generation of smaller and potentially more powerful devices to restore lost vision and hearing functions. The devices collect and transform data into precise electrical signals that are delivered directly to the human nervous system. Degenerative diseases of the retina, such as retinitis pigmentosa or age related macular degeneration, decrease night vision and can progress to diminishing peripheral vision and blindness. These retinal diseases may lead to blindness due to a progressive loss of photoreceptors (rods and cones), the light sensitive cells of the eye. In cases where the neural wiring from the eye to the brain is still intact but the eyes' lack photoreceptor activity, photoreceptor loss could be compensated by bridging or bypassing the destroyed photoreceptors and artificially stimulating the adjacent intact cells. Artificially generated impulses could reach the brain and produce visual perception, thereby restoring some elementary vision. In severe hearing loss, patients typically have absent or malfunctioning sensory cells in the cochlea. In a normal ear, sound energy is converted to mechanical energy by the middle ear, which is then converted to mechanical fluid motion in the cochlea. Within the cochlea, the inner and outer hair sensory cells are sensitive transducers that convert mechanical fluid motion into electrical impulses in the auditory nerve. Cochlear implants are designed to substitute for the function of the middle ear, cochlear mechanical motion, and sensory cells. The implants transform sound energy into electrical energy that will initiate impulses in the auditory nerve. Cochlear implants include an electronic circuit that is surgically placed in the skull behind the ear on the mastoid process of the temporal bone. This circuit is attached to a bundle of tiny wires that are inserted into the cochlea. At the end of the wires are typically 8 to 24 electrodes that cause a different pitch percept when stimulated. The other part of the device is external and has a microphone, a speech processor, and connecting cables. Current cochlear implants have a number of drawbacks. They require major surgery and can eliminate any remaining natural hearing. Because of their large size, current cochlear implants often stimulate several nerve fibers at once. This causes users to experience imprecise or distorted sound, especially with complex sounds like music. Microtechnologies and eventually nanotechnologies are being used in a new generation of implantable devices for improving sensory loss in the following areas.

Retinal implants are in development to restore vision by electrically stimulating functional neurons in the retina. One approach being developed by various groups including a project at Argonne National Laboratory is an artificial retina implanted in the back of the retina. The artificial retina uses a miniature video camera attached to a blind person's eyeglasses to capture visual signals. The signals are processed by a microcomputer worn on the belt and transmitted to an array of electrodes placed in the eye. The array stimulates optical nerves, which then carry a signal to the brain. Another approach by Optobionics makes use of a subretinal implant designed to replace photoreceptors in the retina. The visual system is activated when the membrane potential of overlying neurons is altered by current generated by the implant in response to light stimulation. The implant makes use of a microelectrode array powered by as many as 3,500 microscopic solar cells.

A new generation of smaller and more powerful cochlear implants are intended to be more precise and offer greater sound quality. An implanted transducer is pressure-fitted onto the incus bone in the inner ear. The transducer causes the bones to vibrate and move the fluid in the inner ear, which stimulates the auditory nerve. An array at the tip of the device makes use of up to 128 electrodes, which is five times higher than current devices. The higher number of electrodes provides more precision about where and how nerve fibers are stimulated. This can simulate a fuller range of sounds. The implant is connected to a small microprocessor and a microphone, which are built into a wearable device that clips behind the ear. This captures and translates sounds into electric pulses which are send down a connecting wire through a tiny hole made in the middle ear. Implant electrodes are continuously decreasing in size and in time could enter the nanoscale.

Medical devices that contain nano and micro technologies will allow surgeons to perform familiar tasks with greater precision and safety, monitor physiological and biomechanical parameters more accurately, and perform new tasks that are not currently done. Surgery can be physically demanding on both the surgeon and the patient. Open surgery such as heart operations can require a surgeon to make a large cut into the skin and underlying muscles, saw through breastbones, crack open the ribcage, then make delicate and high precision maneuvers with various surgical instruments. Not only can the patient incur severe pain, scarring, complications and extensive recovery time, the surgeon must be intensely focused for long hours hunched over an operating table. This can increase the chance of a surgical error due to fatigue in the neck and back, along with potentially significant hand tremors. In various procedures, the surgeon's view could be limited. Judgment may be required to tell whether the tissue about to be cut has the density of the intended cartilage, or alternatively, a tendon which was grabbed by mistake. To improve surgical results benefiting both the patient and surgeon, minimally invasive surgical procedures are increasingly being done using laparoscopic techniques. Making use of small entry ports into the area of interest, a rod shaped telescope attached to a camera and other long and narrow surgical instruments are used to perform all of the major maneuvers. Even with its considerable benefits, laparoscopic surgery causes a major shift in surgical skills that have to be mastered as the surgeon's vision and access is severely restricted. Visual cues are obtained from a two dimensional video display instead of a three dimensional operating field. As well, the surgeon must work with long laparoscopic instruments to handle delicate tissues. Nano and micro technologies are being applied to improve the role of the surgeon in the following areas.

Surgical tools such as scalpels, forceps, grippers, retractors and drills are being embedded with miniature sensors to provide real-time information and added functionality to aid surgeons. Surgeons can be given continuous data on the force and performance of their instruments, the tissue type about to be cut (i.e. cartilage, bone, muscle, vascular, etc) and specific tissue properties, such as density, temperature, pressure, and electrical impulses. Verimetra is developing an enhanced version of its Data Knife with logic and surgical micro-electromechanical systems that can provide added information and functionality to assist a surgeon during a procedure by stimulating electrodes, measuring and cutting with ultrasonic elements, and cauterizing. Instruments are being developed with specific functionality like tilt and pressure to allow neurosurgeons to perform operating tasks with greater precision and safety. Nanoparticles are also being investigated for optically guiding surgery. This can potentially allow for better removal of lesioned or diseased sites, including tumors.

Robotic surgical systems are being developed to provide surgeons with unprecedented control over precision instruments. This is particularly useful for minimally invasive surgery. Instead of manipulating surgical instruments, surgeons use their thumbs and fingers to move joystick handles on a control console to maneuver two robot arms containing miniature instruments that are inserted into ports in the patient. The surgeon's movements transform large motions on the remote controls into micro-movements on the robot arms to greatly improve mechanical precision and safety. A third robot arm holds a miniature camera, which is inserted through a small opening into the patient. The camera projects highly magnified 3D images on a console to give a broad view of the interior surgical site. The surgeon controlling the robot is seated at an ergonomically designed console with less physical stress than traditional operating room conditions. UCI Medical Center's da Vinci Surgical System is currently approved for gall bladder, prostrate, colorectal, gynecological, esophageal and gastric bypass procedures. Clinical trials are underway for vascular repair and coronary bypass surgery. The Zeus Robotic Surgical System is presently used for cardiac surgery.

Nano and micro technologies provide new solutions for increasing the speed and accuracy of identifying genes and genetic materials for drug discovery and development, and for treatment-linked disease diagnostics products. Thousands of genes and their products (i.e. RNA and proteins) function in a complicated and orchestrated manner. Genetic testing technologies used for drug discovery and in diagnostic products make it possible to measure gene sequences (i.e. genes, gene mutations) and gene expression levels (abundance), which are important indicators of growth, metabolism, development, behaviour and adaptation of living systems. With these tools, many different genes and expression patterns can be characterized as normal or diseased tissue, along with the various stage of disease. There are many steps and processes involved in genetic testing. While the nature of the tests and nomenclature may vary, some of the key principals and related challenges are described below.

Known probes of genetic sequences are prepared or obtained commercially for a desired test. These probes are typically in the form of cDNA or an array of oligonucleotides. While it is desirable to have as many probes as possible to maximize the search for possible results, there is currently a limited number of known specimens. This limits the range of experimentation for drug discovery and disease diagnostics. With advances in the human genome, gene cloning and molecular biology, a vast amount of new probes are increasingly available which could in time cover the entire genome, and potentially screen multiple targets in a single test. DNA arrays are created by placing probes as individual spots in an orderly array arrangement on a glass microscope slide, a nylon membrane substrate or a silicon chip. The latter case is also referred to as a biochip.

Unknown cells or molecules of interest (called "targets") need to be collected and then compared with the known probes in the DNA array. This begins with tissue, blood or other biological samples where nucleic acids (RNA or DNA) are isolated. The nucleic acids are converted into labeled targets by tagging the DNA marker with an observable fluorescent or radioactive label. The labeled targets are incubated with the probes to permit hybridization and form stable DNA and/or RNA duplex. This is used for locating or identifying nucleotide sequences of interest.

After the incubation, nonhybridized samples are washed away, and measurements are made of the signal produced when hybridization occurs at particular probe locations. The level of hybridization between a specific probe and a target indicates the level of the gene corresponding to that probe in the test solution. Current assay technologies use fluorescent dyes to label molecules and require expensive equipment such as a laser to light up biological interactions, and an optical microscope to detect the binding sites. Fluorescent dyes are not always precise or sufficiently sensitive to detect every gene. Also, genes that are not distinguished separately can bleed together and result in a false or inconsistent result. The hybridization may also require gene amplification or Polymerase Chain Reaction (PCR) which is used to produce millions of copies of a specific piece of DNA in a test tube from a single cell. Amplification could require a repetitive series of cycles making use of various chemical and processing stages. Nano and micro technologies can provide new solutions in the following areas.

Several new technologies are being developed to improve the ability to label and detect unknown target genes. At Genicon, gold nanoparticle probes are being treated with chemicals that cling to target genetic materials and illuminate when the sample is exposed to light. Another approach by Chad Mirkin and Nanosphere uses gold nanoparticle probes coated with a string of nucleotides that complement one end of a target sequence in the sample. Another set of nucleotides, complementing the other end, is attached to a surface between two electrodes. If the target sequence is present, it anchors the nanoprobes to the surface like little balloons, and when treated with a silver solution, they create a bridge between the electrodes and produce a current.

Quantum Dot Corp. uses quantum dots to detect biological material. Because their color can be tailored by changing the size of the dot, the potential for multiple colors increases the number of biological molecules that can be tracked simultaneously. In addition, quantum dots do not fade when exposed to ultraviolet light and the stability of their fluorescence allows longer periods of observation. These technologies are expected to be more sensitive than fluorescent dyes and could more effectively detect low abundance and low-level expressioning genes. They may also make use of smaller and less expensive equipment to light and detect the samples. Without the need for gene amplification, they can also provide results in less time.

Arrays are usually classified by their sample spot sizes with 200 micrometer diameter as the traditional cutoff between macroarrays and microarrays. Microarrays contain thousands of spots. As the sample spot size further decreases in size, more spots can fit onto a substrate. This reduces the area and corresponding time that needs to be sampled. With continued miniaturization beyond micro, the possibility exists to greatly increase the number of spots on a single chip, with the ultimate objective of including the entire genome. At the same time, quicker analyses are being made possible by performing multiple analyses in parallel rather than in series to generate more information in less time. By using nanomaterials as sensing particles, chips could be reduced in size. This would allow scientists to read thousands of molecules with the possibility of using cheaper equipment. Smaller, portable, cheaper and more precise integrated diagnostic kits for diseases such as systemic lupus or other multi-marker diseases may be made possible by using microfluidic and nanofluidic technologies. DNA arrays typically perform one type of analysis thousands of times. In the event that other experiments or processes are also required, micro and nanofluidic devices such as lab-on-a-chip, can integrate mixing, moving, incubation, separation, detection and data processing in a small portable device. At the micro and nanoscale, fluids move through pipes in laminar flow, as opposed to turbulent flow at the macro level. This provides the opportunity for microfluidic devices to exploit certain physical behaviours. For example, two liquids can separately circulate through micro pipelines and valves without mixing each other. The combination of arrays and fluidic capabilities can greatly increase the speed and accuracy of various genetic testing. Within the drug discovery and diagnostic fields, microfluidic devices in the form of nanoarrays or "lab on a chip" technologies could allow the production of more efficient and disposable DNA and protein sequencers for drug discovery and diagnostic kits.

Nano and micro technologies offer new imaging technologies that provide high quality images not possible with current devices, along with new methods of treatment. Malignant tumors are highly localized during the early stage of their development. If detected early, the tumors can often be surgically removed with high success. The longer a patient has a malignant tumor the more likely the cancer will spread to neighboring lymph nodes and other anatomic structures. Aggressive surgery or chemotherapy, or very high doses of radiation may kill the cancer but at the same time severely injure normal tissue. In many patients, particularly those with breast cancer, the cancer can spread to other parts of the body even after the original cancer tumor has been removed. Brain tumors have other challenges. Removing tumors and other cancerous cells in the brain through surgery can risk the loss of cognitive ability, hearing, speech, and other functions. Chemotherapy therapeutics can kill cancerous cells but because of their large physical size may be prevented from entering the brain by the blood-brain barrier. Nanotechnology can offer new solutions for the early detection of cancer and other diseases. Complementary technologies can also destroy cancer cells, which are too small to locate and remove surgically.

Researchers at the University of Michigan are developing nanoprobes that can be used with magnetic resonance imaging (MRI). Nanoparticles with a magnetic core are attached to a cancer antibody that attracts cancer cells. The nanoparticles are also linked with a dye which is highly visible on an MRI. When these nanoprobes latch onto cancer cells they can be detected on the MRI. The cancer cells can then be destroyed by laser or low dosage killing agents that attack only the diseased cells. In brain tumors, the presence of cancer can weaken the blood-brain barrier. Research is being conducted to have nanoprobes cross the weakened barrier to the tumor but not cross into healthy brain tissue. Another group at Washington University is using nanoparticles to attract to proteins emitted from newly forming capillaries that deliver blood to solid tumors. The nanoparticles circulate through the bloodstream and attach to blood vessels containing their complementary protein. Once attached, chemotherapy is released into the capillary membrane. The nanoparticles traveling in the bloodstream would be able to locate additional cancer sites which may have spread to other parts of the body. Nanotechnology can also be used to improve the contrast agents for MRI's. Using fullerenes to encapsulate contrast agents such as gadolinium, Luna Innovations claims that traditional images can be enhanced by 50 times. If successful, this can result in the use of lower magnetic fields through smaller, cheaper and portable imaging equipment. Alternatively, smaller amounts of contrast agents could be used which would reduce the risk of allergic reaction in some patients.

Miniature wireless devices are being developed with nano and micro technologies for providing high quality images not possible with traditional devices. Given Imaging has developed a pill containing a miniature video system. When the pill is swallowed, it moves through the digestive system and takes pictures every few seconds. The entire digestive system can be assessed for tumors, bleeding, and diseases in areas not accessible with colonoscopies and endoscopies. Another company, Medirad is trying to develop a miniature x-ray device that can be inserted into the body. They are attempting to make carbon nanotubes into a needle shape cathode. The cathode would generate electron emissions to create extremely small x-ray doses directly at a target area without damaging surrounding normal tissue.

Nanotechnology, which came into its own in 2000 with the birth of the National Nanotechnology Initiative, is the next stop in the miniaturization path that gave us micro-electronics, microchips and microcircuits. The word nanotechnology derives from nanometer; which is one-thousandth of a micrometer (micron), or the approximate size of a single molecule. Nanotechnology -the study, manipulation and manufacture of ultra-small structures and machines made of as few as one molecule-was made possible by the development of microscopic tools for imaging and manipulating single molecules and measuring the electro-magnetic forces between them. Nanobiotechnology joins the breakthroughs in nanotechnology to those in molecular biology. Molecular biologists help nanotechnologists understand and access the nanostructures and nanomachines designed by 4 billion years of engineering—cell machinery and biological molecules. Exploiting the extraordinary properties of biological molecules and cell processes, nanotechnologists can accomplish many goals that are difficult or impossible to achieve by other means. For example, rather than build silicon scaffolding for nanostructures, DNA's ladder structure provides nanotechnologists with a natural framework for assembling nanostructllres; and its highly specific bonding properties bring atoms together in a predictable pattern to create a nanostructure. Nanotechnologists also rely on the self-assembling properties of biological molecules to create nanostructures, such as lipids that spontaneously form liquid crystals. DNA has been used not only to build nanostructures but also as an essential component of nanomachines. Most appropriately, DNA, the information storage molecule, may serve as the basis of the next generation of computers. As microprocessors and microcircuits shrink to nanoprocessors and nanocircuits, DNA molecules mounted onto silicon chips may replace microchips, with electron flow-channels etched in silicon. Such biochips are DNA-based processors that use DNAs extraordinary information storage capacity. Conceptually, they are very different from the DNA chips discussed below. Biochips exploit the properties of DNA to solve computational problems; in essence, they use DNA to do math. Scientists have shown that 1,000 DNA molecules can solve in four months computational problems that require a centllry for a computer to solve. Other biological molecules are assisting in our continual quest to store and transmit more information in smaller places. For example, some researchers are using light-absorbing molecules, such as those found in our retinas, to increase the storage capacity of CDs a thousandfold. Some more immminent applications of bio-nanotechnology include:

- Increasing the speed and power of disease diagnostics.
- Creating bio-nanostructures for getting functional molecules into cells.

- Improving the specificity and timing of drug delivery.
- Miniaturizing biosensors by integrating the biological and electronic components into a single, minute component. encouraging the development of green manufacturing practices.

Medical Revolution

Imagine having bones woven with a fabric such that one could fall out of a building and walk away. Imagine that in the event of a fire, microscopic vessels just ten billionths of a meter wide, pressurized with 1,000 atmospheres of pure oxygen could sense oxygen levels in the blood and provide hours of respiratory requirements for the body. Imagine medical nanites being injected into the bloodstream, consuming atherosclerotic plaques in the walls of the blood vessels; repairing cell damage caused by cancer. Or imagine nanomouthwashes that could eliminate gum disease and tooth decay—nanomachines acting as security guards and attacking any foreign entity in the body1. Sounds like something from a science fiction movie? Absolutely not. Welcome to the world of nanomedicine.

Despite advances in modern medicine over the last century, the state of current medicine has been limited both by its understanding and by its tools and in many ways is still more of an art than a science. Only within the last 50 years has medical science begun to examine disease pathology on a molecular level; thus, from a molecular viewpoint, modern medicine remains crude. For example, today's drug is essentially a single molecule with an often sophisticated but always limited repertoire. With nanomedicine, tomorrow's "smart pharmaceuticals" could essentially be programable machines with a range of "sensory," "decision-making," and "effector" capabilities. They can avoid side effects and allergic reactions by coming in generic, biocompatible housings; becoming active only upon reaching their ultimate destinations; and attaining almost complete specificity of action. They can check for overdosage before becoming active, thus preventing accidental or intentional poisoning. They may work in concert with three or four "sister" agents that together produce versatility unattainable by one agent alone. Despite the vast increase in complexity over present-day drugs, such agents via nanomedicine can be expected to be totally "pure" and predictable in their behaviour. Safety and efficacy may be inherent in the designs of such "drugs," in which case regulatory issues could be simplified rather than complicated. So what exactly is nanomedicine? Technically it is the application of nanotechnology (engineering of tiny machines) to the prevention and treatment of disease in the human body.

More specifically, it is the use of engineered nanodevices and nanostructures to monitor, repair, construct and control the human biological system on a molecular level. The most elementary of nanomedical devices will be used in the diagnosis of illnesses. A more advanced use of nanotechnology might involve implanted devices to dispense drugs or hormones as needed in people with chronic imbalance or deficiency states. Lastly, the most advanced nanomedicine involves the use of nanorobots as miniature surgeons. Such machines might repair damaged cells, or get inside cells and replace or assist damaged intracellular structures. At the extreme, nanomachines might replicate themselves, or correct genetic deficiencies by altering or replacing DNA (deoxyribonucleic acid) molecules.

In 1959, in a talk at CalTech titled, "There's Plenty of Room at the Bottom," acclaimed physics lecturer and Nobel laureate Richard Feynman suggested that nature could be manipulated at a nanometer scale—atom by atom. Wiith the convergence of human gene sequencing and advances in nantotechnological engineering and a deepening understanding of the function of cellular systems, nanotechnologists believe it will be possible to design medicallyactive microscopic machines to fight disease and affect physiological repairs at the cellular level. Invading bacteria will automatically be destroyed, since they do not "fit" within the physiological blueprint encoded in the nanorobot patrolling the body. Diagnosis of disease will no longer rely upon patient history and the results of laboratory tests, but will result from an ongoing internal examination of deviations from the encoded molecular blueprint and programed repair functions designed to correct anomalies at the molecular level6. Cell damage attributed to aging will be repaired from the inside out. The potential benefits of medical applications of nanotechnology—nanomedicine—are read more like science fiction than science fact. In 1997, a group of health scientists with the Department of Defense concluded that if a breakthrough to a molecular assembler occurs within 10 to 15 years, an entirely new field of nanomedicine will emerge by 20207. These scientists postulated that initial applications would be focused outside the body in diagnostic and pharmaceutical manufacturing, but the most powerful uses would be within the body. Some of the applications they discussed included cell-herding machines to stimulate rapid healing and tissue reconstruction and cell repair machines to perform genetic surgery. Many of these research goals may take 20 or more years to achieve, which is one reason the federal government in US is the major financial supporter of nanotechnology. Researchers have been able to move molecules and even construct nanomotors. The first privately held nanotechnology company, Zyvex, was started in mid 1997 with the goal of building the key tool for creating molecular nanotechnology, the assembler. According the Zyvex Founder, President, and CEO James Von Ehr II, the first assembler could be available within 10 years. However, it will be some time after that before medical nanorobots will reach the market.

The problems with nanotechnology begin with assembly. Recall that nanotechnology is molecular manufacturing—building things one atom or molecule at a time with programed nanoscopic robotic arms. Atoms don't simply let themselves be pushed around. They are constantly moving, jiggling, combining, and reacting, always ready to confound the would-be molecular architect. A design that might seem perfectly reasonable to an engineer in the macroscopic world might, on a nanoscale, fold instantly into a ball of goop. Imagine, anyway, that you have a robot hand that can assemble nanomachines, that you have managed to build a stable structure containing ten atoms, and you now want to add another piece to it. You would need to grab another atom or molecule, all the time making sure that none of the positions of the neighboring atoms are altered. Each atom's position is specified in three dimensions, so you'd have a 3D problem, (at least, if the fourth dimension—timing is not a factor). This is not easy to calculate. The difficulties continue. Because the robot handles one atom at a time, it has to work fast, and with speed come mistakes.

A free radical—a highly reactive molecule—can clip off a single atom, and thereby undo all that the robot has contrived. The robot would have to detect and correct errors as they happened, not only as it assembled a device, but also afterward, in order to maintain its integrity. Another problem is scalability. Unlike the technology for sequencing genes, which began small and only later moved to mass production, carbon nanotubes have been produced by the thousands from the very start. Unfortunately, thousands aren't good enough if the goal is to knit the nanotubes into organs, limbs, or other cellular structure. For that, tons of nanotubes are required with no currently suitable means of mass production. However, on the bright side, the same things were said about combinatorial chemistry, recombinant DNA, decoding the human genome, and landing to the moon.

A nanomedical company with a successful product will reap huge profits much like biotechnology and pharmaceuticals companies do. However, like biotechnology or pharmaceutical companies, they will also experience the tremendous expenditures associated with bringing a product to market. In fact, such companies will incur even higher costs than the traditional biotechnology and pharmaceutical companies because of the diversity of technology required in nanomedicine. Take a pharmaceutical company, for example. The technology that such a company has to deal with is normally confined to molecular technology. A nanomedical company, on the other hand, will require expertise in the same realm of expertise as the pharmaceutical company plus a multitude of expertise in other areas such as computer technology, nanopropullsion technology, and circuitry technology, just to name a few. The cost of developing innovative technology in just one of these areas is expensive enough, let alone multiple areas. Furthermore, there is the basic question of can a company achieve such diversity of technologies? It seems difficult for a starting company to develop all these competencies in all these different areas at once. Companies that have a more realistic chance of pursuing nanomedicine alone are the well-established companies that are already highly diversified—companies such as GE, Cannon, Mitsubishi, IBM. These companies already have technological expertise in a variety of areas which they can leverage.

The difficulty with drug discovery today is that it is part science, part art, and part luck. Because of these elements, the average time to discover and win FDA approval for a prescription drug is 15 years, with 3 out of every 10 drugs failing to meet FDA approval. This long process plus the protection afforded by patents and patent extensions have allowed pharmaceutical products a relatively long product life cycle. For example,

Prozac, launched in January 1988, maintained a monopoly afforded to it by patent protection until only recently when it become generic. Nanomedicine, although still protected by patents, will enjoy shorter periods of monopoly with each product. Why? Because in nanomedicine every molecule is designed—nothing is left to luck. As such, once a general level of assembly technology is understood and common in the industry, the differentiator between companies and products becomes a function of engineering skills. Take digital cameras, for example. When Sony released a 2 megapixel camera, Olympus introduced a 3.2 megapixel camera in less than a year. Sony then countered in 6 months with a 3.2 megapixel camera capable of 10 X optical zoom. Apply this same pattern to nanomedicine and you can see how the product life cycle of nanomedical devices will share a similar fate to consumer electronics. Picture Company A producing a nanopill that when swallowed will measure a variety of parameters such as blood pressure, tissue oxygenation, internal chemistry, etc. Now picture Company B introducing a different nanopill 9 months later that can measure more parameters with greater sensitivity and accuracy. Company A then launches a new version of their original nanopill within 6 months of Company B's launch. Product life cycles have been reduced and prices consequently eroded.

As alluded to in the "financial and competency challenge" section, the ability for a "stand-alone" company to be a serious competitor in nanomedicine will become less likely. Because nanomedicine spans many different industries—materials, computer, circuitry, etc., the role of partnership and cross licensing of technology becomes ever more important. Companies can no longer operate in a vacuum. Imagine Pfizer branching into nanomedicine. If they were to develop a nanobot of some type, Pfizer lacks the engineering capability to develop the sensor, mechanical components or computer guidance system. The company would have to look outward for partners who have the capabilities to develop the necessary components. Development of nanomedical technology will require a cooperative effort and as result, no one company can claim one hundred percent of the profits for itself. The challenge of sole proprietorship does not only apply to new nanomedical companies. It also applies as a challenge for existing companies rich in research and development. Few technologically wealthy companies are actively managing the licensing of their intellectual property as a profit center for the organization. The advancement of nanomedicine, and nanotechnology in general, will require such companies to rethink their business model and role of R&D within the company. As the doors of licensing open, the role of R&D centers cannot be viewed as centers of discovery and thus cost centers, but as potential profit pools and access to other company's technology through cross-licensing. Another interesting question is how the FDA will view/classify medical applications of nanotechnology. Would nanomedicine be regulated as drugs, devices, or something else? Statutes provide that treatments that operate through mechanical means are to be treated as devices; those that rely on chemical effects are considered drugs. This distinction is not very satisfactory in the context of nanotechnology-based treatments.

Like genetic engineering, the question facing nanomedicine is not "Can we?", but "Should we?" Nanomedicine can eradicate biological sources of human suffering and human limitations. For example, advocates of nanomedicine believe that the technology must be used to eliminate human vulnerability to disease. Fully developed nanorobots, in their view, will be able to correct any damage to DNA, eliminate any tumors or infections, and repair most structural damage to the body. Such technology can, in principle, make disease via inheritance or infection a thing of the past. The same technology, however, can be used to prevent aging, remove the inherent limits on the human lifespan. Bones could be made of diamond and lungs rebuilt to breathe poisonous atmospheres. Brain enhancements by means of artificial, improved neurons will mean that limits on memory and intelligence need no longer exist. A single artificial neuron could store the entire Library of Congress, accessible to an individual on demand. Brains could have the ability to link directly via nanoengineered devices with computers, with other brains, or with the Internet. Where is the line drawn between the repair and restoration of lost function and the augmentation of "normal" capabilities? Furthermore, can we even make clear distinctions between what constitutes treatment of a disease or disability and what is "simply" self-improvement? And if some improvement is appropriate, what distinguishes legitimate improvement from manipulative alterations of people for sinister or even benign purposes?

Nanomedical Appliance to Replace Human Blood

The vasculoid is a single, complex, multisegmented nanotechnological medical robotic system capable of duplicating all essential thermal and biochemical transport functions of the blood, including circulation of respiratory gases, glucose, hormones, cytokines, waste products, and cellular components. This nanorobotic system, a very aggressive and physiologically intrusive macroscale nanomedical device comprised of ~500 trillion stored or active individual nanorobots, weighs ~2 kg and consumes from 30-200 watts of power in the basic human model, depending on activity level. The vasculoid system conforms to the shape of existing blood vessels and serves as a complete replacement for natural blood. Here we present a preliminary theoretical scaling analysis including transport capacity, thermal conduction, control and biocompatibility considerations, along with a hypothetical installation scenario and a description of some useful optional equipment. We begin with a very preliminary technical examination of an idea originally referred to as "roboblood". The concept stimulated informal interest in a particularly aggressive nanomedical design and elicited considerable online popular discussion on sci.nanotech and elsewhere Here we examine a set of appropriate design constraints, scaling issues, and reference designs to investigate whether or not the basic idea of a blood replacement appliance might be feasible, and to determine key limitations of such designs.

The idea of the vasculoid originated in the asking of a simple question: Once a mature molecular nanotechnology becomes available, could we replace blood with a single, complex robot? This robot would duplicate all essential thermal and biochemical transport functions of the blood, including circulation of respiratory gases, glucose, hormones, cytokines, waste products, and all necessary cellular components. The device would conform to the shape of existing blood vessels. Ideally, it would replace natural blood so thoroughly that the rest of the body would remain, at least physiochemically, essentially unaffected, but sustained in a cardioplegic state. It is, in effect, a mechanically engineered redesign of the human circulatory system that attempts to integrate itself as an intimate personal appliance with minimal adaptation on the part of the host human body. A robotic device that replaces and extends the human vascular system is properly called a "vasculoid," a vascular-like machine. But the vasculoid is more than just an artificial vascular system. Rather, it is a member of a class of space- or volume-filling nanomedical augmentation devices whose function applies to the human vascular tree. The device is extremely complex, having ~500 trillion independent cooperating nanorobots. In simplest terms, the vasculoid is a watertight coating of nanomachinery distributed across the luminal surface of the entire human vascular tree. This nanomachinery uses a ciliary array to transport important nutrients and biological cells to the tissues, containerized either in "tankers" (for molecules) or "boxcars" (for cells). The basic device weighs ~2 kg and releases ~30 watts of waste heat at a basal activity level and a maximum of ~200 watts of power at peak (e.g., Olympic sprint) activity level. The power dissipation of the human body ranges from ~100 watts (basal) to ~1600 watts (peak), so the device presents no adverse thermogenic consequences to the user. The appliance is powered by glucose and oxygen, as may be common in medical nanorobotic systems. The most important basic structural component of the exemplar vasculoid robot is a ~300 m^2 two-dimensional vascular-surface-conforming array of ~150 trillion "sapphiroid" (i.e., using sapphire-like building materials) basic plates. (Thermal conductivity favors sapphire over diamond) These square plates are nanorobots that cover the entire luminal surface of all blood vessels in the body, to one-plate thickness. Each basic plate is an individual self-contained nanorobot ~1 micron thick and ~2 $micron^2$ in surface area, a size small enough to allow adequate clearance even in the narrowest human capillaries. Molecule-conveying "docking bays" comprise ~24 trillion, or 16 per cent, of all vasculoid basic plates. Tankers containing molecules for distribution can dock at these bays and load or unload their cargo. Cell-conveying "cellulocks" are built on "cellulock plates" which span the area of 30 basic plates, or 60 $micron^2$ each. Boxcars containing biological cells for distribution can dock at these cellulocks and load or unload their cargo. With only 32.6 billion cellulock plates in the entire vasculoid design, cellulocks occupy the area of 0.978 trillion basic plates or only 0.65 per cent of the entire vasculoid surface. The remaining ~125 trillion basic plates are reserved for special equipment and other as-yet undefined applications. All nanomachinery within each plate is of modular design, permitting easy replacement and repair by mobile repair nanorobots called vasculocytes.

Adjacent plates abut through flexible but watertight mechanical interfaces on metamorphic bumpers along the entire perimeter of each plate. Each bumper has controllable variable volume, permitting the vasculoid surface: to slightly expand or contract in area, or to flex, either in response to macroscale body movements or in response to vascular surface corrugations or other irregularities to the same degree or better than the natural endothelium. Thus, plated surfaces readily accommodate the natural cyclical volume changes of various organs such as lung, bladder, or spleen. Rigidity of the plate array is also subject to engineering control and to localized real-time control as well, via the bumpers; diamondoid or sapphire plating may be made substantially stiffer than natural endothelium, if desired. The rupture strength of individual plates is ~40,000 atm, taking plate wall thickness t_{wall} ~ 0.1 mm, effective radius R ~ 0.5 mm, and working stress s_w ~ 10^{10} N/m^2. Bumper actuation power is of order ~0.1 pW, comparable to the power draw of a single plate cilium.

The advantages of installing a vasculoid are potentially numerous. Many of these benefits theoretically could be provided on a temporary or more limited basis using terabot-quantity doses of considerably less aggressive bloodborne nanomedical devices. However, the vasculoid appliance simultaneously provides all benefits on an essentially permanent and whole-body basis. Additionally, some benefits appear unique to the vasculoid and can be achieved in no other way. Whether the entire package is sufficiently attractive to warrant installation will probably be a matter of personal taste rather than of medical necessity, since a molecular nanotechnology capable of building and deploying a complex vasculoid is likely to offer complete non-vasculoid cures for most circulatory and blood-related disorders that plague humanity today, and biological enhancements may also be available. The choice between an augmentation technology that works alongside a natural system (e.g., respirocytes) and an augmentation technology that entirely replaces a natural system (e.g., vasculoid) may involve significant safety, psychological, and even ethical considerations. The most important benefits of vasculoid installation may include:

- Exclusion of parasites, bacteria, viruses, and metastasizing cancer cells from the bloodflow, thus limiting the spread of bloodborne disease. Such microorganisms and cells are easily eliminated from the blood using ~cm^3 doses of appropriately programed nanobiotics, but such individual nanorobotic devices might not normally be deployed on a permanent basis. Intracellular pathogens that can infect motile phagocytic cells (e.g., the tuberculosis *Mycobacterium* or the bacterium *Listeria*, both of which can reside inside macrophages) cannot be directly excluded from the tissues when infected cells are transported by the vasculoid. However, cell surface markers will often reveal such infection, so vasculoid systems can check for the presence of such markers and thus deny these cells re-entry to human tissues. For example, the membrane surface of macrophages infected by *Mycobacterium microti* is antigenically different from that of uninfected macrophages; Listeria, derived peptides are found acting as integral membrane proteins in the plasma membrane of infected macrophages, and other *Listeria*-infected antigen-presenting cells display hsp60 on their plasma membranes only when infected. Note that the natural bacterial inhabitants of the human gastrointestinal tract should be relatively unaffected by the installation of a vasculoid appliance, since gut microbes normally do not enter the bloodstream; similar considerations apply to the natural microbial flora of the skin, mouth, nasal passages, and so forth. The vasculoid could also significantly reduce the prospect of death by cancer by refusing either to transport angiogenesis factors or to transport nonvalidated native stem cells (thus preventing metastasis), or by refusing to install itself in any capillaries arising in locations where angiogenesis is disallowed in an adult body.

- Faster and more reliable trafficking of lymphocytes throughout the secondary lymphoid organs, allowing them to survey for targeted antigens in minutes or hours, rather than days (because both white cells and antigenic sources can be efficiently concentrated), thus greatly speeding the natural immune system response to foreign antigen. This lymphocyte function might also be augmented or replaced using individual histomobile medical nanorobots or biorobots. If biorobots are developed first, many vasculoid installations might take place in patients possessing largely artificial immune systems, thus obviating the need for much of the cellular component trafficking.
- Eradication of most serious circulatory-related pathological conditions including all vascular disease (e.g., aortic dissection, vessel blockages, spasms, aneurysms, phlebitis, varicose veins), heart disease, syncope (including orthostatic hypotension) and shock, strokes, and bleeding, due to the elimination of unconstrained metabolite and fluid circulation. Certain other conditions due to localized prevention of blood flow such as bedsores and subclinical paresthesias (e.g., "pins-and-needles" sensation) can also be ameliorated, since stiffened blood vessels will not be nearly so easy to close via external compression. Again, many of these conditions may already have adequate nanomedical treatments by the time the vasculoid can be built, but other conditions might not yet be readily or as conveniently treatable, such as the dangers of large-scale bleeding (both internal and external).
- Reduced susceptibility to chemical, biochemical, and parasitic poisons of all kinds, including allergenic substances in food, air and water, although bloodborne nanotankers or pharmacytes may be able to partially duplicate this function as well. Note that toxins, novel metabolites or other unknown foreign substances may arrive in the tissues via solvents penetrating the dermis (e.g., DMSO) or by other nonvascular routes, and would be removed from the natural blood in due course by the kidney which extracts all small molecules by default, then actively reabsorbs only useful molecules such as glucose, electrolytes, vitamins, and water. By contrast, in the present design the vasculoid transports only recognized substances. Novel

undesired substances must therefore be removed by employing a small number of previously unallocated applications plates:

(a) as miniature analytical laboratories which can detect foreign analytes and alter a programmable binding site to bind them, allowing transport to the kidney for disposal;

(b) as miniature chemical processing systems, possibly involving well-designed sets of artificial digestive enzymes, which can break down the foreign substance into well-recognized simpler substances; or

(c) as analytical laboratories or chemical processing systems previously described which can receive appropriate new instructions directly from the patient, the physician, or other external sources of information.

The lymphatic system is unmodified in the basic appliance design and may also serve to remove nonvascular toxins. Processing of such toxins may be deferred to the points where the lymph is reabsorbed into the vasculoid system, using either dedicated application plates or specialized processing stations.

- Faster metabolite transport and distribution, significantly improving physical endurance and stamina, including the ability to breathe at low O_2 partial pressures and the ability to flush out unwanted specific biochemicals from the body (a feature which might be duplicated using bloodborne respirocyte-class devices). The architecture would also permit convenient long-term storage of protein, or amino acid recovery and recycling, which could prove nutritionally useful. As another example, an installed vasculoid could mitigate the effects of weightlessness or acceleration on nutrient transport and somatic fluid distribution by decoupling transport from gravity. The appliance can differentially control the release of water and electrolytes in various regions of the body. In addition, by making the transport function independent of gravity, vasculoid avoids the hydrostatic effects of extreme, varying, or absent gravity vectors. This would eliminate the facial puffiness seen in astronauts and the dependence of fighter pilots on external prostheses to force blood out of the legs during high-G maneuvers.
- Direct, rapid user control of many hormonal- and neurochemical-mediated, and all blood-mediated, physiological responses. It would be difficult (though not impossible) to provide equivalent comprehensive whole-body physiological control using individual micron-scale bloodborne nanorobots alone. Additionally, the appliance can continuously monitor the basic physiochemical status of vascularized tissues to near-micron resolution if necessary, since each micron-size plate is initialized with its physical position relative to the vascular tree, and medically relevant information

is communicated back to a central user (or physician) interface. This can permit extremely rapid and specific detection of health problems as will generally be the case in nanomedicine, al though the vasculoid may provide an easier and more thorough method for accomplishing this.

- Voluntary control of capillary conductance and rigidity permitting conscious regulation of thermal energy exchange with the environment and at least limited control of whole-body morphological structure, rigidity (e.g., stiffness, bending modulus, etc.), and volume with ~millisecond response times.
- At least partial protection from various accidents and other physical harm (e.g., insect stings, animal bites, collisions, bullet or shrapnel penetrations, falling from heights). This is perhaps the only specific benefit of the vasculoid appliance that could not be achieved by any less radical means: extreme trauma resistance, especially resistance to exsanguination and cushioning against mechanical shock.

Medically oriented people might properly wonder why anyone would want to discuss replacing a perfectly functional natural fluid transport system with an untested, complex, artificial, dry system with which humans have no experience today. There are several answers to this very good question. First of all, medical skeptics should bear in mind that the vasculoid appliance is clearly a highly sophisticated medical nanosystem. It cannot be built without using a manufacturing system based on a mature molecular nanotechnology. Its use would come only after many decades of previous engineering experience in building, testing, and operating such highly complex systems inside the human body. In the future nanomedicine-rich milieu in which it would be deployed, the vasculoid as a medical intervention may be closer to the typical than to the extreme (as it might appear today). In the nanomedical era, it will bē a matter of personal preference and choice for each patient, in consultation with their physician, whether the aforementioned benefits of the vasculoid appliance are worth the risks. The following chapters are only a preliminary scaling study to investigate whether the machinery involved will fit in the allotted space, perform all required functions fast enough to be useful, operate within an acceptable energy budget even at peak loads, promise a strong likelihood of being safe, reliable, and biocompatible, and so forth.

Recent Advances in NanoBiology Imperative

All living things are nanofoundries. Nature has perfected the artform of biological nanotechnology for billions of years. Now, an emerging technology domain is poised to present a toolkit from which new lifeforms can be crafted, the inner molecular workings of living cells can be directly manipulated, even aging may be treated not as a disease, but as a reversable pathology. The very definition of life itself is perched at the edge of the next great revolution in medicine: nanobiology. Currently existing are technologies and applications in the arenas of biomolecular components and biocompatible surfaces integrated into microscale systems, implantable biochip devices, synthetically engineered quasi-viral components, modified DNA, structured proteomics, pseudoproteins, biomolecular “devices”. What is coming are artificially engineered organelles and cells, technologies which combine organic and inorganic materials and substrates into integrated nanoscale systems, “biomolecular prosthetics”, and intra-cellular modification strategies which will redefine the very essense of what is commonly referred to as life. 3D rendering of a theorectical quasi-viral component, by Charles Ostman, 1993. First patents actually issued for such “viral nano-machinery”, a modified baccilovirus utilized for proteomic targeting and delivery, issued to Onyx Pharmceuticals in 1996 Dendrimer molecules, spherical branching strucutures which can be grown to very precise size and complexity configurations. This example has antibodies attached to its outer extremities, as a biomedical “device”, but dendrimer components can be utilized for a very wide range of applicaitions, including self assembling molecular systems and structures.

Molecular protein delivery to targeted, specific cell types is currently being developed as a method for instigating physiological adjustment or modifications within living cells. In this context, protein can be viewed as being the functional equivilant of “software” which instructs the activities of biomolecular mechanisms and organelles within the cell, somewhat in the same way that machine code instructs computer processor chips to perform various functions. Among many such examples of this process being currently studied, is the utilization of the P53 protein, which can act like a “switch” to literally shut down the metabolism of a living cell.

This has potential, for instance, as a possible cure for various forms of cancer, in that the cancer cells have a genetic identity different than the original cell types from which they emanated, and therefore can be “targeted” for delivery of this P53 protein. Photograph of a colony of live neurons living on the surface of a biochip, from Dr. Gunther Gross, University of N TexasBiolithography, the process of photonically instigating the binding of biomolecules to a variety of target surfaces, such as metals, glass, plastic, even teflon, was orginally patented by BSI (now Surmodics) over a decade ago. Now the procedure has become very robust, enabling the precise “attachment” of proteins, lipids, enzymes, even entire living cells, with molecular precision to almost any type of surface. Above and below: 3D rendered scences from the film Nanotechnology indicating a neural prosthetic biochip, and a “nanobot” repairing a damaged nerve axon.

Charles Ostman, and the production team from AAC, San Francisco. Self assembling organic nanotubes as a strucutural system, and (left) hexagonal nanophotonic structures composed from rhodopsins (retinal proteins). Phodopsins are extremely interesting becuase of their potential application in spatial light modulation, which facilitates optical computing and storage with biological materials—from the Theoretical Biophysics Group, U of Illinois, Urbana DNA, the software for all living things, can also be utilized as a structural proteomic system, and a “logical” component for molecular computing. The example configuration shown here is one of many developed by Ned Seeman at NYU, and others exploring similar concepts Self assembling / self organizing molecular systems, emanating from the development of applied nano-biology, foster the creation of molecular computing platforms.

The biotech industry is extremely IT and computational resource intensive, which in turn benefits directly from the advent of evermore powerful and diverse forms of computing enabled by developments in nanocomputing systems. Biological metaphors in computing, such as genetic and evolutionary computation, currently implemented on reconfigurable computing platforms, further acclerates the pace biotech development via bioinformatics and in-silico biology processes. Various implementations of nanocomputing components will also be able to support related versions of biolgical metaphors in computing. Note the computer generated "virtual protein" model. This complex molecular structure was actually evolved, utilizing genetic algorithms on a reconfigurable computing platform to derive this proteomic solution. This is an example of biological metaphors in computing being utilized to create the physical biological materials of our future evolution.

Dave Fogel, Natural Solutions Inc.3D procedural rendering of a theoretical pseudo-proteomic subunit Charles OstmanThe worlds of biotechnology and nanotechnology currently converge into nanobiology. The first patents for utilizing a modified virus as a proteomic delivery vehicle, that being a molecular scale "device" which can seek out specific cells and deliver various materials to the inner parts of those cells, were issued to Onyx pharmaceuticals in 1996. In 1997, the Nanobiological Systems Group of Searle Labs (a division of Monsanto Chemical) was using dendrimer molecules, a sort of complex nanoscale branching structure which can be "grown" into very specific geometric forms, as another type of cellular targeting and delivery system. These same dendrimer molecules also are part of a growing collection of self-assembling and self organizing "components", the leggo blocks of nanotechnology. Now several companies have emerged into the marketplace with their versions of these dendrimer molecular systems.

Biological materials systems provide a rich collection of options from which self-assembling and self-organizing molecular systems can be "harvested", enabling access to materials to facilitate the construction of 2D structures and 3D manifolds. Many development programs utilizing proteomic components and other forms of biomolecular substrates which can either be used directly as functional components, or as nano-scale manipulation mechanisms for positioning other molecular subsystems, have been underway for a number of years.

The first examples of quasi-viral component molecular "devices" (synthetically engineered viral "entities") utilized for delivery of molecular protein components to specifically targeted cells are already in development. Synthetic organisms and biobots, quasi-viral components and engineered microbes as molecular scale medical "devices";

these are just a few of the technologies residing at the beginning edge of a new domain of development which redefines the very essense of what is traditionally referred to as "medicine".

NeuroChipThis 3D rendering represents the latest development in a series of "biochip" devices, in this case, the precursory example technology which could have applications ranging from neural prosthetics to the next generation of biological computing and sensory system components. This example indicates molecular protein "snippets" induced into bonding with a receptor site, as a dynamic contact/ molecular conduction node embedded into a silicon micro device substrate.

Potential applications—Neural prosthetics, sensory enhancement systems, direct neural interconnect to implantable biochip components and integrated systems. Potential risks—applications which extend beyond benign or "situationally neutral" engineered solutions for human sensory and behavioral capacities

The topic of nanobiology is, in fact, not a singular "science" or technical discipline at all, but rather a concept, as fundamental as life itself, and will eventually affect what can even be described as life, as it is currently defined. And yet, it is the very nature of such a complex and diverse arena of development which is precisely what the latest genre' of computer visualization graphics techniques, incorporating the combination of complex molecular modelling, volumetric topographical scanning of biomolecular and subcellular objects, and 3D rendering and visuallization techniques that enables such a rapidly developing technology domain.

The paradigm of the near future (and even today, in some arenas of development) will be biomedical "systems" which are reconstructed or "assembled" from biomolecular components, and mimic the physiology of living organisms in their operational and behavioral characteristics. As a solution to a highly diverse range of medical conditions, to which the only treatment has been the delivery of drugs or physiologically altering treatments of chemicals or radiation, this is the beginning edge of a radical new design concept. The world, as we currently know it to be, what it will become . . . . and beyond . . .Strategic Synergist—At the forefront of evolving technologies.

There is a state of change at hand, the rate and complexity of which is an acceleration vector never before experienced in human history. The relationships between rapidly evolving technologies, ever exapnding volumes and complexities of of information resources, and socio-economic system substrates are going to reshape almost every aspect of life as it is currently understood to be. To study the nuances and implications of this synergistic codependancy is to become an adaptable entity, capable of evolving and utilizing the tools embedded in this interactive, bi-directional experiential knowledge conveyance system which will be referred to here as the "media matrix".

Examine the details of this conceptual paradigm as a learning experience at the Multimedia Sudies Program, San Francisco State University, and as an ongoing media research and development program manifest in a variety of broadcast and published formats, and also via theInstitute for Global Futures.

The Institute for Global Futures is a strategic technologies research group which engages in trend analysis, technology development stream topology identification, and complex forecasting which examines synergistic thread linkages between various technologies and socio-economic systems. The activities of the institute include consulting services, strategic planning, reports, analysis, media programs, conferences, and strategic alliance building for a variety of global clients.

Complex new technologies are rapidly transforming markets, economies, and industries. The Institute for Global Futures was founded to promote the productive, sustainable, and strategic use of advanced leading edge technologies.

The included imgages are computer generated 3D renderings of theoretical and applied nanocomponents, created by the author as a visualization of various implementations of such molecular substructures and interactive symbiotic components. These are a few examples from a much larger collection of rendered, theoretical and applied nanosubcomponent systems, consisting of various "species" types with specific interactive and symbiotic properties.

The ability to create and exploit devices on the nanoscale is beginning to have a major and practical impact upon biomedicine. For instance, the development of chemical and biological sensors (e.g., the IBM 'NOSE' project) will make real-time diagnostics possible and eventually enable personal therapeutic regimens to be developed. Likewise, the ability to examine biological processes at a scale below that of visible light will yield a new biology—'mesophysiology'—that operates between high-resolution molecular imaging and the bulk biochemical and physiological techniques of traditional biology. The programme of work in the nanobiology section of the Centre will build upon the innovation and expertise of the medical co-applicant and his collaborators in the field. A requirement for success will be the ability to interact with the existing skills of mainstream physical measurement methodologies and the new nanofabrication techniques which will become available as part of this proposal. An interdisciplinary approach is essential to capture the wealth of UCL physics and electrical engineering expertise for the benefit of medicine and nanobiology.

Development of AFM systems for biology. The results are images of surface and intracellular structures of living cells (at left, live cells on glass with prominent cytoskeletal fibres) and maps of physical properties—e.g. receptor binding forces—of cell materials. We have also integrated a confocal microscope with the AFM to accomplish simultaneous fluorescence analysis (at right, a 3D isosurface reconstruction of bone cells on a mineralised substrate (blue) labelled in two colours (red, green) to independently demonstrate the distribution of two membrane proteins). The future lies in developing the bio-AFM further and introducing a range of more sophisticated physical techniques, electrophysiological applications and micromanipulation methods. Each image measures 50 x 50 mm2.

Recently, we have developed several of the technologies that will enable nanoscale analysis of processes in living cells.96-100 We have designed and built a life science optical interface between a commercial AFM and inverted microscope and modified the z-piezo controller to allow imaging and analysis over the large heights required for use with living cells which have a large, irregular and moving surface landscape. Finally, we have combined confocal microscopy, time lapse image capture and patch clamping with the modified AFM to measure real time cell signalling responses by optical and electrophysiological methods. This work has been made possible by 'blue-skies' pharmaceutical industry support and collaboration with instrument manufacturers—we believe that this collaborative approach will be essential for long term exploitation of our findings.

The outcomes from this engineering/development programme have been: (i) the first estimates of single molecule ligand-receptor binding forces in cells using integrin cell adhesion receptors as a paradigm in cells; (ii) generalisation of these techniques to other receptors and binding molecules; (iii) development of mapping techniques to demonstrate the combined location of active receptors with 3-D topography in cells; (iv) micromanipulation and signalling experiments: introduction of a few ligand molecules to a cell using the AFM cantilever at defined positions and under known and controlled mechanical forces with simultaneous recording of signal processes such as intracellular calcium levels, kinase enzyme activation, and induction of apoptosis; and (v) use of an AFM to analyse the material properties of cells and of associated ion channels, some of which are activated by the tip itself.

Based on the above, the first research areas to be pursued in nanobiology in the new Centre will focus on the further development and utilisation of life sciences force microscopy. This will include work on: (i) AFM tips with special shapes and coatings, to be made using the clean room facilities of the Centre, and designed taking advantage of the Centre staff's combined expertise in fabrication and modelling of tip interactions with soft surfaces; (ii) the continuation of our work on the study of cellular signalling events ensuing from receptor activation via reactive ligands on AFM tips; (iii) the analysis of the biomechanics of cellular mechanotransduction; and (iv) the design and manufacture of an AFM-based 'cell scraper' which we will use to study total cellular interaction forces between cells and newly fabricated materials for use in tissue engineering (with members of the UCL Tissue Engineering Centre and the Eastman Dental Institute) and 'lab on a chip' applications within the Centre.

The activities just described will eventually be only a very small part of the nanobiology activities in the Centre. We will be working with a large number of collaborators to further the development of SPM for biology and medicine, as well as to fabricate special purpose chips and even micromachines for assays and eventually even disease treatments. To illustrate this commitment, specific collaborations that we are planning include:

Surface and intracellular properties of cells and their inhomogeneities (e.g., lipid rafts and polarised cell domains) by such methods as surface electrochemical properties (Prof. David Williams, Dept. of Chemistry UCL), ion conductance imaging (Prof. Chris Pitt, Dept. of Electronic & Electrical Engineering UCL), electrophysiology (Prof. Jonathan Ashmore FRS, Dept. of Physiology UCL), cell thermal imaging, and X-ray reflectometry.

Development of new probe techniques such as: single molecule FRET (fluorescence energy transfer) at the tip-cell/macromolecule interface;108 near-IR scanning near-field optical microscopy using microfabricated devices;109-112 magnetic property imaging; probing electron transport in cells (Prof. David Delpy FRS, Dept. of Medical Physics UCL).

Sequence analysis of receptors 'fished' from cells using specific ligand-derivatised AFM tips made in the Centre (Prof. Tony Segal FRS, Dept. of Medicine UCL in the Wellcome Trust proteomic facility).

AFM imaging and structure of 2D crystals and proteins in artificial lipid bilayers (Prof. Helen Saibil, Birkbeck Crystallography and BBSRC Biomolecular Structure Centre); and analysis of structure of process-engineered plasmids, proteins and other biomolecules (Prof. Peter Dunnill, ACBE UCL). The hope here is to learn about proteins occurring in or near membranes, where their conformations may be quite different than in the crystals required for traditional (X-ray) protein structure determination.

Large-scale sensor arrays for diagnostics of biological fluids. The arrays in question can consist of a wide variety of devices, ranging from micromechanical cantilevers, each with a slightly different chemical binding property, to microcalorimeters to measure energy released upon chemical reactions with different molecules resident within each calorimeter (Prof. David Williams).

Electronic properties and exploitation of biomolecules. Although the electrical characteristics of biomolecules may not generally be relevant to their biological function, they do represent a fascinating class of new materials for possible use in electronics. We plan to use scanning tunnelling microscopy (STM) as well as special-purpose substrates prepared in the nanofabrication facility to measure and eventually to make use of these characteristics (Prof. David Delpy FRS).

Our biological research programme is absolutely dependent upon unfettered access not only to the nano-fabrication facility, but also on the presence in a single location of scientists with expertise in theory, hard and soft condensed matter physics, and electrical engineering. All of these will be available in the Centre, and as can be seen from the previous sections of this proposal, are keen to pursue applications of their work in nanobiology. What the Centre will therefore provide is much more than the essential new physical infra-structure, but the unique interdisciplinary environment needed for meeting the challenges of nanobiology.

Rapid advances in nanotechnology are creating new opportunities for interfacing nanotechnology and biology in order to more effectively tackle the nation's biodefense agenda. The advances being made in nanobiology are contributing to new biodefense solutions in detection, diagnostics, therapeutics, and biodefense immunology. Although many of these advances are still in their infancy, the field has progressed with leaps and bounds just within the past year. What's more, the leaders who are pioneering the application of nanobiology to biodefense are enthusiastic about the results to date and the future potential of the technology.

Applications of Nanobiology to Biodefense will provide an excellent opportunity for nanobiologists, bioengineers, microbiologists, and immunologists to get the latest information about the current state of the art and obtain more insight into how they can apply nanobiology technology and nanosystems biology to immunological problems, targeted drug delivery and increased accuracy and sensitivity in analyzing laboratory samples. The goal of the conference is to enable these groups to seize the potential of this powerful new technology. To date, nanobiologists, bioengineers, microbiologists, and immunologists have not engaged in a vigorous interaction, and, therefore, the technology and the bench science have not been sufficiently connected. The conference will bring together these groups so that they can better appreciate the opportunities for the successful application of nanobiology to biodefense and can better assess the opportunities for working together.

Golden Ruler

- Syndicated content from Journal of Nanobiotechnology.
- Syndicated content from Yahoo News.
- Syndicated content from Feedster (different sources).
- Syndicated content from Blogdigger (Different Sources).
- Syndicated content from Topix NanoTechnology.
- Syndicated content from HubMed Nanobiology Alerts from PubMed.
- Syndicated content from NanoBiology on PubMed.
- Syndicated content from AzoNano.
- Syndicated content from NanoVIP.com.
- Submitted by Jameel Ahmad Khan on Mon, 2005-05-23 14:56.

In the current issue of Science (20 MAY 2005) Robert F. Service has reported that in a meeting (NSTI Nanotech 2005, Anaheim, California,) chemist Paul Alivisatos from University of California (UC), Berkeley,has described about a system developed by them, which can be used to measure distances out to 70 nanometers and can keep track of their targets indefinitely.This system uses pairs of gold nanoparticles & works theoratically similar to FRET (fluorescent resonance energy transfer).But neither it has the restriction for the dye molecules to be within the distance of 10 nanometer nor it has problem of photobleaching.

He reports that Alivisatos, Carsten Sönnichsen, Björn Reinhard, and Jan Liphardt—all colleagues at UC Berkeley and Lawrence Berkeley National Laboratory has first immoblised gold nanoparticles on a glass slide. When they shined these particles with white light they found that light scattering from particles changed steadily with distance.On the basis of these results they have attached a pair of nanoparticles at the either end of sigle stranded DNA. When they have added complementry DNA strand they have found a red shift in the wavelength of light scattered by the nanoparticles. Alivisatos hopes that his "longer-lived, longer-range nanoparticle-based FRET will eventually overtake its organic cousin for measurements in which background lightscattering is low".

Nanosize Microtubules

A study in the journal Science is offering new insights into a long-standing mystery about plant growth. The scientists who conducted the experiment say their results could open new avenues of research for developing more effective herbicides and pharmaceuticals. Plant biologists from Stanford University and the Carnegie Institution of Washington report their new findings in the April 24 online edition of Science Express. The researchers are the first to witness the birth and growth of individual "microtubules"—nanosize tubes of protein that form inside living plant cells.

"We have a strong interest in cell growth, form and development, and microtubules play a prominent role in all three," said Stanford scientist Sidney L. Shaw, lead author of the Science study. Shaw is a research associate in the laboratory of Sharon Long, a professor of biological sciences and dean of Stanford's School of Humanities and Sciences.

"Imagine the plant cell as a cylinder with many thin rods, or microtubules, laying along parallel lines just under the surface," noted co-author David W. Ehrhardt, a staff scientist in the Carnegie Institution's Department of Plant Biology located on the Stanford campus. The microtubules form specific patterns that determine the eventual shape of the celi and ultimately the shape of the plant, he explained.

"If plant cells didn't have a particular shape, we wouldn't see trees and grass grow as they do. They'd be squat and lumpy, lacking much of the diversity of form that plays important roles in physiological and ecological function and also makes plants fun to garden with," Ehrhardt added.

A microtubule is only about 25 nanometers in diameter, roughly 2,000 times narrower than a human hair. Found in all plant and animal cells, these hollow rods of protein assemble spontaneously and carry out a number of important functions. Some act as miniature girders that support the cell, while others form tracks that transport vital cellular cargo or guide the separation of chromosomes during cell division.

In animal cells, microtubules lengthen and grow from a fixed site called the "centrosome." But plant cells lack centrosomes, so the origin and development of plant microtubules had been something of a mystery—until now.

"The novelty of this study is that, for the first time, we've been able to watch the birth of microtubules in plants and watch them go streaking across the cell," Shaw explained.

"In plants, microtubules start out in the cell wall in a random pattern as if you'd thrown down a bunch of pick-up sticks," Ehrhardt noted. "But over time, they organize in a really remarkable way to form a three-dimensional array at the periphery of the cell that resembles a helix."

The Science study revealed for the first time how individual microtubules come together to form these large arrays—a discovery that may have ramifications beyond the field of plant biology. That's because the arrays, which look like overgrown DNA helices, are believed to play an important role in the distribution of cellulose, the most abundant organic substance on Earth and a major component of the cell wall.

Cellulose is the principal ingredient in paper, lumber, cotton, rayon and a wide range of other products we take for granted. This tough, water-resistant compound also is found in fruits, vegetables and grains. When doctors recommend a high-fiber diet, they mean a diet rich in cellulose.

"In most plant cells, cellulose is laid down, not randomly, but in a very organized way," Ehrhardt said. "It's thought that microtubules control and guide the machinery that builds the cell wall. Being able to engineer how cellulose is laid down could provide bioengineering opportunities for generating cell walls with different properties for a variety of biomaterials."

In their Science study, the researchers used a confocal microscope to observe single microtubules in the cells of Arabadopsis plants, a member of the mustard family. By decorating the microtubules with green fluorescent protein, the scientists were able to watch them grow and develop.

"One nice thing about our study is that we were looking at the living, intact organism," Ehrhardt said. "Most animal studies are done with isolated cells, so the behavior of their neighboring cells aren't observed, whereas we were looking at live cells in their completely native, multicellular context."

Using time-lapse imagery, the scientists discovered that new microtubules emerged near the outer wall of the plant cell, not in the cell interior where animal microtubules originate. Time-lapse imaging also revealed that plant microtubules appear to move in the cell by a process known as "treadmilling," which occurs when bits of protein (known as "subunits") are added to the leading end of the microtubule and simultaneously removed from the trailing end.

"With treadmilling, the microtubule looks like it's moving in one direction, but in reality, one end is growing while the other is shortening," Ehrhardt noted.

"Treadmilling has only been rarely seen in some animal cells," Shaw added. "What's remarkable about the plant system is that almost all the microtubules we can see are undergoing a treadmilling motility."

The way protein subunits are added and removed from a microtubule also proved interesting, Ehrhardt said: "It isn't the case that they are being smoothly added to the growing end and smoothly removed from the shrinking end. Instead, polymers are quickly added and removed in spurts to one end, while being less quickly removed in spurts on the other.

"The growing end dances back and forth—it grows, it shrinks, it grows, it shrinks. But there is a bias in this dance that results in growth over time. On the other end, the activity is smoother. It shortens slowly, it pauses, then some more gets removed and it pauses again. So the behavior at the two ends is very different, and the combination of the two behaviors 'moves' the assembled microtubule from one location to another in the cell.".

Time-lapse imagery revealed that an individual microtubule starts out in a random "pick-up stick" position, then treadmills across the cell at an average rate of 0.5 microns per minute (one micron equals one-millionth of a meter), encountering other microtubules along the way.

"Then sometimes it bumps into an organized bundle of microtubules, and like a train following a track, it turns direction and follows them, becoming a part of the helical array," Ehrhardt said.

"There's nothing actually physically moving anything, and that makes treadmilling a really unexpected mechanism for assembling higher-ordered structures," Shaw observed.

What guides the microtubule to its new orientation? The researchers hope that subsequent experiments will provide the answer.

"When a microtubule is crawling along the surface of the cell, how does it know it has encountered another microtubule?" Ehrhardt asked. "Are there other proteins that connect microtubules together and guide traveling microtubules into bundles? We suspect that might be the case, but we don't know."

The researchers hope to expand their observations to include other kinds of microtubule arrays, such as those involved in chromosome separation during cell division.

"The drugs that are responsible for slowing cancers, such as taxol, are actually plant molecules than bind to microtubules to prevent a cell from dividing," Shaw explained. "In fact, most of the major microtubule-acting drugs are derived from plants."

Many herbicides are also anti-microtubule, Ehrhardt added: "By understanding more about how microtubules behave, we'll probably get some insights into how these herbicides are actually acting, and that could open up opportunities for developing better herbicides."

The third co-author of the Science study is Roheena Kamyar, a former graduate student in Ehrhardt's laboratory now enrolled at the University of Michigan School of Medicine. Funding for the study was provided by the Carnegie Institution of Washington, the Howard Hughes Medical Institute and the U.S. Department of Energy.

DNA scaffolding

The next frontier for information processing may lie at the interface of nanoelectronics and biotechnology.

An interdisciplinary team led by electrical and computer engineering professor Richard Kiehl is exploring the use of DNA as a programmable scaffolding for the self-assembly of nanoscale electronic components. As a model for fabricating and designing semiconductor devices and circuits, DNA offers two key advantages: size scale and programmability.

Most industry experts believe that within the next 10 to 15 years the ability to scale down conventional technologies will reach its limit. At that point, the operating principles of conventional devices—and the techniques used to fabricate them—will break down. The basic elements of the DNA molecule are at just the right scale, says Kiehl.

Self-assembly uses bio-recognition, a natural process in which one molecule is attracted to and binds with another to form small structures. In the case of DNA, the attraction can be programmed so that the molecules will spontaneously assemble in solution to achieve a desired result.

"It's possible to synthesize small versions of DNA molecules in the laboratory and program in whatever code you want," says Kiehl. "And because the two strands of DNA have complementary codes that match up, you can design one strand of DNA in a certain way so it will match another strand and assemble a nanoscale structure this way."

The matched segments form a scaffolding on which nanoparticles are affixed at highly selective attachment points. It's an approach that offers the programmability and precision needed for assembling electronic circuitry on the nanoscale.

"We have to make a real paradigm shift," Kiehl says. "Not only do we have to keep improving performance, but we also must look at the kinds of devices we can make at those scales and how we want to use them to process information."

To that end, the researchers are turning to the human brain for inspiration. They envision devices whose electrical characteristics resemble those of neuron-like electrical waveforms in the brain. Like certain regions of the brain, the devices would process information based on pattern recognition rather than on individual bits of information. It's a more sophisticated level of information processing than can be achieved using conventional computers.

Kiehl predicts there will be a wide range of applications for this technology, including signal processing, communications systems, and computer systems. "The higher end of this [work] will be things that computers can't do very well today because the operations they use are too restrictive. One is the ability to recognize a pattern, such as identifying a letter as being an 'A' or a 'B', or being able to identify a face.

"It won't be just making things faster and faster in the conventional way," he says. "It will really be opening up new ways to process information in machines."

Micro Manufacturing

The ubiquitous mobile 'phone has had a revolutionising impact on modern life and not just in the social and commercial spheres but also in manufacturing. The need for miniaturisation in the build has spawned developments in the areas of micro and nanotechnology (MNT) and these currently have the potential for exploitation in the manufacture of a wide range of products and not just in cellular communications. At least that is the view of the UK Manufacturing Technologies Association (MTA), an organisation with strong roots in the machine tool field.

Says the MTA, "The mould and die sector and its equipment suppliers will enjoy particular benefits from micro and nano manufacturing through the requirement for high volumes of micro components made from plastics and light alloys. Telecommunications are typical of this trend. Micro and nano manufacturing will be increasingly employed in a huge variety of consumer products and industrial components. Current mobile telephone technology incorporates such aspects: this will permeate not only other product areas but also the tools, machine tools and processes used in their manufacture".

MTA's fear is that the UK will "miss out" on these opportunities unless it increases its awareness and expertise in the micro and nanotechnologies. It was for this reason that it devoted its fifth seminar in the series "Sharing Tomorrow's Technology Today" to the topic of "Micro and Nano Manufacturing", which was held at the University of Birmingham in partnership with the Machining Research Group of its School of Engineering, a team with specialised knowledge and facilities in these technologies. The seminar was supported by demonstrations on these facilities that included high-speed micromachining and micro die sink and wire cut electro discharge machining (EDM).

The status and scope for micro and nanomanufacturing in the UK was explored by Professor David Allen of Cranfield University. He first established some definitions, including "Micromachine, a machine made from micrometer scale components", and "Nanotechnology, the study, development and processing of materials, devices and systems, in which structure definition on a dimension of less than 100?nm is essential to obtain the required functional performance". He also identified the commonly used terms of Microsystems Technology (MST), which is prevalent in Europe, and MicroElectroMechanical Systems (MEMS), widely used in the USA.

Originally, MST or MEMS was a spin-off from the silicon-based microelectronics industry, which relied extensively on photolithography as the manufacturing process. Currently, miniaturised mechanical components are produced and the range of materials has been increased to include metals, polymers and ceramics. In addition, optical components, thermal components and fluidic components are being miniaturised and integrated.

Professor Allen illustrated that the market for MST products has shown a steady rise growing from approximately $25 million in 2000 to an estimated $45 million in 2004, more than half of which is in just two products: disc drive heads and ink-jet printing heads. The former are manufactured using photochemical machining (PCM) mainly in the USA. PCM is also a very active process in the UK with companies such as Precision Micro in Birmingham, which has one of the largest automated PCM facilities certainly in Europe and possibly in the world according to Professor Allen, and Weymouth-based Tecan, who offer "one stop" contract manufacturing of precision mechanical parts and tooling. In Professor Allen's view such facilities are under utilised by the UK MEMS industry.

The other major non-silicon MEMS product, the jet printing head is produced using either laser micromachining or micro-EDM. According to Professor Allen, the

UK is very active in two laser micromachining processes, copper vapour laser machining (CVLM) by Oxford Lasers of Didcot, and excimer laser machining (ELM) by Exitech of Oxford. In the latter, the beam of an excimer excited laser is projected through a mask and focussed on the material to create complex fine detail features.

CVLM involves using pulsed (laser) light energy focussed onto a spot size of less than 5?ìm, to remove material by vapourisation and ablation. It is much faster than micro-EDM due to its high material removal rate (through a combination of high pulse repetition rate and an ability to remove material to depths of 10?ìm in each pulse). Another advantage is the large aspect ratios with up to 40:1 being reported for stainless steel. On the debit side is a variable surface finish, slight degree of taper in holes and lower resolution compared to the EDM process.

Micro-EDM was described in the presentation by Dipl-Ing Axel Demmer of the Fraunhofer Institute of Production Technology (IPT) in Aachen, Germany. He cited some of the advantages of the process, including the ability to machine hard materials and therefore well suited to mould manufacture, the low process forces involved and aspect ratios greater than in micromilling.

The "micro version" of EDM operates at frequencies in the kilohertz to megahertz spectrum in contrast to the kilohertz range in the conventional discharge process, with shorter pulse durations (less than 0.1?ìs vs 100?ìs) and smaller spark gaps (under 10?ìm vs 100?ìm). However, it suffers electrode wear rates over 30 per cent compared to under 5 per cent in conventional EDM.

In his talk, Hanspeter Gruber of Agie in Switzerland differentiated between die sink micro-EDM and the wire cut process. For die sinking, Mr Gruber quoted ribs 20?ìm thick with aspect rations of up to 25, and up to 100 for holes. Electrode sizes were stated to be 10?ìm diameter ×1.5?mm long. Other parameters he listed included, form errors of ±3?ìm, position errors of ±1?ìm and surface qualities of less than 0.1?ìm Ra.

Higher aspect ratios are attainable with wire cut micro-EDM, in the range of 60-80 for grooves and 20-30 for ribs. Other wire cut process parameters also tend to be better, ±1.5?ìm for the form error, position error of ±1?ìm and surface quality under 0.06?ìm Ra, to quote Mr Gruber. Rib sizes are typically 20-40?ìm with inside radii of 20?ìm.

The latest Agie Vertex wire cut EDMs have a significantly improved process performance at speeds eight times that of the earlier models. They are capable of shaping 2D components using wires down to 20?ìm with the smallest thread hole being just 50?ìm larger than the wire diameter. They have all the attributes of modern EDM machines including twin wire cutting and auto threading of the fine wires.

A further possibility is to combine the die sink and wire cut EDM processes, taking the benefits of each to produce intricate micro parts. Mr Gruber cited the example of a four-blade fan measuring 6?mm in diameter and a finished blade leading edge thickness of 15?ìm. The first step was to wire cut the 2D outline shape of the fan with 50?ìm wire and then shape the blade on both surfaces by die sink EDM using a precision milled convex-concave electrode. Mr Gruber went on to list a wide variety of microcomponent for which micro-EDM offers an economic solution. These were:

- mould inserts for electronic devices such as connectors;
- pins and holes in microvalves and nozzles;
- forms for extrusion nozzles;
- punching tools for microgears and lead frames; and
- precision parts for the watch industry.

Moving from the micro to the nanoworld is a three-places-of-decimal step, for which the normal "sliding and rolling" movement of machine tools is totally inadequate. Agie is working with the École Polytechnique Fédérale de Lausanne (EPFL) to solve this problem and has developed the Agie Nano, a prototype nano-EDM machine that uses flexures to create and support nanometre movements. A flexure is simple and robust and has the benefits of zero stick-slip, friction, backlash and wear, a high resonant frequency and a high rigidity. The prototype is a parallel kinematics machine with an 8?mm motion range and a 5?nm static resolution. First tests of this machine, reported by Mr Gruber, show it capable of precise movements with high repeatability and high precision 2D contouring.

Another rapidly developing microcomponent process discussed at the MTA seminar was micromachining, primarily microdrilling and micromilling. IPT Aachen's Mr Demmer itemised some of the recent advances in this process, including ultrafine carbide tools able to mill hardened tool steel up to 55?HRc, 100?ìm diameter ball and square end mills with and without coating, tolerances of ±5?ìm and 10:1 aspect ratios. With these developments, micromilling is able to compete with micro-EDM and offers superior metal removal rates.

Because of the small diameter cutters involved, micromilling requires high speed spindles on machine tools utilising high speed machining (HSM) technology. The design and performance of such machines was described by David Edwards of Matsuura Machinery. He highlighted two machines from the Matsuura range, the LX-1 and the LX-0 series. These are CNC HSM machining centres with large table capacities (700×500?mm on the LX-1) but with micromilling capability. This allows for the machining of microcavities in relatively large tool blocks, for instance.

One of the most important HSM features of the LX-1 for micromilling is the linear motor drive motors for the three major axes. In addition to the high fast traverse and feed rates capability of linear motors – up to 90?m/min for the LX-1 – compared to conventional ball screws, there is no dependency on friction, no backlash or motor reversal errors, no loss of accuracy through component wear and no accumulative errors from the ballscrew-coupling-motor configuration. Also, the same linear motor technology is used for the fourth axis table rotary motion instead of the conventional mechanical worm and gear arrangement.

The other piece of HSM technology on the LX-1 is the spindle, which runs on ceramic bearings lubricated by an oil/air system and is capable of 60,000?rpm. To exploit this potential a shrink fit tool holder is employed with a two-face contact and tool locking system. This ensures both a high precision and long tool life.

Rigidity of the machine structure is also crucial to high speed and micromilling. Matsuura has based the LX-1 on an overhead gantry design optimised using 3D finite element method (FEM) analysis. To overcome the yawing effect symptomatic of the gantry structure, two linear motors are fitted to drive the *Y*-axis, each synchronised with the other and mounted on the outside of the gantry side limbs in isolation from the machining area to eliminate any swarf "pollution" problems.

To further ensure its micromachining attributes, the Matsuura LX machine is fitted with 0.1?ìm resolution linear scales giving a positional feedback to 10?nm. The overall result is a machine able to machine the finest of microcavities with circularity errors of 3?ìm even at the highest feed rate on the LX-1 and under 1?ìm on the LX-0.

For the future, Matsuura is planning an 80,000?rpm spindle machine, which is due in 2005, that will be able to precision grind as well as high speed micromill. Another hybrid machine will be the Photon M, which combines milling and laser machining for mould and die rapid prototyping work, also due in 2005.

A major consideration in micromachining is the CAD data available and its use in programming. As Roger Onions of Delcam emphasised in his presentation, the main priority in CAM is to protect the cutting tool from sudden changes in load, and with micro-sized cutters this is even more important than in conventional machining. Further, because of the fine tolerances, more detailed toolpaths with many more points are needed, which necessitate a sophisticated CAM system, able to complete calculations quickly, and consequently more computing power. He recommended looking out for errors during data exchange as any missing surfaces or gaps can be fatal. Extra time spent in CAD will be rewarded with successful CAM.

Mr Onions suggested, "Model what you intend to machine and don't leave features like internal fillets to be formed implicitly by the cutter as this will cause excessive load and load changes in forming the radii". Other pieces of advice were that holes should be drilled (an all-round load) rather than milled (side load) and when programming, cap over holes or the cutter will try to plunge into material. And, ensure that trimming is accurate as poor trimming is more likely to cause gouging with microcutters.

Delcam has a number of powerful tools within its PowerSHAPE CAM software to minimise tool loads. These include "Raceline machining", a patented Delcam strategy for high-speed roughing that smoothes out tool paths as it moves away from the component form. It minimises sharp direction changes and so allows faster machining and longer cutter life.

Another tool is "Trochoidal machining", which uses a "shaving" approach to machine slots and avoid full-width cuts to protect the cutter at higher speeds. However, it does mean employing a cutter that is smaller than the slot width.

One of the principle aims of the MTA seminar was to encourage "UK Limited" to participate and exploit micro and nanomachining. This is no easy task as in Cranfield Professor Allen's view there was no apparent no national focus less than 4 years ago. He said, "Whilst it was recognised that some centres of excellence existed within the UK, it was felt that the national MEMS effort was un-coordinated and that the links between academia and industry were weak". This became particularly apparent following DTI Trade Missions to study MEMS in France and Switzerland (1999) and Taiwan and Korea (2000). As a result members of these Missions set out to promote national coordination and subsequently formed the Microsystems Manufacturing Association (MMA) under the auspices of the Institution of Mechanical Engineers (IMechE).

The MMA has been lobbying the Government to set-up a National Microsystems Manufacturing Centre (NM2C) and this is currently planned for the South West of England. Also, the Northwest Development Agency is setting up a National Microsystems Packaging Centre (NMPC) in Liverpool.

These moves are an indication that attitudes towards microsystems are changing, a view further confirmed last year when the government announced a new funding initiative. Trevor Fraser of the DTI told the seminar that up to £90 million will be provided to support the development of MNT manufacturing in the UK, in which the emphasis will be on outputs—the products and processes – rather than the inputs – the science and technology.

The funding will be split into two parts; the first of £50 million will be for an "Applied Research" programme, managed by Ernst & Young. It is open to large and small companies and academic institutions and supported on a 50:50 basis. The funds will be released in three annual calls of around £15 million, the first of which closed earlier in 2004.

The second element of the initiative, amounting to £40 million, is to support an MNT network in two areas, the establishment of the "Virtual Hub" directed by Hugh Clare, formerly of Unilever, and the setting up or enhancement of Regional Centres to provide facilities in MNT.

However, the £90 million is not the only source of funds for any organisation wishing to develop MNT. Mr Fraser revealed the wider funding picture in the UK with grants available through other government agencies that could be applied to MNT, including the Research Councils (over £60 million) and the Development Agencies (over £200 million). In addition, £1 billion of EU "Framework 6" money is "up for grabs". All this Europe-wide finance does indicate that the politicians, at least, are taking very seriously the technology necessary to "shrink sizes and tighten tolerances".

Nanocapsules and Dendrimers—

A nanocapsule is any nanoparticle that consists of a shell and a space, in which desired substances may be placed. Technologies for microencapsulating materials have been around for several years, primarily for applications involving minimisation of hygroscopy and chemical interactions, elimination of oxidation, and controlled release of nutraceuticals.

Man-made liposome's have been used in cosmetics for some years to control the release of substances or protect them from the environment. Recently many other materials, such as polymers, have been used to make nanocapsules.

Polymeric nanocapsules can be made in specific sizes, shapes, and in reasonable quantities. Nanocapsules can be made to function in various ways. They can be produced as monodisperse particles with exactly defined biochemical, electrical, optical, and magnetic properties. They can be tailored to suit the complexity of whatever application they are intended for, such causing the release of the contents in response to a particular bimolecular triggering mechanism in targeted drug-delivery systems.

Nanocapsules can be used as smart drugs that have specific chemical receptors and only bind to specific cells. It is this receptor that makes the drug 'smart,' allowing it to target cancer or disease. The advantages of nano-encapsulation technologies for pharmaceutical applications include:

- Higher dose loading with smaller dose volumes
- Longer site-specific dose retention
- More rapid absorption of active drug substances
- Increased bioavailability of the drug
- Higher safety and efficacy
- Improved patient compliance

Beyond the ability to deliver existing drugs to their target, nanocapsules would allow for as much as a 10,000-fold decrease in drug dosages, reducing the harmful side effects of drugs used in chemotherapy. Quite often, drugs don't make it to market is because they have too many unwanted side effects. However, placing the same drug inside a nanocapsule and delivering it directly to its intended target in a reduced dosage, eliminates some of those side effects, or at least reduces them to an acceptable level.

Nanocapsules also have potential applications in agrochemicals, cosmetics, genetic engineering, wastewater treatments, cleaning products, and adhesive component applications. They can be used to encapsulate enzymes, catalysts, oils, adhesives, polymers, inorganic micro- and nanoparticles, latex particles, or even biological cells.

A dendrimer is an artificially manufactured or synthesized large molecule comprised of many smaller ones linked together—built up from branched units called monomers. Technically, dendrimers are a unique class of a polymer, about the size of an average protein, with a compact, tree-like molecular structure, which provides a high degree of surface functionality and versatility. Their shape gives them vast amounts of surface area, making them useful building blocks and carrier molecules at the nanoscale and they come in a variety of forms, with different physical (including optical, electrical and chemical) properties.

Dendrimers can act as biologically active carrier molecules in drug delivery to which can be attached therapeutic agents and as scavengers of metal ions, offering the potential for environmental clean-up operations because their size allows them to be filtered out with ultra-filtration techniques.

A number of projects around the world are exploring the use of nanoparticles on the farm for purposes other than pesticides—from enhanced photosynthesis to better germination and soil management.

Researchers at Kyoto University (Japan) have discovered a method of producing ammonia using buckyballs. Ammonia is a key component of fertiliser, but it is not clear if the resulting product for use in the fields would contain buckyballs.

Scientists at the University of Korea have applied for a patent on a liquid after which time it dissolves in the groundwater and becomes indistinguishable from naturally occurring iron.

The Russian Academy of Sciences reports that they have been able to improve the germination of tomato seeds by spraying a solution of iron nanoparticles on to fields.

In 2003, ETC Group reported on a nanotech-based soil binder called SoilSet developed by Sequoia Pacific Research of Utah (USA). SoilSet is a quick-setting mulch which relies on chemical reactions on the nanoscale to bind the soil together. It was sprayed over 1,400 acres of Encebado mountain in New Mexico to prevent erosion following forest fires, as well as on smaller areas of forest burns in Mendecino County, California.

A number of approaches are being developed to apply nanotechnology and particularly nanoparticles to cleaning up soils contaminated with heavy metals and PCBs. Dr. Wei-Xang Zhang has pioneered a nano clean-up method of injecting nano-scale iron into a contaminated site. The particles flow along with the groundwater and decontaminate en route, which is much less expensive than digging out the soil to treat it. Dr. Zhang's tests with nano-scale iron show significantly lower contaminant levels within a day or two. The tests also show that the nano-scale iron will remain active in the soil for six to eight weeks, after which time it dissolves in the groundwater and becomes indistinguishable from naturally occurring iron.

Case Study 1: Respirocytes—Mechanical Artificial Red Cell

The World Health Organization, as well as most public health organizations, unanimously agree: the world is facing a severe shortage of donor blood. The American Red Cross has 36,000 units per day to satisfy a daily demand of 80,000 units. And unfortunately, like oxygen, blood (more so the red blood cell component of blood) is essential to life and non-substitutable. Furthermore, anemia (hemoglobin—the oxygen carrying molecule of red blood cells—less than 12mg/dL) is a prevalent disorder that causes significant morbidity and mortality, and bears huge economic-costs of treatment. Possible artificial oxygen carriers have been investigated for eight decades, starting with the first documented use of hemoglobin solutions in humans in 1916. With an estimated 4 million patients suffering from anemia as a chronic disease, the commercial potential for successful blood substitutes has been estimated at between $5-10 billion/year. Current blood substitutes include hemoglobin formulations and flurocarbon emulsions, neither of which offers performance comparable to Mother Nature's molecule.

Enter nanomedicine to offer a solution where conventional medicine has failed. Scientists at the Institute for Molecular Manufacturing have purposed a mechanical artificial red cell called a "Respirocyte". The purposed artificial Respirocyte is a spherical nanomedical device 1 micron in diameter made up of a flawless diamond or sapphire shell constructed atom by atom. The device utilizes an active means of conveying gas molecules into, and out of, pressurized microvessels. For this task, engineers have designed molecular sorting rotors with binding site "pockets" along the rim exposed alternately to the blood plasma and interior chamber by the rotation of the disk. Each pocket selectively binds a specific molecule when exposed to the plasma. Once the binding site rotates to expose it to the interior chamber, the bound molecules are forcibly ejected by rods and stored in tanks. Onboard power is provided by a mechanochemical engine that combines glucose and oxygen to generate mechanical energy to drive molecular sorting rotors and other subsystems. Sensors and an onboard computer provide the necessary controls for respiratory gas loading and unloading, rotor field and ballast tank management, glucose engine throttling, power distribution, interpretation of sensor data and commands received from the outside, self-diagnosis and activation of failsafe shutdown protocols, and ongoing revision or correction of protocols in vivo.

Having developed a technology/product is one thing; developing a technology/ product that has commercial demand is a completely different story. The mouse trap is the most highly designed and patented device in the U.S. patent office, and yet the market does not see a flood of mouse traps. Why? Because the market does not command the variety of mouse traps that inventors have created. Market demand is but one of many factors to consider prior to commercialization. No plan is complete without examining other factors such as financing, technology testing and validation, strategy for commercialization, transition from plan to operations, just to name a few. Considering all these elements, then, can mechanical red blood cells "cross the chasm" and become commercially viable?

As alluded to earlier, the treatment of anemia consists of either transfusion of natural red blood cells or red blood substitutes. Natural red blood cells are in short supply, which suggests a market demand. The next logical question is whether red blood substitutes can satisfy the remaining demand.

Analysis of the two forms of blood substitutes—hemoglobin solutions and fluorocarbon emulsions reveal limitations that translate into opportunities for nanomedical red blood cells14. Most of the red cell substitutes under trial at present have far too short a survival time in the circulation to be useful in the treatment of chronic anemia, and are not specifically designed to regulate carbon dioxide or to participate in acid/base buffering. Several cell-free hemoglobin preparations evidently cause vasoconstriction, decreasing tissue oxygenation, and there are reports of increased susceptibility to bacterial infection. Fluosol, a widely available fluorocarbon perfusant, is approved by the FDA for use in patients at high risk of ischemic complications during percutaneous transluminal coronary angioplasty, but its experimental use in the treatment of severe anemia has been disappointing and cannot yet be recommended. The radical innovation associated with Respirocytes addresses the "pain points" where current technology fails. Nanomedical red blood cells thus has the potential to become the standard for anemia management How about the market size? A strong market demand does not necessarily mean a profitable business. For example, demand is present for orphan drugs such as GENOTROPIN because it treats a rare disease (Prader-Will Syndrome); however the drug is not profitable because fewer than 200,000 Americans are affected by the disease. Because of the level of technology involved with nanomedical red cells, the market size must be sufficient to warrant the capital investment necessary to bring such a product to market. In this case, the target market of Respirocytes—anemia —is certainly attractive. With 4 million people in the U.S. alone suffering from chronic anemia and an additional 1 million units (minimum) of blood required for transfusion annually, the cost of treating anemia has been estimated to be in excess of $5 billion per year.

Development is one of the key challenges to the commercialization of Respirocyte. The challenge is not in manufacturing the components of the nanodevice, but in assembling the components. Molecular nanotechnology violates no physical laws. In

1959, the Nobel physicist Richard P. Feynman observed that: "The principles of physics, as far as I can see, do not speak against the possibility of maneuvering things atom by atom. It is not an attempt to violate any laws; it is something, in principle, that can be done; but, in practice, it has not been done because we are too big".

It is generally accepted that molecular nanotechnology will be developed; the proper question is no longer "if" but "when." Developmental success leading towards molecular nanotechnology has been achieved. For example, A Scanning Tunneling Microscope was used in 1989 to spell out "IBM" using 35 individual xenon atoms on a nickel surface. Atomic Force Microscopes (AFMs) have performed nanomachining operations: two rectangular slots and a 50-nm rectangular sliding member were milled from a crystal, and then the member was slid repeatedly from one slot to the other, making a reversible mechanical latch. Chemists have built a large number of potentially useful rigid nanoparts including molecular-scale rods, rings, springs, cubes, spheres, tetrahedrons, hollow tubes, propellers, and tongs, and wire-frame nanostructures of many shapes made of polymerized DNA. Micro Eléctro-Mechanical Systems (MEMS) has produced 100-micron-scale accelerometers, flow valves, pistons, gear trains, and piezo-driven motors that are available today off-the shelf in mass quantities; microgrippers used to manipulate individual 2.7-micron polystyrene spheres, dried red blood cells, and various protozoa; and a/1000th scale working electric car just 2.8 millimeters wide, complete with motor, wheels, body, spare tire, bumpers, and even a 10-micron thick license plate.

As discussed earlier, market research suggests a commercial application of IMM's Respiorcyte design. However, the analysis has also demonstrated the constraints surrounding the commercialization of this product, namely development. Developmental challenges aside, the technology also faces other obstacles, such as financing, regulatory, and marketing. Furthermore, since nanotechnology has been achieved along so many independent pathways, the real question for scientists at the Institute of Molecular Manufacturing looking to commercialize this design is whether or not to partner with or license to other companies involved in nanotechnology. Several compelling reasons speak to the need for IMM to consider licensing its design to other companies for development. First, the financial burden associated with commercializing the product is unlike any other consumer product on the market. Commercializing a medical product is a multi-tiered process, with each step being more expensive than the last. According to a recent study conducted by Tufts University, the average cost for a biotechnology/pharmaceutical company to develop a prescription drug is $800 million (the breakdown of these costs is approximately: basic/investigative, 10 per cent; preclinical, 15 per cent; manufacturing/process, 15 per cent; clinical, 55 per cent; post-marketing, 5 per cent). Using these numbers as a benchmark, it is reasonable to assume that the present cost of developing Respirocytes would be in excess of the average $800 million given the immaturity of technology and process. In addition, because of the novelty of this technology, regulatory bodies such as the FDA will require strong evidence of the product's efficacy and safety. In clinical trial terms this translates into bigger sample size, more metrics, and longer duration of study—all of which exponentially increases the cost of development. Secondly, Respirocytes, unlike a drug molecule, is a combination of material science, biological science, mechanical engineering, electrical engineering, and computer science. IMM may have the competency for material design and assembly, but where will they get the expertise for computer design necessary for control of the device? Where will they gèt the technical know-how to build a nanomotor? Lastly, again relating to competency, IMM's competency resides in R&D. Assuming that Respirocytes can be manufactured and does get FDA approval, the company is now faced with the unfamiliar task of marketing. Granted that the product will probably sell itself because of the promise associated with artificial mechanical red blood cells, radical technologies such as these always faced the first mover disadvantage.

Case Study 2: Using DNA as a Construction Medium

DNA is thought of as the building blocks of all living organisms—the chemical basis of heredity, the blue print of all living things. But DNA is not normally thought of as a building material. A group of chemists at NYU have developed a means of utilizing strands of DNA to construct 3-dimentional geometric structures. Utilizing the specific base paring relationship as "the glue", the group has been able to link strands of DNA together. From this, the research group has introduced bends, kinks, twists and other shapes, which serve as the building blocks for even more complex structures. The use of DNA has several distinct advantages. The first advantage is the specificity of DNA interactions (i.e. "sticky ends") combined with branched molecules results in a system whereby the DNA intermolecular interactions are both readily programable and predictable.

Therefore, it is possible to gain large amounts of control over both linking and branching topology. The second advantage of using strands of DNA as a molecular scaffold is the availability of arbitrary sequences. In other words, for a segment of DNA to "make sense" and code for a protein, a specific sequence of bases is required. For example, a sequence ATTCGT may code for a simple amino acid chain. Any deviation from this code will not be understood and produce nothing. As such, the length and therefore size of the structure can be controlled without any adverse consequences. The third advantage is that DNA can be manipulated and modified by a large battery of enzymes, including DNA ligase, restriction endonucleases, kinases and exonucleases. Lastly, the double helical DNA is a stiff polymer, stable molecule, and has an external code that can be read by proteins and nucleic acids.

Unlike the CASE 1, the technology discussed here represents a more immediately viable commercial technology. As previously illustrated, early iterations of nanomedicine will not be in the form of nanobots with self-replicating capabilities, but in simpler forms such as diagnostic and monitoring devices or drug delivery systems. In fact, the commercial application of this technology can be immediately applied to issues such as drug delivery.

Drug delivery techniques were established to deliver or control the amount, rate and, sometimes, location of a drug in the body to optimize its therapeutic effect, convenience and dose. Developing a well established formulation of a drug with a new delivery system is a relatively low risk activity and can be used to enhance a company's product portfolio by extending the drug's commercial life-cycle. New, therapeutic biological drugs may also depend on a novel technology to enable them to reach their target within the body. Any drug delivery system that: results in reduced side effects; is simple to use (i.e. does not require administration by a trained professional); is pain-free; is pleasant tasting; and is infrequent in its dosing schedule, will increase patients' quality of life, meaning that more patients are likely to use it and respond to the treatment. Likewise, a promising product in vitro will have limited success in vivo and on the market if it cannot: be delivered into the body with proven high bioavailability, high absorption by the body; be effective in small enough doses that cause minimal side effects; manufactured simply and costefficiently; and be tailored to the needs of the disease type being treated or prevented.

Current studies have shown that existing drug delivery systems are less than perfect. In the year 2000, 15 per cent of all hospital admissions were due to adverse drug effects with 100,000 deaths. Adverse drug effects have resulted in $136 billion in health care costs. Furthermore, from a commercial view point, drug delivery is a rapidly expanding subsection of the market for therapeutic drugs. The company MediJect estimates the current global drug delivery market to be worth $20 billion. Drug delivery is the fastest growing healthcare sector, currently growing at a rate twice that of pharmaceuticals. Drug delivery industry sales are expected to increase at an annual rate of 15 per cent, and this growth is thought likely to be sustained throughout the 21st century. There are, therefore, significant opportunities for growth and development within this market. The US drug delivery market alone is projected to grow from $14.3 billion in 1998 to $24 billion in 2003 and it is estimated that by the end of 2005, 20 per cent of pharmaceutical expenditure will be used on products using drug delivery technologies.

Over the past 10 years there has been an increase in R&D into drug delivery technologies. This growth has been, and will continue to be, fueled by a number of factors:

- Controlling overall healthcare costs—the cost of developing and launching a new chemical entity can be $300–600 million, whereas the reformulation of an existing product may cost only $50 million and still achieve peak sales of $100–200 million.

- Enhancing a product's life-cycle—pharmaceutical companies need to enhance older products reaching the end of their commercial life-cycle in order to optimize the company portfolio and maintain sales against competition.
- Improving patient compliance—this can be achieved by improving the choice of delivery system with pain-free, pleasant tasting, convenient alternatives. Patients are also more likely to take a drug which has reduced side effects and which works quickly.
- Rapid growth in the worldwide biotechnology industry—this has increased drug discovery and subsequently resulted in promising compounds such as proteins, peptides, hormones and therapeutic biologicals which do not penetrate the body easily.
- The need to reduce side effects—this has lead to the development of targeted drug delivery systems so that treatment is localized and systemic side effects can be reduced.
- The need to deliver peptides and proteins—there is a demand for an alternative delivery system to injectable or infusion-delivered insulin for the estimated 18 million sufferers of Type 1 diabetes worldwide. A growing number of the 15 million Type 2 sufferers worldwide are also now administering insulin themselves. At present, peptide and protein pharmaceuticals are worth approximately $10 billion worldwide and is expected to increase two to three times over the next decade.

What makes this technology highly viable in terms of commercialization is the ability to satisfy such market needs. The most obvious benefit of utilizing DNA as a construction material is its ability to avoid aggravating the body's immune system and thus reducing side effects. Drug molecules, which would normally elicit immune reactions, are in effected shielded from antibodies by the inert structure of the DNA. Furthermore, aside from a protective effect, the DNA shell can also be functional. After all, the DNA shell can "code" for proteins and peptides. Again using insulin as an example, a DNA constructed structure containing molecules of insulin within it could be injected into diabetic patients28. As enzymes within the body begin reading the DNA coding on the shell, and manufacturing nascent insulin molecules, the exogenous molecules of insulin are gradually being released either through simple diffusion or a sustained release mechanism. This process has resulted in the release and production of enough insulin to last the patient 24 hours, reducing the normal dosing of insulin from three times a day to once a day, and greatly improving patient compliance. Furthermore, this process is has not mounted any immune reaction because the porousness of the engineered DNA shell is big enough for insulin molecules to escape, but small enough so that antibodies cannot enter.

So far, the analysis has shown that the technology is viable from an engineering and scientific perspective. That a profitable market exists for this technology, and that the technology can address "pain points" and market needs. The next question to consider is how to bring the technology to market. Unlike CASE 1, the financing needs with this technology are more feasible. Average costs to develop and bring to market a drug delivery system are approximately $50 million. Given the novelty of this technology, a more liberal estimate of the cost of development should be applied, say $200 million. Is this company and technology worth $200 million? A simple calculation suggests that a company/technology such as this has a NPV well in excess of $200 million. In fact, a conservative calculation shows that this company's technology is worth $1.3 billion. Funding for this technology should not be an issue. Where the company should turn for capital is another issue that depends on multiple factors such as: how soon the company needs the funds, corporate goals, degree of control the company wishes to relinquish, among others.

Although the challenge of financing is less prominent for the commercialization of DNA construction medium technology, the company and technology face other challenges that must be considered as the technology is brought to market. One challenge is that of manufacturing and the scaling from laboratory to industrial. To date, the success of the experiments have been on the scale of laboratory experiments; however, it is unknown if the manufacturing process when scaled up will function as expected. The other challenge facing the company is that of management. Who will lead the company from the commercial end? The current organization is comprised primarily of scientists and engineers who may not have the necessary business skills or experience to make the company and technology a commercial success. This issue spills over into the issue of organizational structure. As the company begins to grow, roles and responsibilities will require better definition. What degree of control will the founders be willing to concede? How will the environment and culture of the organization continue to foster innovation while maintaining competitive from a business end? Lastly, the company must be aware of marketing risks. The marketing risk described here is not only with advertising or distribution channels, but also with the risk of myopia. It is all too easy for the company to become complacent through the capture of profits from "low hanging fruit" and miss the greater opportunity29. For example, the technology has been discussed in the context of drug delivery systems, but the company should not limit its applications to drug delivery alone. This is not to say that the company must pursue other markets outside of drug delivery—its area of expertise, but that it should always consider the option of licensing or crosslicensing. It should be cognizant of the fact that the business of licensing technology and intellectual property can be an equally profitable business proposition.

Although the two cases presented here represent nanomedical technologies that are on different planes of development, they do demonstrate fundamental difficulties common to all technology commercialization: the problem of translating ideas into practice; the problem of evolving from prototype to mass market; and the problem of environmental factor conditions that are outside the control of the entrepreneur. Both technologies were to face regulatory and legal scrutiny as well as market adoption uncertainty. The cases suggest the disadvantage of a first mover strategy and that "the first application is not the killer application". The cases have also shown the differences in obstacles facing technologies at different stages of development. With the mechanical red blood cells, one can appreciate the magnitude of challenges facing the commercialization of the technology compared to that of the DNA construction medium. In the first case, financing and even being able to build a prototype was an issue, but in the second case, the salient issue revolved around marketing and product launch; transitioning from plan to product.

The field of nanomedicine will certainly take decades to develop. Perhaps before we let our imaginations run wild dreaming of the promises that nanomedicine can deliver, our first task is to decide whether the capabilities are indeed possible. Bear in mind that as we look at how to solve various problems, we notice that some that look difficult today will become easy, while others that might seem easier turn out to be more difficult. The seeming difficulty of treating disorders is always changing: Once polio was frequent and incurable, today it is easily prevented. Syphilis once caused steady physical decline leading to insanity and death; now it is cured with a shot. In the longer term, though, the story of nanotechnology in medicine will be the story of extending surgical control to the molecular level.

REFERENCES

Akis, R., D. Ferry, C. Musgrave, "Kinetic Lattice Monte Carlo Simulations of Processes on the Silicon (100) Surface," *Physica E-Low Dimensional Systems and Nanostructures* 19, 183-187 (2003).

Albrecht, T.R., Grütter, P., Horne, D., and Rugar, D. (1991) Frequency modulation detection using high-Q cantilevers for enhanced force microscope sensitivity. J. Appl. Phys. 69(2), 668-673

Alexander, S., Hellemans, L., Marti, O., Schneir, J., Elings, V., Hansma, P.K., Longmiro, M., and Gurley, J. (1989) An atomic-resolution atomic-force microscope implemented using an optical lever. J. Appl. Phys. 65(1), 164-167

B. Ding, R. Sha & N.C. Seeman, Pseudohexagonal 2D DNA Crystals from Double Crossover Cohesion, J. Am. Chem. Soc. 126, 10230-10231 (2004).

B.H. Robinson and N.C. Seeman, The Design of a Biochip: A Self-Assembling Molecular-Scale Memory Device. Protein Engineering 1, 295-300 (1987).

C. Mao, W. Sun and N.C. Seeman, Designed Two-Dimensional DNA Holliday Junction Arrays Visualized by Atomic Force Microscopy, Journal of the American Chemical Society 121, 5437-5443 (1999).

F. Mathieu, S. Liao, C. Mao, J. Kopatsch, T. Wang, and N.C. Seeman, Six-Helix Bundles Designed from DNA, NanoLetters 5, 661-665 (2005).

G. Serrano & N.C. Seeman, Nanotecnologia basada en ADN, Revista de Química 19, 11-20 (2004), in Spanish.

Hoh, J.H. and Hansma, P.K. (1992) Atomic force microscopy for high-resolution imaging in cell biology. Trends Cell Bio. 2, 208-213

J.H. Reif, T.H. LaBean & N.C. Seeman, Challenges and Applications for Self-Assembled DNA Nanostructures, Sixth International Workshop on DNA-Based Computers, DNA 2000, Leiden, The Netherlands, (June, 2000) ed. A. Condon, G. Rozenberg. Springer-Verlag, Berlin Heidelberg, Lecture Notes in Computer Science 2054, 173-198, (2001).

Kang, J. and C. Musgrave, "The Mechanism of Atomic Layer Deposition of SiO 2 on the Silicon (100)-2x1 Surface Using SiC1 4 and H 2 O as Precursors," *Journal of Applied Physics* 91, 3408-3414 (2002).

L. Zhu, P.S. Lukeman, J. Canary & N.C. Seeman, Nylon/DNA: Single-Stranded DNA with Covalently Stitched Nylon Lining, J. Am. Chem. Soc. 125, 10178-10179 (2003).

Mui, C. and C. Musgrave, "The Hydroxylation and Oxidation of the Ge(100)-2x1 Surface by Hydrogen Peroxide and Water," *Langmuir*, 20, 7604-7609 (2004).

N. Jonoska, P. Sa-Ardyen and N.C. Seeman, Computation by Self-Assembly of DNA Graphs, Journal of Genetic Programming and Evolvable Machines 4, 123-137 (2003).

N.C. Seeman and N.R. Kallenbach, Nucleic Acid Junctions, The Tensors of Life? In: Nucleic Acids: The Vectors of Life, ed. by B. Pullman and J. Jortner, D. Reidel, Dordrecht, 183-200 (1983).

N.C. Seeman, De Novo Design of Sequences for Nucleic Acid Structure Engineering, Journal of Biomolecular Structure and Dynamics 8, 573-581 (1990).

P.S. Lukeman, A. Mittal, & N.C. Seeman, Two Dimensional PNA/DNA Arrays: Estimating the Helicity of Unusual Nucleic Acid Polymers, Chemical Communications 2004, 1694-1695 (2004).

Senosiain, J., C. Musgrave and D. Golden, "Temperature and Pressure Dependence of the Reaction of OH and CO: Master Equation Modeling on a High Level Potential Energy Surface," *International Journal of Chemical Kinetics* 35, 464-474 (2003).

W. Shen, M. Bruist, S. Goodman & N.C. Seeman, A Nanomechanical Device for Measuring the Excess Binding Energy of Proteins that Distort DNA, Angew. Chem. Int. Ed. 43, 4750-4752 (2004); Angew.Chem. 116, 4854-4856 (2004).

Weisenhorn, A.L., Hansma, P.K., Albrecht, T.R., and Quate, C.F. (1989) Forces in atomic force microscopy in air and water. Appl. Phys. Lett. 54(26), 2651-2653

Widjaja, Y. and C. Musgrave, "Quantum Chemical Study of the Elementary Reactions in Zirconium Oxide Atomic Layer Deposition," *Applied Physics Letters* 81, 304-306 (2002).

X. Zhang, H. Yan, Z. Shen and N.C. Seeman, Paranemic Cohesion of Topologically-Closed DNA Molecules, J Am. Chem. Soc.124, 12940-12941 (2002).

Xu, Y. and C. Musgrave, "A DFT Study of the Atomic Layer Deposition of Al 2 O 3 on SAMs: The Effect of SAM Termination," *Chemistry of Materials,* 16, 646 (2004).

5

Impact of Nanotechnology on Food Agriculture, Fishing, Veterinary and Environment

Nanomaterials and Nanoparticles for Agriculture and Environment

Livestock tracking has been a problem for farmers since before Little Bo Peep lost her sheep. Nano-Bo-Peep, however, would have no such problems. Just as converging technologies in crop production will use nanosensor networks to continuously monitor the health of plants, so, too, will sensors monitor livestock. The USDA envisions the rise of 'smart herds'—cows, sheep and pigs fitted with sensors and locators relaying data about their health and geographical location to a central computer.

This is a vision of precision agriculture on the hoof. The long-term aim is not merely to monitor, but also to automatically and autonomously intervene with pharmaceuticals using small drug delivery devices that can be implanted into the animal in advance of illness. The notion of linking in-built sensors to in-built smart delivery systems has been called "the fuel injection principle" since it mimics the way modern cars use sensors to time fuel-delivery to the engine. The closest applications to market are implantable insulin-delivery devices or "drug chips" that will be linked with glucose sensors for (human) diabetics to automatically regulate blood sugar levels. Over time, this could become the model for all drug delivery, in both humans and animals.

One of the current barriers to implantable medical devices is that their composite materials (e.g., metal or plastics) are often incompatible with living tissue. New materials, engineered at the nano-scale to be biocompatible, seek to address this problem.

Implanting tracking devices in animals is nothing new—either in pets, valuable farm animals or for wildlife conservation. Injectable microchips are already used in a variety of ways with the aim of improving animal welfare and safety—to study animal behaviour in the wild, to track meat products back to their source or to reunite strays with their human guardians. In the nanotech era, however, retrofitting farm animals with sensors, drug chips and nanocapsules will further extend the vision of animals as industrial production units.

Although the challenge of financing is less prominent for the commercialization of DNA construction medium technology, the company and technology face other challenges that must be considered as the technology is brought to market. One challenge is that of manufacturing and the scaling from laboratory to industrial. To date, the success of the experiments have been on the scale of laboratory experiments; however, it is unknown if the manufacturing process when scaled up will function as expected. The other challenge facing the company is that of management. Who will lead the company from the commercial end? The current organization is comprised primarily of scientists and engineers who may not have the necessary business skills or experience to make the company and technology a commercial success. This issue spills over into the issue of organizational structure. As the company begins to grow, roles and responsibilities will require better definition. What degree of control will the founders be willing to concede? How will the environment and culture of the organization continue to foster innovation while maintaining competitive from a business end? Lastly, the company must be aware of marketing risks. The marketing risk described here is not only with advertising or distribution channels, but also with the risk of myopia. It is all too easy for the company to become complacent through the capture of profits from "low hanging fruit" and miss the greater opportunity29. For example, the technology has been discussed in the context of drug delivery systems, but the company should not limit its applications to drug delivery alone. This is not to say that the company must pursue other markets outside of drug delivery—its area of expertise, but that it should always consider the option of licensing or crosslicensing. It should be cognizant of the fact that the business of licensing technology and intellectual property can be an equally profitable business proposition.

Although the two cases presented here represent nanomedical technologies that are on different planes of development, they do demonstrate fundamental difficulties common to all technology commercialization: the problem of translating ideas into practice; the problem of evolving from prototype to mass market; and the problem of environmental factor conditions that are outside the control of the entrepreneur. Both technologies were to face regulatory and legal scrutiny as well as market adoption uncertainty. The cases suggest the disadvantage of a first mover strategy and that "the first application is not the killer application". The cases have also shown the differences in obstacles facing technologies at different stages of development. With the mechanical red blood cells, one can appreciate the magnitude of challenges facing the commercialization of the technology compared to that of the DNA construction medium. In the first case, financing and even being able to build a prototype was an issue, but in the second case, the salient issue revolved around marketing and product launch; transitioning from plan to product.

The field of nanomedicine will certainly take decades to develop. Perhaps before we let our imaginations run wild dreaming of the promises that nanomedicine can deliver, our first task is to decide whether the capabilities are indeed possible. Bear in mind that as we look at how to solve various problems, we notice that some that look difficult today will become easy, while others that might seem easier turn out to be more difficult. The seeming difficulty of treating disorders is always changing: Once polio was frequent and incurable, today it is easily prevented. Syphilis once caused steady physical decline leading to insanity and death; now it is cured with a shot. In the longer term, though, the story of nanotechnology in medicine will be the story of extending surgical control to the molecular level.Animals also are likely to be used as the testing ground for less savoury or more risky applications that could later be extended to human beings. Using microfluidics for breeding is likely to accelerate genetic uniformity within livestock species and also opens the possibility of applying new nano-eugenic technologies to humans in the future. The ability to remotely regulate animals may have adverse affects as livestock go longer periods without direct human care.

The same technologies transferred to humans raises profound concerns about quality of life and civil liberties. In October 2004, the US Food and Drug Administration approved the use of implantable microchips in humans to provide easy access to an individual's medical records—the first approval of microchips for medical uses in the United States.

As healthcare is driven more and more by the bottom line, the future use of implantable chips for automated drug delivery may become economically preferable to nursing. When dealing with the elderly or those with different cognitive abilities or with any condition requiring regular treatment, ethical questions may arise about who decides to make an individual 'fuel injected.' Automated drug delivery could allow some people to live independently who would otherwise be institutionalised. However, the absence of human caretakers is also a factor.

The world's fastest growing area of animal production is the farming of fish, crustaceans and molluscs, particularly in Asia. According to the FAO there were 45.7 million tonnes of aquaculture production in 2000 and it is growing at a rate of more than 9% per year. With a strong history of adopting new technologies, the highly integrated fish farming industry may be among the first to incorporate and commercialise nanotech products.

Nevada-based Altair Nanotechnologies makes a water cleaning product for swimming pools and fishponds called 'NanoCheck.' It uses 40 nm particles of a lanthanum-based compound which absorbs phosphates from the water and prevents algae growth.

NanoCheck is currently undergoing large-scale testing in swimming pools and Altair is expected to launch a swimming pool cleaner in early 2005. Altair has its eye on a potentially large demand for NanoCheck for use in thousands of commercial fish farms worldwide where algae removal and prevention is costly at present. According to Altair, the company plans to expand its tests to confirm that its nanoparticles will not harm fish, but no mention is made of the tests that will be undertaken to examine the impacts of nanoparticle-laden run-off on human health or on the environment.

The USDA is completing trials on a system for mass vaccination of fish using ultrasound. Nanocapsules containing short strands of DNA are added to a fishpond where they are absorbed into the cells of the fish. Ultrasound is then used to rupture the capsules, releasing the DNA and eliciting an immune response from the fish. This technology has so far been tested on rainbow trout by Clear Springs Foods (Idaho, US)—a major aquaculture company that produces about one third of all US farmed trout.

Scientists from the Russian Academy of Sciences have reported that young carp and sturgeon exhibited a faster rate of growth (30% and 24% respectively) when they were fed nanoparticles of iron.

Food Production Using Nanotechnologies—Recommendations for Regulation by Environmentalist Groups

By allowing nanotech products to come to market in the absence of public debate and regulatory oversight, governments, agribusiness and scientific institutions have already jeopardised the potential benefits of nano-scale technologies. First and foremost, society—including farmers, civil society organisations and social movements—must engage in a wide debate about nanotechnology and its multiple economic, health and environmental implications.

In keeping with the Precautionary Principle, all food, feed and beverage products (including nutritional supplements) that incorporate manufactured nanoparticles should be removed from the shelves and new ones prohibited from commercialisation until such time as laboratory protocols and regulatory regimes are in place that take into account the special characteristics of these materials, and until they are shown to be safe. Similarly, nano-scale formulations of agricultural input products such as pesticides, fertilisers and soil treatments should be prohibited from environmental release until a new regulatory regime specifically designed to examine these products finds them safe.

Governments must also move immediately to establish a moratorium on lab experimentation with—and the release of—"synthetic biology" materials until society can engage in a thorough analysis of the health, environmental and socio-economic implications. Any efforts by governments or industry to confine discussions to meetings of experts or to focus debate solely on the health and safety aspects of nano-scale technologies will be a mistake. The broader social and ethical issues must also be addressed.

At the intergovernmental level, the Food and Agriculture Organization's (FAO) standing committees and commissions on agriculture, fisheries, forestry and genetic resources should be monitoring and debating the new technologies—with active input and feedback from peasant and small farmers' organisations. FAO's Committee on Commodity Problems should immediately begin to examine the socio-economic implications for farmers, food safety and national governments. The UN/FAO Committee on World Food Security should be discussing the implications for agro-terrorism as well as food sovereignty. Additionally, the UN Convention on Biological Diversity should review nanobiotech's potential impact, especially on bio-safety.

Other UN agencies such as the United Nations Conference on Trade and Development (UNCTAD) and International Labour Organization (ILO) should join with FAO to examine the impact of nanotech on the ownership and control of the world's food supply, commodities and labour. The international community should establish a body dedicated to tracking, evaluating and monitoring new technologies and their products through an International Convention for the Evaluatión of New Technologies (ICENT).

Thai physicists are looking to nanotechnology to develop a new rice strain and improve the quality of silk. They will alter the rice by making a nanosize hole through a rice cell. A nitrogen atom shot through the hole will rearrange rice DNA to produce a new rice strain. Their goal is a fragrant rice plant with short stems that isn't overly sensitive to sunlight. Other engineers are planning to use plasma to coat the surface of silk so it doesn't get wet easily and attracts less dust. The plasma causes a nanolevel change to the silk surface. The surface treatment causes water to bead on the surface and simply roll off.

Molecular Food Manufacturing

Some people claim that in the future, molecular engineering will enable us to "grow" unlimited quantities of food without soil, seed, farms or farmers—and that it will wipe out global hunger in the process. Consider the views outlined below.

What Are the Experts Saying About Molecular Food Manufacturing?

- "Nanomachines could create unlimited amounts of food by synthesis at the atomic level, which would eradicate hunger."—Carmen I. Moraru et al., Professor of Food Science, Cornell University (USA), on nanotech's potential impact on food science.
- "Molecular biosynthesis and robotic replenishment may allow quick replacement of production, so we wouldn't have to depend on centralized

systems to grow and deliver our food. In the first, primitive stages of molecular assembly, we'd build packaged greenhouses, radically different from those today, that would allow local or individualized production by millions who know nothing about farming...At the next stage of molecular manufacturing, food synthesis could occur directly, without growing crops or livestock."—Douglas Mulhall, 'Our Molecular Future'.

- "Why can't human beings imitate nature's methodology? Instead of harvesting grain and cattle for carbohydrates and protein, nanomachines (nanobots) could assemble the desired steak or flour from carbon, hydrogen, and oxygen atoms present in the air as water and carbon dioxide. Nanobots present in foods could circulate through the blood system, cleaning out fat deposits and killing pathogens."—Dr. Marvin J. Rudolph, Director, DuPont Food Industry Solutions, in 'Food Technology', January 2004.

Producing food by molecular manufacturing is the most ambitious goal of nanotech—and the least likely to materialize any time soon. To those who have followed the biotech debate over the past two decades, enthusiastic claims that a new technology will feed hungry people is a tired and empty refrain. Nano-optimists see the future through the biotech industry's rose- (and green-) coloured glasses: now it's nanotech, they claim, that will eradicate hunger by increasing agricultural yields, enhancing the nutritional content of food and eliminating the risk of food allergens.

ETC Group concludes that present-day "nanotech for tummies" is following the same trajectory as other nano-scale R&D, with the earliest applications in the area of "smart" materials and sensors. More revolutionary applications, such as the atomic modification of food, are perhaps more distant. But it's worth noting that a few ambitious scientists are trying to create food in the lab.

Tissue engineers at Touro College (New York City) and at the Medical University of South Carolina (USA) are experimenting with growing meat by "marinating" fish myoblast (muscle) cells in liquid nutrients to encourage the cells to divide and multiply on their own. The first goal is to keep astronauts in space from going hungry.

REFERENCES

Baker, R.T.K. Synthesis, properties and applications of graphite nanofibers. In *R&D status and trends*, ed. Siegel et al.

Bate, R., Frazier, G., Frensley, W., Reed., M., "An Overview of Nanoelectronics," *Texas Instruments Technical Journal*, July-August 1989, pp. 13-20.

Beck, J.S., J.C. Vartuli, W.J. Roth, M.E. Leonowicz, C.T. Kresge, K.D. Schmitt, C.T.-W. Chu, D.H. Olsen, E.W. Shepard, S.B. McCullen, J.B. Higgins, and J.L. Schlenker. 1992. *J. Am. Chem. Soc.* 114:10834.

Bensaude-Vincent, B.: 2001, 'The Construction of a Discipline: Materials Science in the United States', *Historical Studies in the Physical and Biological Sciences*, 31, 223-248.

Braun, T.; Schubert, A. & Zsindely, S.: 1997, 'Nanoscience and nanotechnology on the balance', *Scientometrics*, 38, 321-325.

Brinker, C.J. 1996. *Curr. Opin. Solid State Mater. Sci.* 1:798.

Brown, N.: 2004, 'Needle on the Real: Technoscience and Poetry at the Limits of Fabrication', in: N.K. Hayles (ed.), *Nanoculture: Implications of the New Technoscience*, Bristol, UK: Intellect Books, pp. 173-190.

Bueno, O.: 2004, 'The Drexler-Smalley Debated on Nanotechnology: Incommensurability at Work?', *Hyle: International Journal for Philosophy of Chemistry*, 10(2), 83-98.

Bueno, O.: 2004, 'Von Neumann, Self-Reproduction and the Constitution of Nanophenomena', in: D. Baird, A. Nordmann & J. Schummer (eds.), *Discovering the Nanoscale*, Amsterdam: IOS Press, pp. 101-115.

Crandall, B.C. and Lewis, J., *Nanotechnology: Research and Perspectives*, MIT Press, Cambridge, 1992.

Crow, M. & Sarewitz, D.: 2001, 'Nanotechnology and Societal Transformation', in: M.C. Roco and W.S. Bainbridge (eds.), *Societal Implications of Nanoscience and Nanotechnology*, Dordrecht: Kluwer, pp. 45-54.

Einsiedel, E.F.; Goldenberg, L.: 2004, 'Dwarfing the Social? Nanotechnology Lessons from the Biotechnology Front', *Bulletin of Science, Technology & Society*, 24, 28-33.

European Comission: *European Workshop on Social and Economic Research on Nanotechnologies and Nanosciences, Brussels, 14-15 April 2004* [www.stage-research.net/STAGE/PAGES/Nano.html].

European Commission (Community Health and Consumer Protection): 2004, *Nanotechnologies: A Preliminary Risk Analysis on theBasis of a Workshop, Brussels, 1-2 March 2004* [www.europa.eu.int/comm/health/ ph_risk/documents/ev_20040301_en.pdf].

European Commission: 2004, *Converging Technologies: Shaping the Future of European Societies (Report of the High Level Expert Group "Foresighting the New Technology Wave")*, Brussels: European Commission Research [http://europa.eu.int/comm/research/conferences/2004/ntw/index_en.html].

F. Buot, "Mesoscopic Physics and Nanoelectronics: Nanoscience and Nanotechnology," *Physics Reports*, pp.73-174, 1993.

F. Capasso and S. Datta, "Quantum Electron Devices," *Physics Today*, pp.74-82, February 1990.

Feynman, R., "There's Plenty of Room at the Bottom: An invitation to Enter a New Field of Physics," Talk at the Annual Meeting of the American Physical Society, 29 December 1959. Reprinted in Appendix B of Crandall and Lewis. See citation above.

Fielder, F.A.; Reynolds, G.H.: 1994, 'Legal Problems of Nanotechnology: An Overview', *Southern California Interdisciplinary Law Journal*, 3, 593-629.

Fleischer, Th.; Decker, M. & Fiedeler, U. (eds.): 2004, *Große Aufmerksamkeit für kleine Welten ' Nanotechnologie und ihre Folgen*, special issue of *Technikfolgenabschätzung ' Theorie und Praxis*, 13 (2), 5-85 [www.itas.fzk.de/tatup/042/inhalt.htm].

Fogelberg, H. & Glimell, H.: 2003, 'Molecular Matters: In Search of the Real Stuff', in: H. Fogelberg & H. Glimell, *Bringing Visibility to the Invisible: Towards A Social Understanding of Nanotechnology*, Göteborg: Göteborg University, pp. 5-32 [www.sts.gu.se/publications/STS_report_6.pdf].

Fogelberg, H. & Glimell, H.: 2003, *Bringing Visibility to the Invisible: Towards A Social Understanding of Nanotechnology*, Göteborg: Göteborg University [www.sts.gu.se/publications/STS_report_6.pdf].

Fogelberg, H.: 2003, 'The Grand Politics of Technoscience: Contextualizing Nanotechnology', in: H. Fogelberg & H. Glimell, *Bringing Visibility to the Invisible: Towards A Social Understanding of Nanotechnology*, Göteborg: Göteborg University, pp. 33-54 [www.sts.gu.se/publications/STS_report_6.pdf].

Fogelberg, H.: 2003, 'The Material Culture of Nanotechnology', in: H. Fogelberg & H. Glimell, *Bringing Visibility to the Invisible: Towards A Social Understanding of Nanotechnology*, Göteborg: Göteborg University, pp. 99-114.

Frazier, G., "An Ideology For Nanoelectronics," in *Concurrent Computations: Algorithms, Architecture, and Technology*, Plenum Press, New York, 1988.

Goldhaber-Gordon, D., Montemerlo, M.S., Love, J.C., Opiteck, G.J., and Ellenbogen, J.C., "Overview of Nanoelectronic Devices," submitted to the *Proceedings of the IEEE*, February 1997. For more information, please send e-mail to nanotech@mitre.org.

Gorman, M.E.; Groves, J.F. & Shrager, J.: 2004, 'Societal Dimensions of Nanotechnology as a Trading Zone: Results from a Pilot Project', in: D. Baird, A. Nordmann & J. Schummer (eds.), *Discovering the Nanoscale*, Amsterdam: IOS Press, pp. 63-73.

Grinbaum, A. & Dupuy, J.-P.: 2004, 'Living with Uncertainty: Toward the Ongoing Normative Assessment of Nanotechnology', *Techne: Research in Philosophy and Technology*, 8(3) (forthcoming).

Grinbaum, A.: 2004, 'La condition de l'homme moderne et les nanotechnologies', in: G. Nivat (ed.), *Les limites de l´humain. 39èmes Rencontres Internationales de Genève*, L´Age d´Homme, Genève, p. 141.

Grunwald, A.: 2004, 'Ethische Aspekte der Nanotechnologie. Eine Felderkundung', *Technikfolgenabschätzung ' Theorie und Praxis*, 13 (2), 71-78.

Gupta, V.K. & Pangannaya, N.B.: 2000, 'Carbon nanotubes: bibliometric analysis of patents', *World Patent Information*, 22, 185-189.

Hansson, S.O.: 2004, 'Great Uncertainty about Small Things', *Techne: Research in Philosophy and Technology*, 8(3) (forthcoming).

Haruta, M. 1997. *Catalyst surveys of Japan* 1:61 and references therein.

Hayles, N.K. (ed.): 2004, *Nanoculture: Implications of the New Technoscience*, Bristol, UK: Intellect Books.

Hayles, N.K.: 2004, 'Connecting the Quantum Dots: Nanotechscience and Culture', in: N.K. Hayles (ed.), *Nanoculture: Implications of the New Technoscience*, Bristol, UK: Intellect Books, pp. 11-26.

Hemerén, G.: 2004, 'Nano Ethics Primer', in: European Commission (Community Health and Consumer Protection): *Nanotechnologies: A Preliminary Risk Analysis on theBasis of a Workshop, Brussels, 1-2 March 2004* (www.europa.eu.int/comm/health/ ph_risk/documents/ev_20040301_en.pdf), pp. 95-102.

Hennig, J.: 2004, 'Changes in the Design of Scanning Tunneling Microscopic Images from 1980 to 1990', *Techne: Research in Philosophy and Technology*, 8(3) (forthcoming).

Hessenbruch, A.: 2004, 'Nanotechnology and the Negotiation of Novelty', in: D. Baird, A. Nordmann & J. Schummer (eds.), *Discovering the Nanoscale*, Amsterdam: IOS Press, pp. 135-144.

Hullmann, A. & Meyer, M.: 2003, 'Publications and Patents in Nanotechnology. An overview of previous studies and the state of the art', *Scientometrics*, 58 (3) 507-527.

Huo, Q., D.I. Margolese, U. Ciesla, P. Feng, T.E. Gier, P. Sieger, R. Leon, P.M. Petroff, F. Schuth, and G.D. Stucky. 1994. *Nature* 368:317.

Jena, P., S.N. Khanna, and B.K. Rao. 1996. In *Science and technology of atomically engineered materials*, ed. P. Jena. River Edge, NJ: World Scientific.

Johansson, M.: 2003, 'Plenty of room at the bottom: Towards an anthropology of nanoscience', *Anthropology Today*, 19 (No. 6), 3-6.

Johnson, A.: 2004, 'The End of Pure Science: Science Policy from Bayh-Dole to the NNI', in: D. Baird, A. Nordmann & J. Schummer (eds.), *Discovering the Nanoscale*, Amsterdam: IOS Press, pp. 217-230.

Kresge, C.T., M.E. Leonowicz, W.J. Roth, J.C. Vartuli, and J.S. Beck. 1992. *Nature* 359:710.

Kupperman, A., S. Nadimi, S. Oliver, G. Ozin, J. Garcés, and M. Olken. 1993. *Nature* 365:239.

Kuusi, O.; Meyer, M.: 2002, 'Technological generalizations and leitbilder ' the anticipation of technological opportunities', *Technological Forecasting & Social Change*, 69, 625-639.

Landon, B.: 2004, 'Less is More: Much Less is Much More: The Insistent Allure of Nanotechnology Narratives in Science Fiction', in: N.K. Hayles (ed.), *Nanoculture: Implications of the New Technoscience*, Bristol, UK: Intellect Books, pp. 131-146.

Laszlo, P.: 2004, 'Is There Life After Partington?', *Hyle: International Journal for Philosophy of Chemistry*, 10(2), 169-178.

Lenhard, J.: 2004, 'Nanoscience and the Janus-Faced Character of Simulations', in: D. Baird, A. Nordmann & J. Schummer (eds.), *Discovering the Nanoscale*, Amsterdam: IOS Press, pp. 93-100.

Lent, C.S., Tougaw, P.D., Porod, W., Bernstein, G.H., "Quantum Cellular Automata," *Nanotechnology*, Vol. 4, p. 49, 1993.

Paschen, H.; Coenen, C.; Fleischer, T.; Grünwald, R.; Oertel, D. & Revermann, C.: 2004, *Nanotechnologie: Forschung, Entwicklung, Anwendung*, Berlin: Springer (*Nanotechnologie, TAB-Arbeitsbericht 92*, Berlin: Büro für Technikfolgen-Abschätzung beim Deutschen Bundestag).

Roberts, J.A.: 2004, 'Deciding the Future of Nanotechnologies: Legal Perspectives on Issues of Democracy and Technology', in: D. Baird, A. Nordmann & J. Schummer (eds.), *Discovering the Nanoscale*, Amsterdam: IOS Press, pp. 247-255.

Robinson, C.: 2004, 'Images in NanoScience/Technology', in: D. Baird, A. Nordmann & J. Schummer (eds.), *Discovering the Nanoscale*, Amsterdam: IOS Press, pp. 165-169.

Robison, W.L.: 2004, 'Nano-Ethics', in: D. Baird, A. Nordmann & J. Schummer (eds.), *Discovering the Nanoscale*, Amsterdam: IOS Press, pp. 285-300.

Roco, M.C. & Bainbridge, W.S. (eds.): 2001, *Societal implications of nanoscience and nanotechnology*, (Proceedings of a workshop organized by the National Science Foundation, September 28-29, 2000), Kluwer: Dordrecht [available online at http://itri.loyola.edu/nano/societalimpact/nanosi.pdf]

Roco, M.C. & Tomellini, R. (eds.): 2002, *Nanotechnology: Revolutionary Opportunities and Societal Implications, Workshop, Lecce (Italy), 31 January - 1 February 2002*, Luxemburg: Office for Official Publications of the European Communities, [200 pp.]

Roco, M.C.; Bainbridge, W.S. (eds.): 2002, *Converging Technologies for Improving Human Performance: Nanotechnology, Biotechnology, Information Technology and the Cognitive Science*, Arlington, VA: National Science Foundation.

Roher, H. 1993. *Jpn. J. Appl. Phys.* 32:1335.

Rohlfing, E.A., D.M. Cox, and A. Kaldor. 1984. *J. Chem. Phys.* 81:3846.

Rosen, A. 1998. A periodic table in three dimensions: A sightseeing tour in the nanometer world. In *Advances in quantum chemistry*. In press.

Ruthven, D.M., S. Farooq, K.S. Knaebel. 1994. *Pressure swing adsorption*. New York: VCH Publishers.

Sarewitz, D.; Woodhouse, E.: 2003, 'Small is Powerful', in: A. Lightman, D. Sarewitz & Chr. Desser, (eds.), *Living with the Genie: Essays on Technology and the Quest for Human Mastery*, Washington, DC: Island Press, pp. 63-83.

Schummer, J.: 2001, 'Ethics of Chemical Synthesis', *Hyle: International Journal for Philosophy of Chemistry*, 7, 103-124 [online: www.hyle.org/journal/issues/7/schummer.htm].

Schummer, J.: 2003, 'The Notion of Nature in Chemistry', *Studies in History and Philosophy of Science*, 34 (2003), 705-736.

Schummer, J.: 2004, 'Naturverhältnisse in der modernen Wirkstoff-Forschung', in: K. Kornwachs (ed.), *Technik ' System ' Verantwortung*, Münster: LIT, pp. 629-638.

Schummer, J.: 2004, 'Why do Chemists Perform Experiments?', in: D. Sobczynska, P. Zeidler, E. Zielonacka-Lis (eds.), *Chemistry in the Philosophical Melting Pot*, Peter Lang (Frankfurt/M.), 2004, pp. 395-410.

Schummer, J.; Baird, D. (eds.): 2004-5, *Nanotech Challenges*, joint special issue of *Hyle: International Journal for Philosophy of Chemistry & Techne: Research in Philosophy and Technology* (forthcoming).

Schwarz, A.E.: 2004, 'Shrinking the 'Ecological Footprint' with NanoTechnoScience?', in: D. Baird, A. Nordmann & J. Schummer (eds.), *Discovering the Nanoscale*, Amsterdam: IOS Press, pp. 203-208.

Glossary

Acrosomal process: Long, thin, actin-containing spike produced from the head of certain sperm when they make contact with the egg. Seen in sea urchins and other marine invertebrates whose eggs are surrounded by a thick gelatinous coat.

Actin: Abundant protein that forms actin filaments in all eucaryotic cells. The monomeric form is sometimes called globular or G-actin; the polymeric form is filamentous or F-actin. A protein found in many cells. Actin assembles into long polymer rods called microfilaments. These microfilaments form part of the cell's cytoskeleton and they interact with myosin to cause movement in muscles. In muscle contraction, filaments of actin and myosin alternately unlink and chemically link in a sliding action. The energy for this reaction is supplied by adenosine triphosphate (ATP).

Active shield: A defensive system with built-in constraints to limit or prevent its offensive use.

Active site: Region of an enzyme surface to which a substrate molecule binds in order to undergo a catalyzed reaction.

Avogadro's number: The number of atoms in exactly 12 grams of pure ^{12}C, equal to 6.022×10^{23}.

Bacteriophage (phage): Any virus that infects bacteria. They were the first organisms used for the study of molecular genetics and are now widely used as cloning vectors. (From Greek *phagein*, to eat.)

Binding: The process by which a molecule (or ligand) becomes bound, that is, confined in position (and often orientation) with respect to a receptor. Confinement occurs because structural features of the receptor create a potential well for the ligand; van der Waals and electrostatic interactions commonly contribute.

Bio-assemblies or Biomolecular Assemblies: Containing several protein units, DNA loops, lipids, various ligands, etc.

Biochauvinism: The prejudice that biological systems have an intrinsic superiority that will always give them a monopoly on self-reproduction and intelligence.

Biocompatible coated materials: Biocompatible materials usually used in dental and bone implants that enhance biologic fixation, thereby increasing the bond strength between the coated material and bone, and minimize possible biological effects that may result from the implant itself. MeSH, 1999

Biocompatible materials: Synthetic or natural materials, other than drugs, that are used to replace or repair any body tissue or bodily function. MeSH, 1973

Biovorous: From "biovore;" an organism capable of converting biological material into energy for sustenance.

Bipolar-junction transistor: Transistor with n-type and p-type semiconductors having base-emitter and collector-base junctions.

Bit: Numbering system based on powers of 2 using only the digits 0 and 1, called "bits".

Blue Goo: Opposite of Grey goo. Benificial tech, or "police" nanobots.

Body-centered cubic (BCC): A common crystal structure found in some elemental metals. Within the cubic unit cell, atoms are located at corner and cell center positions.

Bogosity Filter: A mechanism for detecting bogus ideas and propositions.

Bond: Two atoms are said to be bonded when the energy required to separate them is substantially larger than the van der Waals attraction energy. Ionic bonds result from the electrostatic attraction between ions; covalent and metallic bonds result from the sharing of electrons among atoms; hydrogen bonds are weaker and result from dipole interactions and limited electron sharing. When used without modification, "bond" usually refers to a covalent bond.

Bone substitutes: Synthetic or natural materials for the replacement of bones or bone tissue. They include hard tissue replacement polymers, natural coral, hydroxyapatite, beta- tricalcium phosphate, and various other biomaterials. The bone substitutes as inert materials can be incorporated into surrounding tissue or gradually replaced by original tissue.

Breakdown: Failure of a material resulting from an electrical overload. The resulting damage may be in the form of thermal damage (melting or burning) or electrical damage (loss of polarization in piezoelectric materials). The failing of a physical principle or unit of measurement as the scale is changed.

Brownian Assembly: Brownian motion in a fluid brings molecules together in various position and orientations. If molecules have suitable complementary surfaces, they can bind, assembling to form a specific structure. Brownian assembly is a less paradoxical name for self-assembly (how can a structure assemble itself, or do anything, when it does not yet exist?).

Brownian Motion: Motion of a particle in a fluid owing to thermal agitation, observed in 1827 by Robert Brown. (Originally thought to be caused by vital force, Brownian motion in fact plays a vital role in the assembly and activity of the molecular structures of life).

Bus: Transmission medium for electrical or optical signals that perform a particular function, such as computer control.

Calibration: A process of adapting a sensor output to a known physical quantity to improve sensor output accuracy. Ubiquitous Ca^{2+}-binding protein whose binding to other proteins is governed by changes in intracellular Ca^{2+} concentration. Its binding modifies the activity of many target enzymes and membrane transport proteins.

Carbonyl: A chemical moiety consisting of O with a double bond to C. If the C is bonded to N, the resulting structure is termed an amide; if it is bonded to O, it is termed a carboxylic acid or an ester linkage.

Carboxylic acid: A molecule that includes a C having a double bond to O and a single bond to OH.

Compliance: The reciprocal of stiffness; in a linear elastic system, displacement equals force times compliance.

Copolymer: A polymer that consists of two or more dissimilar monomer units in combination in its molecular chains. Also a polymer formed from the polymerization of more than one type of monomer.

Corrosion: Deteriorative loss of a metal as a result of dissolution environmental reactions.

Coulomb [C]: Measure of electrical charge: 1 C is an amount of charge equal to that of about $6.24x10^{18}$ electrons.

Coupling: A connection between more than one object or energy pathway so that together they function as a single unit.

Covalent bond: A bond formed by sharing a pair of electrons between two atoms. A primary interatomic bond that is formed by the sharing of electrons between neighboring atoms. An attachment between two atoms resulting from the sharing of a pair of electrons

Covalent radius: Given a set of N elements that can form covalent single bonds in molecules, with $N(N - 1)$ possible elemental pairings, it has proved possible to define a covalent radius for each element such that the actual bond length between any two elements that form a covalent single bond is roughly equal to the sums of their covalent radii.

CPU: The central processing unit of a computer, responsible for executing instructions to process information.

Creep: The time-dependent permanent deformation that occurs under stress; for most materials it is important only at elevated temperatures.

Crosslinked polymer: A polymer in which adjacent linear molecular chains are joined at various positions by covalent bonds.

Cross-linking: A process forming chemical bonds between two separate molecular chains.

Cross-sensitivity: The influence of one measurand on the sensitivity of a sensor, another measurand.

Crosstalk: Electromagnetic noise transmitted between leads or circuits in close proximity to each other.

Crybiology: The science of biology at low temperatures; research in cryobiology has made possible the freezing and storing of sperm and blood for later use.

Crystal Lattice: The regular three-dimensional pattern of atoms in a crystal.

Crystal structure: For crystalline materials, the manner in which atoms or ions are arrayed in space. It is defined in terms of the unit cell geometry and the atom positions within the unit cell.

Crystallescence: In medical nanorobotics, the crystallization of solid solute that is offloaded by nanorobot sorting rotors at a concentration that exceeds the solvation capacity of the surrounding solvent.

Curie temperature (also Curie point) (T_c): The temperature above which a ferromagnetic or ferrimagnetic material becomes paramagnetic. For iron the Curie point is 760oC and for nickel 356°C.

Current [A]: Measure of rate of flow of electric charge: a one-ampere current is a flow of 1 C of charge per second.

Cutoff: Condition in a diode or bipolar junction transistor in which the potential across a p-n junction prevents current flow.

Cyclic: A structure is termed cyclic if its covalent bonds form one or more rings.

Cycloaddition: A reaction in which two unsaturated molecules (or moieties within a molecule) join, forming a ring.

Cyclotron: A type of particle accelerator in which an ion introduced at the center is accelerated in an expanding spiral path by use of alternating electrical fields in the presence of a magnetic field.

Cytocarriage: In medical nanorobotics, the commandeering of a natural motile cell, by a medical nanorobot, for the purposes of in vivo transport (of the nanorobot), or to perform a herding function (of the affected cell), or for other purposes.

Cytocide: The killing of living cells.

Cytography: A physical description (and mapping) of the living cell.

Cytoidentification: Identification of cell type.

Cytometrics: The quantitative measurement of cell sizes, shapes, structures, and numbers.

Cytopenetration: In medical nanorobotics, entry into cells by penetrating the plasma membrane.

Deoxyribonucleic acid (DNA): A huge nucleotide polymer having a double helical structure with complementary bases on two strands. Its major functions are protein synthesis and the storage and transport of genetic information.

Deoxyribonucleic acid: Deoxyribonucleic acid are long chains consisting of four kinds of nucleotides; the order of these nucleotides encodes the information needed to construct protein molecules. These in turn make up much of the molecular machinery of the cell. DNA is the genetic material of cells.

Dermal Zippers: In medical nanorobotics, a theorized medical nanorobot that can rapidly perform incision-wound repairs to the dermis and epidermis; dermal zippers.

Design Ahead: The use of known principles of science and engineering to design systems that can only be built with tools not yet available; this permits faster exploitation of the abilities of new tools.

Design Diversity: A form of redundancy in which components of different design serve the same purpose; this can enable systems to function properly despite design flaws.

Design: To plan and delineate with an end in mind and subject to constraints.

Desmosome: Specialized cell-cell junction, usually formed between two epithelial cells, characterized by dense plaques of protein into which intermediate filaments in the two adjoining cells insert.

Detergent: Type of small amphipathic molecule that tends to coalesce in water, with its hydrophobic tails buried and its hydrophilic heads exposed; widely used to solubilize membrane proteins.

Determination: Commitment by an embryonic cell to a particular specialized path of development; it reflects a change in the internal character of the cell.

Development: Succession of changes that take place in an organism as a fertilized egg gives rise to an adult plant or animal.

Devitrification: The process in which a glass (noncrystalline or vitreous solid) transforms to a crystalline solid.

Dialysis: A phenomenon in which a semipermeable membrane allows transfer of both solvent molecules and small solute molecules and ions.

Diode: two-terminal device that conducts current well in one direction and poorly in the other.

Dip: Dual In-line Package - common ceramic or plastic enclosure for an integrated circuit.

Dipole (electric): A pair of equal yet opposite electrical charges that are separated by a small distance.

Directed-Assembler: A specific type of assembler that makes use of directed-assembly, such that the assembly process requires external energy or information input.

Disassembler: A system of nanomachines able to take an object apart a few atoms at a time, while recording its structure at the molecular level. (2) An instrument able to take apart structures a few atoms at a time, recording structural information at each step. (3) In molecular nanotechnology, a nanomachine or system of nanomachines able to take an object apart while at each step recording the structure and composition of that object at the molecular level.

Disasterbation: Idly fantasizing about possible catastrophes (ecological collapse, full-blown totalitarianism) without considering their likelihood or considering their possible solutions and preventions. [David Krieger, 1993]

Disequilibration: In medical nanorobotics, maintenance or inducement of a state of perpetual ionic, chemical, or energetic disequilibrium in a living cell by a medical nanorobot, usually for the purpose of inducing cytocide.

Dislocation: A linear crystalline defect around which there is atomic misalignment. Plastic deformation corresponds to the motion of dislocations in response to an applied shear stress. Edge, screw, and mixed dislocations are possible.

Disruptive Technology: Technology that is significantly cheaper than current, is much higher performing, has greater functionality, and is frequently more convenient to use. Will revolutionize markets by superseding existing technology. "Paradigm shifting" is a well-worn connotation. Although the term may sound negative to some, it is in fact neutral. It is only negative when businesses who are unprepared for change fail to adapt, only to fall behind and fail. The results are not *evolutionary*, they are *revolutionary*.

Disruptive techonlogies: In his book, *The Innovator's Dilemma*, Harvard professor Clayton Christensen used the term "disruptive technologies" to describe developments capable of displacing established market leaders.

Dissolution: Deterioration in an organism such that its original structure cannot be determined from its current state.

Distributed Intelligence: An intelligent entity which is distributed over a large volume (or inside another system, like a computer network) with no distinct center. This is the opposite to the strategy of Concentrated intelligences. Distributed intelligences have much longer communications lags, but are more flexible in their structure and can survive damage to their parts.

DNA (Deoxyriobonucleic Acid): DNA molecules are long chains consisting of four kinds of nucleotides; the order of these nucleotides encodes the information needed to construct protein molecules. These in turn make up much of the molecular machinery of the cell. DNA is the genetic material of cells.

DSP: Digital Signal Processing; a process by which a sampled and digitized data stream (real-time data such as sound or images) is modified in order to extract relevant information. Also, a digital signal processor.

Ductility: A measure of a material's ability to undergo plastic deformation before fracture; expressed as percent elongation (%EL) or percent area reduction (%AR) from a tensile test.

DumbSizing: Apealing to the least common denominator by explaining difficult concepts in such a manner so they loose meaning. Also, talking down to someone less informed or learned. [uhf]

Dynamic characteristics: A description of an instrument's behavior between the time a measured quantity changes value and the time the instrument obtains a steady response.

Dynamic error: The error that occurs when the output does not precisely follow the transient response of the measured quantity.

Dynamic range: The ratio of the largest to the smallest values of a range, often expressed in decibels.

Electrical breakdown: Condition in which, particularly with high electric field, a nominal insulator becomes electrically conducting.

Electromotive force (emf) series: A series of chemical elements arranged in order of their electromotive force. The electromotive force is the greatest potential difference that can be generated by a particular source of electric current. In practice this potential may be observable only when the source is not supplying current, because of its internal resistance.

Enabling science and technologies: Areas of research relevant to a particular goal, such as nanotechnology. Also, technology that "enables" other technology to advance, such as the transistor enabled the computer chip revolution, as did photolithography.

Endocrine cell: Specialized animal cell that secretes a hormone into the blood; usually part of a gland, such as the thyroid or pituitary gland.

Endocytosis: Uptake of material into a cell by an invagination of the plasma membrane and its internalization in a membrane-bounded vesicle.

Endoplasmic reticulum (ER): Labyrinthine, membrane-bounded compartment in the cytoplasm of eucaryotic cells, where lipids are synthesized and membrane-bound proteins are made.

Endosome: Membrane-bounded organelle in animal cells that carries materials newly ingested by endocytosis and passes many of them on to lysosomes for degradation.

Endothelium: Single sheet of highly flattened cells (endothelial cells) that forms the lining of all blood vessels. Regulates exchanges between the bloodstream and surrounding tissues and is usually surrounded by a basal lamina.

Energy [J]: Capacity for performing work or to cause heat flow. Like work itself, it is measured in Joules.

Energy Conversion: Changing energy from one form to another.

Engineering: The use of scientific knowledge and trial-and-error to design systems.

Engulf formation: In medical nanorobotics, a configuration that may be adopted by a metamorphic nanorobot, in which the nanorobot reshapes itself to create an interior cavity capable of trapping a living cell, virion, or other biological particle.

Exponential Growth: Growth that proceeds in a manner characterized by periodic doublings. Inaccurately referred to as "self-replication," exponential growth refers to the process of growth or replication involving doubling within a given period.

Extrinsic: Characterizes doped, rather than pure, semiconductor.

Fabricator: A small nano-robotic device that can use supplied chemicals to manufacture nanoscale products under external control. Fabricators could work together to build macroscale products by convergent assembly. Similar to assemblers, but less complex, easier to build, and probably more efficient.

Face-centered cubic: A crystal structure found in common elemental metals also FCC. Within the cubic unit cell, atoms are located at all corner and face-centered positions.

Fail-stop: Describes a component or subsystem that, in the event of a failure, produces no output (e.g., of material or data) rather than producing a damaged or incorrect output.

Faraday: A constant representing the charge on one mole of electrons; 96, 485 coulombs.

Far-Field Diffraction Limit: A physical limit to how focused a beam of light can be.

Fatty acid: Compound such as palmitic acid that has a carboxylic acid attached to a long hydrocarbon chain. Used as a major source of energy during metabolism and as a starting point for the synthesis of phospholipids.

Fault-tolerant: Describes a system that can suffer failure in a component or subsystem, yet continue to function correctly.

Femtoengineering: Will involve engineering using mechanisms within a quark.

Femtometer: [abbr: fm] A unit suitable to express the size of atomic nuclei. One quadrillionth (10 to minus 15) of a meter.

Femtosecond: Femtosecond is one quadrillionth of a second, and is to a second what a second is to 32,700,000 years. At 186,000 miles per second, in one femtosecond light travels only far enough to traverse about 1,000 silicon atoms. When used to time a laser pulse, it allows for ultra-precise micromachining, with virtually no damage to surrounding material.

Femtotechnology: Femtotechnology the art of manipulating materials on the scale of elementary particles (leptons, hadrons, and quarks). [CA-B] The next step smaller after picotechnology, which is the next step smaller after nanotechnology.

Fermi energy: The energy level in a solid at which the probability of finding an electron is 1/2. For a metal, the energy corresponding to the highest filled electron state in the valence band at 0 K.

Fermi Paradox: "If there are other intelligent beings in the Universe, why aren't they here?" Since it appears to be quite possible for a technological species to spread across the galaxy in less than 10 million years (using von Neumann machines) or otherwise change things on such a large scale that it would be very visible, the lack of such evidence is puzzling or implies that other technological civilizations doen't exist. There have been many attempts to explain this, for example the "Wildlife Preserve" idea (the aliens doesn't want to interfere with younger civilizations), that they transcend and become incomprehensible, that they hide or that they are actually here, hidden on the nanoscale, but the problem with these attempts is that most of them just explain why some aliens would not be apparent. [E. Fermi]

Ferroelectric material: A dielectric material such as Rochelle salt and barium titanate with a domain structure containing dipoles (asymmetric distributions of electrical charge) which spontaneously align. Their domain structure makes them analogous to ferromagnetic materials. They exhibit hysteresis and usually the piezoelectric effect.

Ferromagnetism: Permanent and large magnetizations found in some metals (e.g. Fe, Ni and Co), resulting from the parallel alignment of neighboring magnetic moments.

Fertilization: Fusion of a male and a female gamete (both haploid) to form a diploid zygote, which develops into a new individual.

FET: Field-Effect Transistor — semiconductor device whose insulated gate electrode controls current flow.

Fiber-optic: Relates to transmission of information as modulated light in tiny transparent fibers instead of copper wires.

Fibrin sealant: A biological tissue glue derived from human plasma. Sometimes referred to as fibrin glue, it contains the components necessary for the last step of coagulation: i) fibrinogen and other coagulation factors and ii) thrombin. It is used in surgical procedures to arrest bleeding and as an adjunct to wound healing.

Fibrin: A protein derived from FIBRINOGEN in the presence of THROMBIN, which forms part of the blood clot. [MeSH]

Foglet: A mesoscale machine. A discreet component of utility fog. [J. S. Hall 1994]

Follicle cell: One of the cell types that surround a developing oocyte or egg.

Follower: A component in a cam system that is driven through a pattern of displacements as it rests against a moving contoured surface.

Free Radicial: A molecule containing an unpaired electron, typically highly unstable and reactive. Free radicals can damage the molecular machinery of biological systems, leading to cross-linking and mutation.

Free-energy change (DG): Change in the free energy during a reaction: the free energy of the product molecules minus the free energy of the starting molecules. A large negative value of DG indicates that the reaction has a strong tendency to occur.

Frequency response: Two relations between sets of inputs and outputs. One relates frequencies to the output-input amplitude ratio; the other relates frequencies to the phase difference between the output and input.

Frequency: Number of times per second that a quantity representing a signal, such as a voltage, changes state. Also, the number of waves (cycles) per second that pass a given point in space.

FUD: Fear, Uncertainty, Doubt.

Fullerene: A class of cage-like carbon compounds composed of fused, pentagonal and/or hexagonal sp^2 carbon rings. A new allotrope of carbon characterized by a closed cage structure consisting of an even number of three coordinate carbon atoms devoid of hydrogen atoms. This class was originally limited to closed-cage structures with twelve isolated five- membered rings, the rest being six- membered rings. [IUPAC Provisional Recommendations for the Nomenclature for the C_{60}-I_h and C_{70}-$D_{5h(6)}$ Fullerenes, 2001]

Fullerenes: Fullerenes are a molecular form of pure carbon discovered in 1985. They are cage-like structures of carbon atoms, the most abundant form produced is buckminsterfullerene (C60), with 60 carbon atoms arranged in a spherical structure. There are larger fullerenes containing from 70 to 500 carbon atoms. [Wid] What are fullerenes? A very good explanation, even though some of the translation is a bit rough.

Functional Group: Part of a molecule (group) with the potential to take part in a chemical reaction

Functional navigation: In medical nanorobotics, a form of nanorobotic navigation in which nanodevices seek to detect subtle variations in their environment, comparing sensor readings with target tissue/cell profiles and then congregating wherever a precisely defined set of preconditions exists.

G protein: One of a large family of heterotrimeric GTP-binding proteins that are important intermediaries in cell-signaling pathways. Usually activated by the binding of a hormone or other signaling ligand to a seven-pass transmembrane receptor protein.

G0 phase: (G-"zero" phase) State of withdrawal from the eucaroytic cell-division cycle by entry into a quiescent G1 phase; often seen in differentiated cells.

G1 phase: Gap 1 phase of the eucaryotic cell-division cycle, between the end of cytokinesis and the start of DNA synthesis.

G2 phase: Gap 2 phase of the eucaryotic cell-division cycle, between the end of DNA synthesis and the beginning of mitosis.

GAG (glycosaminoglycan): Long, linear, highly charged polysaccharide composed of a repeating pair of sugars, one of which is always an amino sugar. Mainly found covalently linked to a protein core in extracellular matrix proteoglycans. Examples include chondroitin sulfate, hyaluronic acid, and heparin.

Gain: The ratio of the amplitude of an output to input signal.

Galvanic corrosion: The preferential corrosion of the more chemically active of two metals electrically coupled and exposed to an electrolyte.

Gamete: Specialized haploid cell, either a sperm or an egg, serving for sexual reproduction.

Ganglion (plural ganglia): Cluster of nerve cells and associated glial cells located outside the central nervous system.

Ganglioside: Any glycolipid having one or more sialic acid residues in its structure. Found in the plasma membrane of eucaryotic cells and especially abundant in nerve cells.

Gap junction: Communicating cell-cell junction that allows ions and small molecules to pass from the cytoplasm of one cell to the cytoplasm of the next.

Gastrula: Animal embryo at an early stage of development where cells are invaginating to form the rudiment of a gut cavity. (From Greek *gaster*, belly.)

Gate: Circuit whose logical output variables are determined by its inputs. In digital logic, a component that can switch the state of an output dependent on the states of one or more inputs.

Gauss: The cgs unit used in measuring magnetic induction.

Gene: A given segment of the DNA molecule that contains the code for a specific protein.

Genegeneering: Genetic engineering.

General-purpose molecular manufacturing: A manufacturing technology that will find many applications across many segments of society. Its extreme flexibility, precision, high capacity, and low cost will cause rapid adoption almost everywhere, and therefore will have disruptive effects in many industries.

Genetic Algorithm: Any algorithm which seeks to solve a problem by considering numerous possibilities at once, ranking them according to some standard of fitness, and then combining ("breeding") the fittest in some way. In other words, any algorithm which imitates natural selection. [AS]

GENIE: An AI combined with an assembler or other universal constructor, programmed to build anything the owner wishes. Sometimes called a Santa Machine. This assumes a very high level of AI and nanotechnology. [AS]

Genomics: "The study of the full complement of genes that make up an organism."

Giant Magnetoresistance (GMR): It results from subtle electron-spin effects in ultra-thin 'multilayers' of magnetic materials, which cause huge changes in their electrical resistance when a magnetic field is applied. GMR is 200 times stronger than ordinary magnetoresistance.

Gibbs free energy: The Gibbs free energy is the Helmholtz free energy plus the product of the system volume and the external pressure. Changes in the Gibbs free energy at a constant pressure thus include work done against external pressure as a system undergoes volumetric changes. This proves convenient for describing equilibria in gases and liquids at a constant pressure (e.g., at one atmosphere), but is of little use in describing machine-phase chemical processes. Changes in the Gibbs free energy caused by a change in the applied pressure (at constant volume) have no direct physical significance.

Glass transition temperature (T_g): The temperature at which, upon cooling, a noncrystalline ceramic or polymer transforms from a supercooled liquid to a rigid glass.

Glass: An amorphous solid obtained when silica is mixed with other compounds, heated above its melting point, and then cooled rapidly.

GNR technologies: (Genetic Engineering, Nanotechnology, and Robotics)

Golden Goo: Another member of the grey goo family of nanotechnology disaster scenarios. The idea is to use nanomachines to filter gold from seawater. If this process got out of control we would get piles of golden goo (the "Wizard's Apprentice Problem"). This scenario demonstrates the need of keeping populations of self-replicating machines under control; it is much more likely than grey goo, but also more manageable. [AS - Originated on sci.nanotech 1996]

Grain boundary: The interface separating two adjoining grains having different crystallographic orientations.

Grain growth: The increase in average grain size of a polycrystalline material: for most materials, an elevated temperature heat treatment is necessary.

Grain size: The average grain diameter as determined from a random cross section.

Gray Goo or Grey Goo - destructive nanobots [AKA: "gray dust"]: Opposite of Blue Goo. See Star Trek scenario. Vast legions of *destructive* nanites. Typically, created by accident. Left unchecked, they will basically convert everthing they contact into more of themselves, or consume and digest it for energy. Either way, its pretty much bad news. Also - Self-replicating (von Neumann) nanomachines spreading uncontrolably, building copies of themselves using all available material. This is a commonly mentioned nanotechnology disaster scenario, although it is rather unlikely due to energy constraints and elemental abundances. More probable disaster scenarios are the green goo, golden goo, red goo, khaki goo scenarios. As a protection blue goo has been proposed.

Gray goo problem: Nanotech Now's Goo and Paste Glossary. The following quote was taken from Eric Drexler's book Engines of creation. The early transistorized computers soon beat the most advanced vacuum-tube computers because they were based on superior devices. For the same reason, early assembler-based replicators could beat the most advanced modern organisms. "Plants" with "leaves" no more efficient than today's solar cells could out-compete real plants, crowding the biosphere with an inedible foliage. Tough, omnivorous "bacteria" could out-compete real bacteria: they could spread like blowing pollen, replicate swiftly, and reduce the biosphere to dust in a matter of days. Dangerous replicators could easily be too tough, small, and rapidly spreading to stop - at least if we made no preparation. We have trouble enough controlling viruses and fruit flies.

Gray goo: The name given to free-range self-replicating miniature machines that could, in theory, run out of control and cause severe damage to the biosphere. The actual threat is generally overrated, as we explain here.

Green ceramic body: A ceramic piece, formed as a particulate aggregate, that has been dried but not fired.

Green Goo: Nanomachines or bio-engineered organisms used for population control of humans, either by governments or eco-terrorist groups. Would most probably work by sterilizing people through otherwise harmless infections. See Nick Szabo's essay Green Goo — Life in the Era of Humane Genocide. [AS] "A scary concept dreamed up by Erik K Drexler whereby tiny *assemblers*, or molecular machines, that are capable of making copies of themselves, are let loose and proceed to replicate uncontrollably, consuming everything in their path and turning it into a grey goo."

Ground state: The lowest-energy state of a system. The electronic ground state of a system cannot reduce its energy by an electronic transition, but may contain vibrational energy (kinetic and potential energy associated with the motions and positions of its atoms); extended systems at ordinary temperatures are always vibrationally excited, and so "ground state" is often taken to mean "electronic ground state."

Ground: To make electrical connection to the earth or to the chassis of a device (verb); the connection point so used (noun).

Group velocity: In wave propagation, the speed of the waveform (e.g., of a peak) can be different from the speed of a group of waves (e.g., of a set of ripples in water). The latter is the *group velocity*, and is the speed of propagation of information and wave energy. The waveform speed is the *phase velocity*.

Group: A set of linked atoms in a molecule; a defined substructure. Typically, a set that is usefully regarded as a unit in chemical reactions of interest.

GUI: Graphical User Interface — hardware, software, and firmware that produces the display on modern personal computers.

Guy Fawkes Scenario: If nanotechnology becomes widely available, it might become trivial for anyone to committ acts of terrorism (such as making nanomachines build a large amount of explosives under government buildings a la Guy Fawkes). This would either force strict control over nanotechnology (hard) or a decentralized mode of organization.

Hair cell: Specialized sensory epithelial cell in the ear with bundles of giant microvilli (stereocilia) protruding from its apical surface. Sound vibrations tilt the stereocilia, evoking an electrical change in the hair cell, which thus acts as a sound detector.

Hall effect: The phenomenon whereby a force is applied to a moving electron or hole by a magnetic field that is applied perpendicular to the direction of motion. The force direction is perpendicular to both the magnetic field and the particle motion directions.

Haploid: Having only one set of chromosomes, as in a sperm cell or a bacterium, as distinct from diploid (having two sets of chromosomes).

Hardness: The measure of a material's resistance to deformation by surface indentation or by abrasion. There are various scales in use to express hardness. The Mohs scale is qualitative and somewhat arbitrary and ranges from 1 on the soft end for talc to 10 for diamond. Quantitative scales are the Rockwell (HR), Brinell (indicated by HB), Knoop (HK) and Vickers (HV). Knoop and Vickers are referred to as microhardness testing methods on the basis of load and indenter size.

Harmonic oscillator: A system in which a mass is subject to a linear restoring force, like an ideal spring. A harmonic oscillator vibrates at a fixed frequency, independent of amplitude.

Heat capacity (C_v at constant volume and C_p at constant pressure): The quantity of heat required to produce a unit temperature rise per mole of material. The ratio of the heat input to the temperature increase in a system. Note that this definition does not imply that a system *contains* heat, despite the name *heat capacity*.

Heat shock protein (stress-response protein): Protein synthesized in response to an elevated temperature or other stressful treatment; usually helps the cell to survive the stress.

Heat: As defined in thermodynamics, heat is the energy that flows between two systems as a result of temperature differences (a system *contains* neither heat nor work, but can *produce* heat or *do* work). Heat thus differs from thermal energy.

Heisenberg Uncertainty Principle: A quantum-mechanical principle with the consequence that the position and momentum of an object cannot be precisely determined. The Heisenberg principle helps determine the size of electron clouds, and hence the size of atoms.

HeLa cell: Line of human epithelial cells that grows vigorously in culture. Derived from a human cervical carcinoma.

Heme: An iron complex. Cyclic organic molecule containing an iron atom that carries oxygen in hemoglobin and carries an electron in cytochromes.

Hemidesmosome: Specialized cell junction between an epithelial cell and the underlying basal lamina.

Hemoglobin: A biomolecule composed of four myoglobine-like units (proteins plus heme) that can bind and transport four oxygen molecules in the blood.

The major protein in red blood cells that associates with O_2 in the lungs by means of a bound heme group.

Hemopoiesis (hematopoiesis): Generation of blood cells, mainly in the bone marrow.

HEMT: High electron mobility transistor.

Henry (H): Unit of inductance (see"inductance"). One henry (H) is the inductance of a closed circuit in which an electromotive force of 1 volt is produced when the electric current in the circuit varies uniformly at the rate of 1 ampere per second.

Henry's law: The amount of gas dissolved in a solution is directly proportional to the pressure of the gas above the solution.

Heterocaryon: Cell with two or more nuclei produced by the fusion of two or more different cells.

Heterochromatin: Region of a chromosome that remains unusually condensed and transcriptionally inactive during interphase.

Heterodimer: Protein complex composed of two different polypeptide chains.

Heteronuclear: Consisting of different elements. [TNTWeekly]

Heterozygote: Diploid cell or individual having two different alleles of a specified gene.

Heuristics: Rules of thumb used to guide one in the direction of probable solutions to a problem.

Highest Occupied Molecular Orbital (HOMO): The highest energy molecular orbital of an atom or molecule that contains an electron.

Histone: A protein present in the eukaryotic nucleus that is bound to the DNA at regularly spaced intervals.

Homogeneous and Heterogeneous Assays: A homogeneous assay does not require a separation step to remove free antigen from bound antigen and relies upon the fact that the function of the label is modified upon binding, leading to a change in signal intensity. Because of high background signal a heterogeneous approach incorporating a separation step of bound and unbound makes the detection limit lower, approaching the values obtained by RIA. The homogeneous assay is less technically demanding.

Human Genome Project: A research initiative that has mapped the entire human genome.

Human tissue engineering No universally agreed definition exists. But a Commission working hypothesis suggests a human tissue engineered product means any autologous (emanating from the patient himself) or allogeneic (coming from another human being) product which: contains, consists of, or results in engineered human cells or tissues; and has properties for, or is presented as having properties for, the regeneration, repair or replacement of tissue, where the new tissue or cells, in whole or in part, are structurally and functionally analogous to the original tissue that is being regenerated, repaired or replaced.

Hydrogen Bond: A bond formed between hydrogen and two other atoms. A hydrogen atom covalently bound to an electronegative atom (e.g., nitrogen, oxygen) has a significant positive charge and can form a weak bond to another electronegative atom; this is termed a hydrogen bond.

Hydrophobic force: Water molecules are linked by a network of hydrogen bonds. A nonpolar, nonwetting, surface (e.g., wax) cannot form hydrogen bonds. To form their full complement of hydrogen bonds, the nearby water molecules must form a more orderly (hence lower entropy) network. This both increases free energy and causes forces that tend to draw hydrophobic surfaces together across distances of several nanometers.

Hydrophobic: Literally - water fearing, from the Greek *hydro* - "water" and *phobo* - "fear".

Hypertext: A computer-based system for linking text and other information with cross-references, making access fast and criticisms easy to publish and find.

Hypsithermal limit: The maximum amount of energy that may be released at Earth's surface, as a result of human technological activities, without significantly altering the natural global energy balance; estimated as 10^13-10^15 watts.

Hysteresis: The difference in the output when a specific input value is approached first with an increasing and then with a decreasing input. This phenomenon occurs in ferroelectric materials and results in irreversible loss of energy through heat dissipation.

Imaging Contrast Agent: A molecule or molecular complex that increases the intensity of the signal detected by an imaging technique, including MRI and ultrasound. An MRI contrast agent, for example, might contain gadolinium attached to a targeting antibody. The antibody would bind to a specific target—a metastatic melanoma cell, for example – while the gadolinium would increase the magnetic signal detected by the MRI scanner.

Imaging Contrast Agent: A molecule or molecular complex that increases the intensity of the signal detected by an imaging technique, including MRI and ultrasound. An MRI contrast agent, for example, might contain gadolinium attached to a targeting antibody. The antibody would bind to a specific target—a metastatic melanoma cell, for example – while the gadolinium would increase the magnetic signal detected by the MRI scanner.

Immortalization: Production of a cell line capable of an unlimited number of cell divisions. Can be the result of a chemical or viral transformation or of fusion with cells of a tumor line.

Immune Machines: Medical nanomachines designed for internal use, especially in the bloodstream and digestive tract, able to identify and disable intruders such as bacteria and viruses.

Immune response: Response made by the immune system of a vertebrate when a foreign substance or microorganism enters its body.

Immune system: Population of lymphocytes and other white blood cells in the vertebrate body that defends it against infection.

Immunoglobulin (Ig): An antibody molecule. Higher vertebrates have five classes of immunoglobulin - IgA, IgD, IgE, IgG, and IgM - each with a different role in the immune response.

Immunoglobulin like (Ig-like) domain: Characteristic protein domain of about 100 amino acids that is found in antibody molecules and in many other proteins that form the Ig superfamily.

IMP: Electronic implant, especially in the brain. [Ron Hale Evans]

Impedance: The complex ratio of a force-like quantity (force, pressure, voltage, temperature, or electric field) to a corresponding related velocity-like quantity (velocity, volume velocity, current, heat flow, or magnetic field strength).

***In situ* hybridization:** Technique in which a single-stranded RNA or DNA probe is used to locate a gene or an mRNA molecule in a cell or tissue.

***In vitro*:** Term used by biochemists to describe a process taking place in an isolated cell-free extract. Also used by cell biologists to refer to cells growing in culture (*in vitro*), as opposed to in an organism (*in vivo*). (Latin for "in glass.")

Inductance [in Henry, H]: That property of an electric circuit which tends to oppose change in current in the circuit. One henry (H) is the inductance of a closed circuit in which an electromotive force of 1 volt is produced when the electric current in the circuit varies uniformly at the rate of 1 ampere per second.

Induction (embryonic): Change in the developmental fate of one tissue caused by an interaction with another tissue.

Inductor: Energy storage circuit component consisting of a coil of wire and possibly a magnetic material.

Inflammatory response: Local response of a tissue to injury or infection. Caused by invasion of white blood cells, which release various local mediators such as histamine.

Infrared: Invisible electromagnetic radiation having a longer wavelength, and lower frequency, than visible red light.

Inhibitor: A chemical substance that, when added in relatively low concentrations, slows down a chemical reaction.

Inline Universities: (As opposed to online universities), nanocomputer implants serving to increase intelligence and education of their owners, essentially turning them into walking universities.

Inmessaging: In medical nanorobotics, conveyance of information from a source external to the human body, or external to working nanodevices, to a receiver located inside the human body.

Insertion point (in lithography context): Adaptation of a new lithography technique is referred to as the insertion point of that technique.

Insulator: A substance, object or material that does not conduct electricity. Material that conducts electricity very poorly.

Integrated Circuit (IC): An electronic circuit consisting of many interconnected devices on one piece of semiconductor, typically into 10 millimeters on a side. ICs are the major building blocks of today's computers. Semiconductor circuit, typically on a very small silicon chip, containing microfabricated transistors, diodes, resistors, capacitors, etc.

Interagency Working Group on Nanoscience, Engineering and Technology IWGN, National Science and Technology Council NSTC, US

Interface: In physical terms, an interface is the boundry between two phases, for instance between a solid and liquid or between a liquid and gas.

Intermolecular: Between more than one molecule. Describes an interaction (e.g., a chemical reaction) between different molecules.

Internal energy: The sum of the kinetic and potential energies (including electromagnetic field energies) of the particles that make up a system.

Internet: Worldwide digital communication network in which packets of information travel between senders and recipients.

Interstitial diffusion: a diffusion mechanism that causes atomic motion from interstitial site to interstitial site.

Intramolecular: Describes an interaction (e.g., a chemical reaction) within a single molecule. Intramolecular interactions between widely separated parts of a molecule resemble intermolecular interactions in most respects. (2) Within a single molecule.

Intrinsic: Characterizes pure undoped semiconductor; electrical conductivity depends only on temperature and the band gap energy.

Ion: An atom or molecule with a net charge. **(2)** An atom with more or fewer electrons than those needed to cancel the electronic charge of the nucleus. An ion is an atom with a net electric charge.

Ionic bond: A chemical bond resulting chiefly from the electrostatic attraction between positive and negative ions. A coulombic interatomic bond existing between two adjacent and oppositely charged ions.

Ionophore: A macro-organic molecule capable of specifically solubilizing an inorganic ion of suitable size in organic mediums.

Junna: Minor Negentropy Alliance world. During the period from 3267 to 3421 was center for the Institute of Baseline Psychotypology. Suffered during the Version War and became an MPA garrison. Since the age of separative empires the locals have drifted increasingly towards Zoeticism

Jupiter-Brain: A posthuman being of extremely high computational power and size. This is the archetypal concentrated intelligence. The term originated due to an idea by Keith Henson that nanomachines could be used to turn the mass of Jupiter into computers running an upgraded version of himself.

Karyotype: Full set of chromosomes of a cell arranged with respect to size, shape, and number.

Keratin (cytokeratin): Member of the family of proteins that form keratin intermediate filaments, mainly in epithelial cells. Some specialized keratins are found in hair, nails, and feathers.

Ketone: Organic molecule containing a carbonyl group linked to two alkyl groups.

Kevlar (TM): A synthetic fiber made by E. I. du Pont de Nemours & Co., Inc. Stronger than most steels, Kevlar is among the strongest commercially available materials and is used in aerospace construction, bulletproof vests, and other applications requiring a high strength-to-weight ratio.

Kilobyte (kB): 2^{10} (= 1024, or about one thousand) bytes of information.

Kilocalorie (kcal): Unit of heat energy equal to 1000 calories. Often used to express the energy content of food or molecules: bond strengths, for example, are measured in kcal/mole. An alternative unit in wide use is the kilojoule, equal to 0.24 kcal.

Kilohertz (kHz): One thousand cycles per second.

Kilojoule: Standard unit of energy equal to 1000 joules, or 0.24 kilocalories.

Kinesin: One type of motor protein that uses the energy of ATP hydrolysis to move along a microtubule.

Kinetic energy: Energy resulting from the motion of masses.

Kinetic molecular theory: A model that assumes that an ideal gas is composed of tiny particles (molecules) in constant motion.

Kinetochore: Complex structure formed from proteins on a mitotic chromosome to which microtubules attach and which plays an active part in the movement of chromosomes to the pole. The kinetochore forms on the part of the chromosome known as the centromere.

Knowbots: Knowledge robots, first developed Vinton G. Cref and Robert E. Kahn for National Research Initiatives. Knowbots are programmed by users to scan networks for various kinds of related information, regardless of the language or form in which it expressed. "Knowbots support parallel computations at different sites. They communicate with one another, and with various servers in the network and with users."

Label: Chemical group or radioactive atom added to a molecule in order to follow its progress through a biochemical reaction or to locate it spatially. Also, as a verb, to add such a group or atom to a cell or molecule.

Lagging strand: One of the two newly made strands of DNA found at a replication fork. The lagging strand is made in discontinuous lengths that are later joined covalently.

Lambda bacteriophage (l bacteriophage): Virus that infects *E. coli;* widely used as a DNA cloning vector.

Lamin: Extracellular matrix protein found in basal laminae.

Lamins: Intermediate filament proteins that form the fibrous matrix (nuclear lamina) on the inner surface of the nuclear envelope.

Langmuir-Blodgett (LB) Film: A film of organic material (often surfactant molecules) assembled at the liquid-gas interface and subsequently transferred onto a solid substrate.

Langmuir-Blodgett: The name of a nanofabrication technique used to create ultrathin films (monolayers and isolated molecular layers), the end result of which is called a "Langmuir-Blodgett film".

Laser trimming: A method for adjusting the value of thin- or thick-film resistors by using a computer-controlled laser system.

Laser tweezers: Laser tweezers and related approaches are important for manipulation and isolation of subcellular *organelles* and structural components. [National Center for Research Resources "Integrated Genomics Technologies Workshop Report" Jan 1999]

Laser: Light Amplification by the Stimulated Emission of Radiation — quantum device that produces coherent light.

LCD (Liquid Crystal Display): LCD is the predominant technology used in flat panel displays. The principle that makes the display work is this: A crystalís alignment can be altered with an electric current. If the crystal is lined up one way ñ it will allow the light waves to pass through a polarized filter, but if the electric current alters the crystalís alignment, it will guide light so that the polarized filter blocks the light. By densely packing red, blue and green light emitting crystals next to each other on a sheet (ìcalled a substrateî), one can create a full color display. The great thing about LCD is that the crystals can be packed together closely, allowing for a higher-resolution, finer-detail display. The con is that LCDs are somewhat fragile, require a lot of power and are relatively less bright. Liquid Crystal Display — display device employing light source and electrically alterable optically active thin film.

Leading strand: One of the two newly made strands of DNA found at a replication fork. The leading strand is made by continuous synthesis in the 5′-to-3′ direction.

Leakage: The loss of all or parts of a useful agent, as of the electric current that flows through an insulator or the magnetic flux that passes outside useful flux circuits.

Lectin: Protein that binds tightly to a specific sugar. Abundant lectins derived from plant seeds are often used as affinity reagents to purify glycoproteins or to detect them on the surface of cells.

LED: Light-Emitting diode — semiconducting diode that produces visible or infrared radiation.

LEDs (Light Emitting Diodes): LEDs work on a completely different concept. Traditionally LEDs are created from two semiconductors. By running current in one direction across the semiconductor the LED emits light of a particular frequency (hence a particular color) depending on the physical characteristics of the semiconductor used. The semiconductor is covered with a piece of plastic that focuses the light and increases the brightness. These semiconductors are very durable, there is no filament, they donít require much power, theyíre brighter and they last a long time. By densely packing red, blue and green LEDs next to each other on a substrate one can create a display. The disadvantage of LEDs is that they are much larger ñ therefore the resolution is not nearly as good as LCD displays. Thatís why most LED displays are large, outdoor displays, not smaller devices, like monitors.

Leucine zipper: Structural motif seen in many DNA-binding proteins in which two a helices from separate proteins are joined together in a coiled-coil, forming a protein dimer.

Life (lifetime): The length of time the sensor can be used before its performance changes.

Life Science Applications of Deep Red T2-MP EviTag Quantum Dots: The new T2-MP EviTags are available in the deep red wavelengths used in life science research – Macoun Red (660nm) and Jonamac Red (680nm) – with carboxyl, amine, biotin and non-functionalized surfaces. Having a biologically compatible quantum dot label like the T2-MP EviTags allows further exploration and development of in vivo assays, following cells inside living organism to get a new understanding of life. In particular, it offers the possibility to create new classes of assays for use where current testing materials could produce cytotoxicity or cell death. Potential applications include live cell and even whole animal imaging, blood cancer assays, and numerous other applications.

Ligand: In protein chemistry, a small molecule that is (or can be) bound by a larger molecule is termed a ligand. In organometallic chemistry, a moiety bonded to a central metal atom is also termed a ligand; the latter definition is more common in general chemistry.

Light Emitting Diodes (LEDs): Light Emitting Diodes work on a completely different concept. Traditionally LEDs are created from two semiconductors. By running current in one direction across the semiconductor the LED emits light of a particular frequency (hence a particular color) depending on the physical characteristics of the semiconductor used. The semiconductor is covered with a piece of plastic that focuses the light and increases the brightness. These semiconductors are very durable, there is no filament, they donít require much power, theyíre brighter and they last a long time. By densely packing red, blue and green LEDs next to each other on a substrate one can create a display.

Lightsail: A spacecraft propulsion system that gains thrust from the pressure of light striking a thin metal film.

Limit of detection: The smallest measurable input. This differs from resolution, which defines the smallest measurable change in input. For a temperature measurement, this would provide an indication of the lowest temperature a sensor could generate an output in response to.

Limited Assembler: An assembler with built-in limits that constrain its use (for example, to make hazardous uses difficult or impossible, or to build just one thing). Assembler capable of making only certain products; faster, more efficient, and less liable to abuse than a general-purpose assembler.

Linde Scenario: A scenario for indefinite survival of intelligent life. It assumes it is possible to either create basement universes connected to the original universe with a wormhole or the existence of other cosmological domains. Intelligent life continually migrates to the new domains as the old grow too entropic to sustain life. [AS/Mitch Porter, 1997. The name refers to Linde's chaotic inflation cosmology, where new universes are continually spawned.]

Linear: Aside from its geometric meaning, *linear* describes systems in which an output is directly proportional to an input. In particular, a linear elastic system is one in which the internal displacements are (at equilibrium) directly proportional to applied forces.

Linearity: The degree to which the calibration curve of a device conforms to a straight line.

Lipids: Water-insoluble substances than can be extracted from cells by nonpolar organic solvents.

Liposome: A type of nanoparticle made of lipids, or fat molecules, surrounding a water core. Liposomes, several of which are widely used to treat infectious diseases and cancer, were the first type of nanoparticle to be used to create therapeutic agents with novel characteristics.

Liposome: A type of nanoparticle made of lipids, or fat molecules, surrounding a water core. Liposomes, several of which are widely used to treat infectious diseases and cancer, were the first type of nanoparticle to be used to create therapeutic agents with novel characteristics.

Liquid Crystal Display: LCD is the predominant technology used in flat panel displays. The principle that makes the display work is this: A crystalís alignment can be altered with an electric current. If the crystal is lined up one way ñ it will allow the light waves to pass through a polarized filter, but if the electric current alters the crystalís alignment, it will guide light so that the polarized filter blocks the light. By densely packing red, blue and green light emitting crystals next to each other on a sheet (ìcalled a substrateî), one can create a full color display. The great thing about LCD is that the crystals can be packed together closely, allowing for a higher-resolution, finer-detail display. The con is that LCDs are somewhat fragile, require a lot of power and are relatively less bright. Liquid Crystal Display — display device employing light source and electrically alterable optically active thin film.

Lithography: A process for creating chemical patterns on a surface.

LMNT: An abbreviation for *limited* molecular nanotechnology; a narrowly specified type of MNT, using only diamondoid reactions; much easier to achieve than general MNT, but with nearly equivalent appeal and impact.

Lofstrom Loop: An beanstalk-like megaconstruction based on a stream of magnetically accelerated bars linked together. The stream is sent into space, where a station rides it using magnetic hooks, redirects it horizontally to another station, which sends it downwards to a receiving station on the ground. From this station the stream is then sent back to the launch station (a purely vertical version is called a space fountain). This structure would contain a large amount of kinetic energy but could be built gradually and would only require enough energy to compensate for losses when finished. Elevators could be run along the streams, and geostationary installations could be placed along the horizontal top. [Named after Keith Lofstrom, who did the first detailed calculations on it in: Lofstrom, Keith H., "The launch loop — a low cost Earth-to-high orbit launch system," AIAA Paper 85-1368, 1985].

London dispersion force: An attractive force caused by quantum-mechanical electron correlation. For example, a neutral spherical molecule (such as a single argon atom) has no charge and produces no external electric field, yet a pair of molecules has a distribution of electron configurations weighted toward those with lesser electron-electron repulsions; this creates a small net attraction. (2) An attractive force between atoms or molecules caused by the numerous transient dipoles resulting from electronic superposition.

Lone pair: Two valence electrons of an atom that share an orbital but do not participate in a bond.

Low-dimension Structures: Quantum wells, quantum wire and quantum dots.

Lowest Unocupied Molecular Orbital (LUMO): The lowest energy molecular orbital of an atom or molecule that does not contain an electron.

Machine-phase chemistry: The chemistry of systems in which all potentially reactive moieties follow controlled trajectories (e.g., guided by molecular machines working in vacuum).

Macroscale: Larger than nanoscale; often implies a design that humans can directly interact with; too large to be built by a single assembler (one cubic micron of diamond contains 176 billion atoms).

Macrosensing: In medical nanorobotics, the detection of global somatic states (inside the human body) and extrasomatic states (sensory data originating outside of the human body) by in vivo nanorobots.

Magnetic field strength (designated by H) [A/m]: Magnetic field produced by a current, independent of the presence of magnetic material. The units of H are ampere-turns per meter, or just amperes per meter.

Magnetic flux density or magnetic induction (designated by B): The magnetic field produced in a substance by an external magnetic field. The units of B are tesla (T). One tesla is the magnetic flux density given by a magnetic flux of 1 weber per square meter. One weber is a magnetic flux that, linking a circuit of 1 turn, would produce in it an electromotive force of 1 volt if it were reduced to zero at a uniform rate in 1 second. Both B and H are field vectors. One henry (H) is the inductance of a closed circuit in which an electromotive force of 1 volt is produced when the electric current in the circuit varies uniformly at the rate of 1 ampere per second. The magnetic field strength and flux density are related according to: $B = \mu H$, where μ is the permeability (see under permeability).

Magnetic Force Microscopy (MFM): A method for observing local magnetic fields near a surface by scanning the surface with a magnetic probe.

Magnetic susceptibility ($\div_m$): The proportionality constant between the magnetization M (see under”magnetization”) and the magnetic field strength H. The magnetic susceptibility is unitless.

Magnetization (M): The total magnetic moment per unit volume of material. Also, a measure of the contribution to the magnetic flux by some material within an H field. The magnitude of M is proportional to the applied field as: $M = \div_m \times H$, with $\div_m$ the magnetic susceptibility.

Magnetostrictive material: A material that changes dimension in the presence of a magnetic field or generates a magnetic field when mechanically deformed.

Manufacturing: The ability to make products, in this case ranging from clothing, to electronics, to medical devices, to books, to building materials, and much more.

Martensite: A metastable iron phase supersaturated in carbon that is the product of a diffusionless (athermal) transformation from austenite.

Mask: Pattern on glass, like a photographic negative, for producing integrated-circuit elements on semiconductor wafer.

Mass to Charge Ratio: A number defining how a particle will respond to an electric or magnetic field that can be calculated by dividing the mass of a particle by its charge.

Massometer: In medical nanorobotics, a nanosensor device for measuring the mass of individual molecules or small physical objects to single-proton resolution.

Materials science: Science of ceramics, glass, metals, plastics, semiconductors.

MatML Materials Markup Language: Materials property data distributed on the World Wide Web in documents using hypertext markup language.

Matter as Software: “Autonomous, motile microdevices clearly are on the horizon. They may be regarded as the first step in the evolution of a technology for “programming” the structure and properties of material objects at the microscopic and the submicroscopic levels. As this evolution progresses, *the physical and economic properties of such* *programmable matter are likely to become much like those of present day software."* [MITRE Corporation]

MCM: MultiChip Module; the interconnection of two or more semiconductor chips in a semiconductor-type package.

Measurand: A physical quantity, condition, or property that is to be measured.

Meat Machine: AKA Cabinet Beast. A box containing assemblers and raw material, within which is formed meat [or whatever else it was programmed to make].

Mechanical: Pertaining to the positions and motions of atoms, as defined by the positions of their nuclei; see electronic. A purely mechanical device can be described in terms of atomic positions and motions without reference to electronic properties, save through their effect on the potential energy function.

Mechanochemistry: Chemistry accomplished by mechanical systems directly controlling the reactant molecules; the formation or breaking of chemical bonds under direct mechanical control. In this volume, the chemistry of processes in which mechanical systems operating with atomic-scale precision either guide, drive, or are driven by chemical transformations. In general usage, the chemistry of processes in which energy is converted from mechanical to chemical form, or vice versa.

Mechanosynthesis: Chemical synthesis controlled by mechanical systems operating with atomic-scale precision, enabling direct positional selection of reaction sites; synthetic applications of mechanochemistry. Suitable mechanical systems include AFM mechanisms, molecular manipulators, and molecular mill systems. Processes that fall outside the intended scope of this definition include reactions guided by the incorporation of reactive moieties into a shared covalent framework (i.e., conventional intramolecular reactions), or by the binding of reagents to enzymes or enzymelike catalysts. Molecular tools with chemically specific tip structures can be used, sequentially, to modify a work piece and build a wide range of molecular structures.

Mechatronics: The study of the melding of AI and electromechanical machines to make machines that are greater than the sum of their parts. The synergistic combination of precision mechanical engineering with electronic control.

Megabyte (MB): 2^{20} (= 1,048,576, or about one million) bytes of information.

Megahertz (Mhz): One million cycles per second

Meme: An idea that replicates through a society as it is propagated through person-to-person interaction, both direct and indirect. Memetics is a field of study that focuses on memes' role in the evolution of a culture. Self-reproducing idea or other information pattern which is propagated in ways similar to that of a gene. [Richard Dawkins, 1976]

MEMS MicroElectro Mechanical Systems: MEMS is less an application area in itself than a manufacturing or fabrication technique that enables other application areas. Many authors use MEMS as shorthand to imply a number of particular application areas. As it is used here, MEMS is a "top- down" fabrication technology that is especially useful for integrating mechanical and electrical systems together on the same chip. It is grouped in the category of integrated microsystems because these same MEMS techniques can be extended in the future to also help integrate biological and chemical components on the same chip, as discussed below. Thus far, MEMS techniques have been used to make some functional commercial devices such as sensors and single- chip measurement devices. Many researchers have used MEMS technologies as analytical tools in other areas of nanotechnology [Central Intelligence Agency, US The Global Technology Revolution, Chapter Two Technology Trends, Genomics, 2001] Stood originally for Micro-ElectroMechanical System — microscopic mechanical elements, fabricated on silicon chips by techniques similar to those used in integrated circuit manufacture, for use as sensors, actuators, and other devices. Today almost any miniaturized device (based on Si technology or traditional precision engineering, chemical or mechanical) is referred to as a MEMS device generic term to describe micron scale electrical/mechanical devices.

MEMS: Stood originally for Micro-ElectroMechanical System — microscopic mechanical elements, fabricated on silicon chips by techniques similar to those used in integrated circuit manufacture, for use as sensors, actuators, and other devices. Today almost any miniaturized device (based on Si technology or traditional precision engineering, chemical or mechanical) is referred to as a MEMS device.

Mesoscale: A device or structure larger than the nanoscale (10^-9 m) and smaller than the megascale; the exact size depends heavily on the context and usually ranges between very large nanodevices (10^-7 m) and the human scale (1 m). [AS]

Messenger molecule: A chemically recognizable molecule which can convey information after it is received and decoded by an appropriate chemical sensor.

Metal nanoshells: A new type of *nanoparticle* composed of a *semiconductor* or dielectric core coated with an ultrathin conductive layer.. By adjusting the relative core and shell thicknesses, metal nanoshells can be fabricated that will absorb or scatter light at any wavelength across the entire visible and infrared range of the electromagnetic spectrum.

Metamorphic: In medical nanorobotics, capable of adopting multiple physical configurations via smooth changes from one configuration to another.

Micellar nanocontainers: Block copolymer micelles are water- soluble biocompatible nanocontainers with great potential for delivering hydrophobic drugs. An understanding of their cellular distribution is essential to achieving selective delivery of drugs at the subcellular level. R. Savic, L. Luo, A. Eisenberg, D. Maysinger, Micellar nanocontainers distribute to defined cytoplasmic organelles Science 300 (5619): 615- 618, Apr. 25, 2003

Micelles: Micelles are small, spherical structures composed of molecules that attract one another to reduce surface tension. The head of the molecule is hydrophilic, meaning it likes water, while the interior portion is hydrophobic, meaning it avoids water.

Microbiotagraphics: Mapping the microbiotic populations present in the human body.

Microbubbles: Very small encapsulated gas bubbles (diameters of micrometers) that can be used in diagnostic and therapeutic applications. Upon exposure to sufficiently intense ultrasound, microbubbles will cavitate, rupture, disappear, release gas content, etc. Such characteristics of the microbubbles can be used to enhance diagnostic tests, dissolve blood clots, and deliver drugs or genes for therapy. MeSH 2004

Microchemistry: The development and use of techniques and equipment to study or perform chemical reactions, with small quantities of materials, frequently less than a milligram or a milliliter. MeSH 2003

Microdevices: Narrower terms: integrated microdevices, nanodevices, SED Single Electron Devices. Related terms microelectronics, microfluidics, micro- TAS.

Micro-Electromechanical Systems (MEMS): "The fabrication or micro-machining of materials to make stationary and moving structures, devices and systems of a nominal size scale from a few centimeters to a few micrometers."

Microelectronics: Narrower terms: MEMS, nanoelectronics, optoelectronics, SED Single Electron Devices. Related terms: molecular electronics, semiconductors

Microencapsulation: Individually encapsulated small particles.

Microfabrication: This technology includes techniques used to manufacture integrated circuits (ICs), discrete microelectronic devices, MEMS devices such as sensors and actuators, and various electro-optic devices. [University of Louisville Microfabrication Lab] is currently serving as a center for research activity in the areas of micromachined sensors and actuators, electro- optic devices, special- purpose microelectronic devices, planar waveguides, chemical transducers, microstrip and microgap radiation detectors, micromachined nozzles, and micromachined ink- jet printheads. Microfabrication holds the key to developing monolithically integrated devices capable of extremely high throughput. ... Microfabrication of surface characteristics provides a means of controlling the cell-environment interaction. Microfabricated structures also present new opportunities for cell, protein and nucleic acid separations and analysis- including high throughput single protein molecule sequencing. New approaches to analysis utilizing conductance techniques or tunneling probe microscopy merely emphasize the breadth of the potential of nanofabrication for future technology development relevant to biological issues. [National Center for Research Resources "Integrated Genomics Technologies Workshop Report" Jan 1999]

Microfluidics: A multidisciplinary field comprising physics, chemistry, engineering and biotechnology that studies the behavior of fluids at volumes thousands of times smaller than a common droplet. Microfluidic components form the basis of so-called "lab-on-a-chip" devices that can process microliter and nanoliter volumes and conduct highly sensitive analytical measurements. The fabrications techniques used to construct microfluidic devices are relatively inexpensive and are amenable both to highly elaborate, multiplexed devices and also to mass production. In a manner similar to that for microelectronics, microfluidic technologies enable the fabrication of highly integrated devices for performing several different functions on the same substrate chip. Microfluidics is a critical component in gene chip and protein chip development efforts. The study of fluid channels and chambers of tiny dimensions of tens to hundreds of micrometers and volumes of nanoliters or picoliters. This is of interest in biological MICROCIRCULATION and used in MICROCHEMISTRY and INVESTIGATIVE TECHNIQUES [MeSH 2004] Enables the fabrication of networks of channels, chambers, and valves for the flow of liquids as minute as one picoliter. These systems have no moving parts and require little assembly. [Coventor "About Coventor" 2002] Within the microelectromechanical systems (MEMS) and biological and chemical detection communities, *microfluidics* refers to the research and development of microscale devices that handle very small volumes of fluids (nano- and picoliter volumes). These devices may perform tasks such as DNA analysis or the separation of human blood cells. [Josh Molho "What is microfluidics?" Stanford Univ., US, 2000]

Microinjection: The insertion of a substance into a cell through a microelectrode. Typical applications include the injection of drugs, histochemical markers (such as horseradish peroxidase or lucifer yellow) and RNA or DNA in molecular biological studies. To extrude the substances through the very fine electrode tips, either hydrostatic pressure (pressure injection) or electric currents (ionophoresis) is employed. [OMD] A technique for introducing a solution of DNA, protein, or other soluble material into a cell using a fine microcapillary pipet. [Life Sciences Dictionary]

Micromachines: Mechanical objects that are fabricated in the same general manner as integrated circuits. They are generally considered to be between 100 nanometres to 100 micrometres in size, though that is debatable. The applications of micromachines include accelerometers that detect when a car has hit an object and trigger an airbag. Complex systems of gears and levers are another application. The fabrications of these devices is usually done by one or both of two techniques: surface micromachining and bulk micromachining.

Micromachining: Techniques used to fabricate **MEMS**. **Broader term: microengineering**

Micron: 10^{-6} Symbol is υ. One millionth of a meter, or about 1/25,000 of an inch.

Microparticles: Applications include calibration of *flow cytometers*, particle and hematology analyzers, *confocal laser scanning microscopes* and zetapotential measuring instruments; flow measurements in gases and liquids like Laser Doppler Anemometry (LDA); Particle Dynamics Analysis (PDA), and Particle Image Velocimetry (PIV); medical diagnostics; separation phases for *chromatography*; support for immobilized enzymes; spacer in liquid crystal displays (LCD's); peptide synthesis; cell separation; *tracers* in environmental science; model systems in medicine, biochemistry, colloid chemistry, and aerosol research. [Microparticles GmbH, Berlin Germany]

Microphone: Device that produces voltage or current in response to a sound wave.

Microprocessor: Chip containing the logical elements for performing calculations and carrying out stored instructions.

Microspectrophotometry: Analytical technique for studying substances present at enzyme concentrations in single cells, in situ, by measuring light absorption. Light from a tungsten strip lamp or xenon arc dispersed by a grating monochromator illuminates the optical system of a microscope. The absorbance of light is measured (in nanometers) by comparing the difference between the image of the sample and a reference image. MeSH 1990

Microspheres: Microspheres are generally defined as small spheres made of any material and sized from about 0.5 µm to 100 µm. Similar, but smaller spheres sized 10 to 500 nm are called *nanospheres*. Ideally, microspheres are completely spherical and homogeneous in size, although less perfect particles are often termed microspheres as well. Depending on the preparation method and material used, microspheres show a typical size distribution which often deviates from the mono-sized ideal.

Microstructure: In material engineering the structural features of a material such as grain boundaries, grain size and structure, subject to observation under a microscope, selective etching etc. In MEMS microstructure unfortunately is also used to designate a micromachined feature. The last decade has seen rapid developments in the fabrication, characterization and conceptual understanding of synthetic microstructures in many different material systems including silicon, III-V and II-VI semiconductors, metals, ceramics and organics. The objective of this journal [Superlattices and Microstructures] is to provide a common interdisciplinary platform for the publication of the latest research results on all such "nanostructures" with dimensions in the range of 1 - 100 nm; the unifying theme here being the dimensions of these artificial structures rather than the material system in which they are fabricated.

Microsystem: A microscale machine that can sense information from the environment and act accordingly. Outside the U.S., it can also refer to microelectromechanical systems (MEMS). [smalltimes glossary, 2002]

MicroTas, micro Total Analysis Systems, uTAS: Although initial research dates back to the early 1970's, the field of micro- TAS formally started in 1990, when Manz et al described the possibility of creating microsystems that would take care of many or all the traditional analytical steps involved in a biochemical analysis (sample introduction, handling, extraction, purification, concentration, filtration, analysis, detection) .... Micro-TAS offer many advantages over traditional analysis systems. Low power consumption and small reaction volumes, faster analysis, ultrasensitive detection, and minimal human intervention are key parameters in the development of micro- TAS. Most biochemical reactions take place in liquid environments. Hence, the development of MicroTAS is intrinsically linked to the design of liquid handling micro- devices. [Biomedical Applications Group (GAB) Centro Nacional de Microelectronica (CNM- IMB) Bellaterra, Spain, 2000]

Microtubule: A hollow cylindrical tube used for the transport of materials inside cells.

Miller indices: A set of 3 integers (4 for hexagonal) that designate crystallographic planes, as determined from reciprocals of fractional axial intercepts.

Millimeter: One thousandth of a meter, or about 1/26 of an inch.

MIMIC: [Micromoulding in capillaries] one-step rapid prototyping technique.

Miniaturization: Desirable for many technologies for overall cost reduction (including reduction in the amount of reagents and analytes). Important to remember that building space is often the least available and most expensive component of an overall laboratory budget.

MIPS: Millions of Instructions per Second — a measure of computing power.

Misreaction: A chemical reaction that fails by yielding an unwanted product.

MM2/CSC: A molecular mechanics program developed by Cambridge Scientific Computing that closely follows the MM2 model, adding a graphical user interface and other features.

MM2: A molecular mechanics program developed by Norman Allinger and coworkers; the "MM2 model" is the molecular potential energy function described by the equations, rules, and parameters embodied in that program.

MNT: An abbreviation for molecular nanotechnology; refers to the concept of building complicated machines out of precisely designed molecules.

Mobility (electron, and hole): The proportionality constant between the carrier drift velocity and applied electric field.

Modulus of elasticity (E) : The ratio of stress to strain when deformation is totally elastic. Also the Young's modulus.

MOEMS MicroOpticalElectroMechanical systems: In addition to mechanical and electrical components, integrate waveguides or other optical features into the body of the silicon chip. Intellisense Corp. MEMS The Next Small Thing Google = about 10 Aug. 8, 2002; about 12 June 23, 2004; about 10 May 2, 2005

Moiety: A portion of a molecular structure having some property of interest.

Molality: The molality or molal concentration (symbol m) is the amount of substance per unit mass of solvent or mol kg^{-1}. Concentration in a liquid solution (symbol c), in terms of the number of moles of a solute dissolved in 10^6 mm^3 (10^3 cm^3) of solution in mol l^{-1}.

Molding (plastics): Shaping a plastic material by forcing it, under pressure at a high temperature, into a mold cavity.

Mole: A number of instances of something (typically a molecular species) equaling ~6.022 1023. *Mole* ordinarily means gram-mole; a kilogram-mole is ~6.022 1026.

Molecular - Nanotechnology: The technology of precisely- constructed molecular-scale machines; from nanometer: a billionth of a meter. For a good introduction, see What is Nanotechnology? or Ralph C. Merkle's nanotechnology page.

Molecular assembler: A general-purpose device for molecular manufacturing, able to guide chemical reactions by positioning individual molecules to atomic accuracy (e.g. mechanosynthesis) and to construct a wide range of useful and stable molecular structures according to precise specifications. Also known as an assembler, a molecular assembler is a molecular machine that can build a molecular structure from its component building blocks. [ZY]

Molecular Beam Epitaxy: [MBE] Process used to make compound (multi-layer) semiconductors. Consists of depositing alternating layers of materials, layer by layer, one type after another (such as the semiconductors gallium arsenide and aluminum gallium arsenide).

Molecular Integrated Microsystems (MIMS): microsystems in which functions found in biological and nanoscale systems are combined with manufacturable materials. See Molecular Integrated Microsystems

Molecular machine: A mechanical device that performs a useful function using components of nanometer scale and defined molecular structure; includes both artificial nanomachines and naturally occurring devices found in biological systems. Any machine with atomically precise parts of nanometer dimensions; can be used to describe molecular devices found in nature.

Molecular Recognition: A chemical term referring to processes in which molecules adhere in a highly specific way, forming a larger structure; an enabling technology for nanotechnology.

Molecular robotics: Ralph Merkle, A New Family of Six Degree of Freedom Positional Devices, Zyvex.

Molecular self-assembly: The spontaneous formation of molecules into covalently bonded, well- defined, stable structures - is a very important concept in biological systems and has increasingly become a focus of non- biological research.

Molecular sorting rotor: A class of nanomechanical device capable of selectively binding (or releasing) molecules from (or to) solution, and of transporting these bound molecules against significant concentration gradients.

Molecular surgery (molecular repair): In medical nanorobotics, the analysis and physical correction of molecular structures in the body using medical nanomachines. Analysis and physical correction of molecular structures in the body using medical nanomachines.

Molecular Systems Engineering: Design, analysis, and construction of systems of molecular parts working together to carry out a useful purpose.

Molecular Wire: A molecular wire - the simplest electronic component - is a quasi-one-dimensional molecule that can transport charge carriers (electrons or holes) between its ends. [Michael D Ward]

Molecularly imprinted polymers MIPs: A new class of materials that have artificially created receptor structures. Since their discovery in 1972, MIPs have attracted considerable interest from scientists and 3engineers involved with the development of chromatographic absorbents, membranes, sensors and enzyme and receptor mimics.

MUX: Device for combining several signals or data streams into a single flow.

Myosin: Myosin is one of the two major proteins responsible for contraction of muscle. In muscle cells, myosin is arranged in long filaments that lie parallel to the microfilaments of actin. In muscle contraction, filaments of actin and myosin alternately unlink and chemically link in a sliding action. The energy for this reaction is supplied by adenosine triphosphate.

NAD+ (nicotine adenine dinucleotide)**:** Coenzyme that participates in an oxidation reaction by accepting a hydride ion (H-) from a donor molecule. The NADH formed is an important carrier of electrons for oxidative phosphorylation. Coenzyme closely related to NAD+ that is used extensively in biosynthetic, rather than catabolic, pathways.

Nanapheresis: In medical nanorobotics, the removal of bloodborne medical nanorobots from the body using aphersis-like processes.

Nanarchist: Someone who circumvents government control to use nanotechnology, or someone who advocates this. [Eli Brandt, October 1991]

Nanite: Machines with atomic-scale components. (Popularized by the Star Trek episode "Evolution") As to their weight, a popular question: "Do you 'feel' heavier after you drink a mouthful of water? A mouthful of water, roughly 5 cm^3, would have the same mass as a ~2 terabot (2 trillion nanites) dose of 1 micron^3 nanorobots. You'll never feel it."

Robert A. Freitas Jr. “Nanobot” and “Nanorobot” usually mean the same thing. Slang term for a nanomachine (see nanotechnology), esp. a machine able to replicate itself. [Popularized by the Star Trek episode “Evolution”]

Nano Cubic Technology: An ultra-thin layer coating that results in higher resolution for recording digital data, ultra-low noise and high signal-to-noise ratios that are ideal for magneto-resistive (MR) heads. It is capable of catapulting data cartridge and digital videotape to one-terabyte native (uncompressed) capacities and floppy disk capacities to three gigabytes. To help visualize the potential, 1TB can store up to 200 two-hour movies.

Nanoarray: an ultra-sensitve, ultra-miniaturized array for biomolecular analysis. BioForce Nanosciences' Nanoarrays utilize approximately 1/10,000th of the surface area occupied by a conventional microarray, and over 1,500 nanoarray spots can be placed in the area occupied by a single microarray domain.

Nanoassembler: the Holy Grail of nanotechnology; once a perfected nanoassembler is availble, building anything becomes possible, with physics and the imagination the only limitation (of course each item would have to be designed first, which is another small hurdle). See Bootstrapping a Nanofactory “So how hard is it to build a nanofactory? You need to start with a working fabricator, a nanoscale device that can combine individual molecules into useful shapes. But once you have that, the rest is pretty straightforward.” —CRN

Nanobarcode: SurroMed's Nanobarcode™ technology uses cylindrically-shaped colloidal metal nanoparticles, in which the metal composition can be alternated along the length and the size of each metal segment can be controlled. Intrinsic differences in reflectivity between the metal segments allow individual particles to be identified by conventional optical microscopy.

Nanobeads: Polymer beads with diameters of between 0.1 to 10 micrometers. Also called nanodots, nanocrystals and quantum beads. *Impregnating fluorescent crystal chips into these beads allows simultaneous measurement of thousands of biological interactions, a stepping stone for breakthroughs in the diagnosis and treatment of disease. ... with the potential to accelerate drug discovery and clinical diagnostics.”*

Nanobot: Machines with atomic-scale components. (Popularized by the Star Trek episode “Evolution”) As to their weight, a popular question: “Do you ‘feel’ heavier after you drink a mouthful of water? A mouthful of water, roughly 5 cm^3, would have the same mass as a ~2 terabot (2 trillion nanites) dose of 1 micron^3 nanorobots. You'll never feel it.” Robert A. Freitas Jr. “Nanobot” and “Nanorobot” usually mean the same thing. Slang term for a nanomachine.

Nanobubbles: tiny air bubbles on colloid surfaces. Thought to reduce drag, such as would be of benefit to swimmers wearing a suit coverd in them.

Nanocantilever: The simplest micro-electro-mechanical system (MEMS) that can be easily machined and mass-produced via the same techniques used to make computer chips. The ability to detect extremely small displacements make nanocantilever beams an ideal device for detecting extremely small forces, stresses and masses. Nanocantilevers coated with antibodies, for example, will bend from the mass added when substrate binds to its antibody, providing a detector capable of sensing the presence of single molecules of clinical importance.

Nanocell Surgery: Modifying cellular structures by the use of medical nanomachines.

Nanocentrifuge: In medical nanorobotics, a proposed nanodevice that can spin materials at very high speed, imparting rotational accelerations of up to one trillion gravities (g's) thus permitting rapid sortation.

Nanocentrifuge: In medical nanorobotics, a proposed nanodevice that can spin materials at very high speed, imparting rotational accelerations of up to one trillion gravities (g's), thus permitting rapid sortation.

Nanochemistry: The goal of the Nanochemistry group is to develop chemistry beyond the realm of individual molecules by exploring how systems of several molecules or other electronically confined entities self-organize into distinct structures on the nanometer scale. The new insight is used to study and manipulate biological nanosystems and to create novel nanoarchitectures with unique structural, electronic, optical and magnetic properties. Specific research areas include bionanotechnology and self- assembling molecular electronics.

Nanochip: In Friday's issue [Aug. 1, 2001] of the journal *Science,* physicists from IBM's Thomas J. Watson Research Center [Philip G. Collins, Michael S. Arnold and Phaedon Avouris] announce their fabrication of the world's first array of transistors made from carbon nanotube. We are approaching the limits of standard microchip technology; thus, the "nanochip" — a next-smaller microchip. [ed] They are also a next-gen device for mass storage, of significantly higher density, with greater speed, and much lower cost.

Nanochondria: Nanomachines existing inside living cells, participating in their biochemistry (like mitochondria) and/or assembling various structures.

Nanochronometer: In medical nanorobotics, a proposed clock or timing mechanism constructed of nanoscale components.

Nanocircles: Stanford scientists have synthesized a molecule of DNA that is capable of shutting off specific genes in living bacteria. Dubbed the "nanocircle," the new nanometer-size molecule might one day give researchers the ability to target harmful genes that cause cancer and other diseases in humans. A nanocluster or nanocrystal is a fragment of solid comprising somewhere between a few atoms to a few tens of thousands of atoms. Nanoclusters are therefore a novel state of matter with properties that are neither those of a bulk crystal nor those of individual atoms and molecules. Over the past 10 years huge advances have been made both in the synthesis of size- tunable, monodisperse (i.e. size- selected) nanoclusters of various chemical compositions and in the development of techniques for their assembly into nanostructured solids (facilitating the synthesis of what have been termed "designer materials"). Nanoparticles, Nanoclusters, Nanoscience Group, Nottingham Univ., UK, 2002

Nanocomputer: A computer made from components (mechanical, electronic, or otherwise) built at the nanometer scale. These computers could be many orders-of-magnititude faster than today's, which enables software to take proportional leaps. (2) A computer whose fundamental components measure only a few nanometers in size. State of the art current computer components are no smaller than about 350 nm. [Mnemosyne Mnews 21 (3) January 2001 " The Nanotechnology Initiative and Future Electronics" Presentation by Gail J. Brown, Air Force Research Laboratory, Wright- Patterson Air Force Base, November 16, 2000] (3) A computer with parts built on a molecular scale.

Nanocones: Nonplanar graphitic structures. Carbon-based structures with five-fold symmetry that form due to disclination defects in two-dimensional graphene sheets. They have been observed as nanotube caps and as freestanding structures. [North Carolina State University].

Nanocontainers: "Micellar nanocontainers" or "Micelles," these are nanoscale polymeric containers that could be used to selectively deliver hydrophobic drugs to specific sites within individual cells.

Nanocrit (Nct): In medical nanorobotics, volume-fraction or bloodstream concentration of medical nanorobots, expressed as a percentage.

Nanomachining: like traditional machining, where portions of the structure are removed or modified, nanomachining involves changing the structure of nano-scale materials or molecules.

Nanomanipulation: The process of manipulating items at an atomic or molecular scale in order to produce precise structures.

Nanosieving: In medical nanorobotics, a nanodevice that can sort molecules or other nanoscale objects by physical sieving.

Nanosieving: In medical nanorobotics, a nanodevice that can sort molecules or other nanoscale objects by physical sieving.

Nanosome: Nanodevices existing symbiotically inside biological cells, doing mechanosynthesis and disassembly for it and replicating with the cell. Similar to nanochondria. [AS January 1998]

Nanosome: Nanodevices existing symbiotically inside biological cells, doing mechanosynthesis and disassembly for it and replicating with the cell. Similar to nanochondria. [Anders Sandberg January 1998]

Nanoswarm: UFog and Goo

Nanosystem: A eutactic set of nanoscale components working together to serve a set of purposes; complex nanosystems can be of macroscopic size.

Nanotech: Slang for nanotechnology.

Nanotechism: the religion of nanotech, as opposed to the science of nanotech

Nanoterrorism: using MNT derived nanites to do damage to people or places.

Nernst equation: An equation relating the potential of an electrochemical cell to the concentrations of the cell components: $E = Eo + RT/zF \ln C_1/C_2$ with z the charge exchanged at the electrode and C_1 and C_2 concentrations of two electro-active compounds.

Neroprosthesis: Implanted cybernetic brain augmentation.

Neural Simulation: Imitating the functions of a neural system—such as the brain—by simulating the function of each cell.

Neurocomputation: The study of how natural and artificial neural networks process information.

Neuron Star: A neutron star used as a basis for a densely packed mind (similar to a jupiter brain or a omegon) exploiting neutronium or quark matter for computa on [Damien Broderick, 1996]

Neuron: A nerve cell, such as those found in the brain.

Neuronaut: A person who explores their own neural functioning and internal mentational processes by various means, including deep introspection and meditation, psychoactive drugs, mind machines, and neuroscientific understanding.

Neurosuspension: Cryonically suspending only the head or brain of a person.

NMOS: An acronym for n-*channel metal-oxide-semiconductor*, as in NMOS transistor and NMOS logic.

Nono: A prefix meaning ten to the minus ninth power, or one billionth.

Nootropic: A cognition-enhancing drug that has no significant side-effects.

NOR: NOT-OR — logic gate whose output is the negation of that of the OR gate.

NOT: Logic gate whose output is binary 1 when its input is 0, and whose output is a 0 when its input is a 1.

Now shock: Being shocked or confused by the rapid changes that has already taken place, a kind of future shock before the present. [Michael Rothschild 1995, "Cornucopia Or Black Hole?", *Upside*]

NRAM™ - Nanotube-based/Nonvolatile RAM, developed by Nantero, using proprietary concepts and methods derived from leading-edge research in nanotechnology.

n-type: Characterizes a semiconductor containing predominantly mobile electrons (also see"p-type").

Nuclear Magnetic Resonance (NMR): An analytical method that can detect subatomic and structural information of molecules by measuring the adsorption of radio-frequency electromagnetic radiation by nuclei under the influence of a magnetic field.

Nucleation: The initial stage in a phase transformation, evidenced by the formation of small particles (nuclei) of the new phase, which are capable of growing.

Nucleotide: A monomer of the nucleic acids composed of a five-carbon sugar, a nitrogen-containing base, and phosphoric acid. A small molecule composed of three parts: a nitrogen base (a purine or pyrimidine), a sugar (ribose or deoxyribose), and phosphate. Nucleotides serve as the building blocks of nucleic acids (DNA and RNA).

Nucleus: In biology, a structure in advanced cells that contains the chromosomes and apparatus to transcribe DNA into RNA. In physics, the small, dense core of an atom. The positively charged core of an atom, an object of ~0.00001 atomic diameters containing >99.9% of the atomic mass. Nuclear positions define atomic positions.

Nutraceutical: A foodstuff or parts of foodstuff that offers medical or health benefits related to the prevention and treatment of disease.

Ohmmeter: Tool for measuring electrical resistance.

Okazaki fragments: Short lengths of DNA produced on the lagging strand during DNA replication, discovered by R. Okazaki. They are rapidly joined by DNA ligase to form a continuous DNA strand.

Peptide nanotubes: NANOTUBES formed from cyclic peptides (PEPTIDES, CYCLIC). Alternating D and L linkages create planar rings that self assemble by stacking into nanotubes. They can form pores through CELL MEMBRANE causing damage. MeSH 2004

Permeability [µ]: From the relation between magnetic induction and magnetic field (B = µH); for free space, $\mu_0 = 1.26 \times 10^{-6}$ H/m.

Permitivity [å]: From the relation between polarization charge and electric field; for free space, $å_0 = 8.85 \times 10^{-12}$ F/m.

Peroxisome: Small membrane-bounded organelle that uses molecular oxygen to oxidize organic molecules. Contains some enzymes that produce and others that degrade hydrogen peroxide (H2O2)

Phagocyte: General term for a professional phagocytic cell - that is, a cell such as a macrophage or neutrophil that is specialized to take up particles and microorganisms by phagocytosis.

Phagocytosis: Process by which particulate material is endocytosed ("eaten") by a cell. Prominent in carnivorous cells, such as *Amoeba proteus*, and in vertebrate macrophages and neutrophils. (From Greek *phagein*, to eat.)

Pharmacyte: In medical nanorobotics, a theorized (nanorobotic) device capable of delivering precise doses of biologically active chemicals to individually-addressed human body tissue cells (e.g. cell-by-cell drug delivery).

Phase shift: A time difference between the input and output signals.

Phase space: A classical system of *N* particles can be described by its 3*N* position and 3*N* momentum coordinates. The phase space associated with the system is the 6*N* dimensional space defined by these coordinates.

Phase transformation: A change in the number and/or character of the phases that make up the microstructure of an alloy.

Photolithography: Writing or creating patterns by means of light.

Photon: A particle of light. The energy E of a photon is given by the Einstein relation, $E = hf$. Here f is the frequency of the light and h is Planck's constant. The momentum p of a photon is given by the de Broglie relation, $p = h/ë$. Here ë is the wavelength of the light.

Photonic Band Gap (PBG): An energy range (and corresponding wavelength range) for which a material neither absorbs light nor allows light propagation.

Phototransistor: Transistor that, when powered, produces amplified voltage (current) in response to illumination.

Pico Technology: (trillionth of a meter) — The next step smaller, after Nano-technology. The art of manipulating materials on a quantum scale.

Piezoelectric material: A ferroelectric material in which an electrical potential difference is created due to mechanical deformation, or conversely, in which the application of a voltage causes dimensional changes in the material.

Piezoelectric: A piezoelectric material expands or contracts when a voltage is applied across it. Piezoelectric tubes are often used to move the tip in an AFM or STM. If a piezoelectric crystal is compressed, a voltage appears across it. This effect is sometimes used to measure displacements. A commonly used piezoelectric is PZT (Pb,Zr)TiO_3.

Pinhole: The term pinhole embraces a wide variety of oxide defects and is used in a broad sense today. Listed in this category are cracks caused by thermal contraction after oxidation or by handling, and regions of oxide with low dielectric strength caused by dust particles, inadequate masking, contamination, or poor resist adhesion.

Pin-out: Diagram showing for electronic components the relations between connecting pins and internal components.

Pitting: A form of very localized corrosion wherein small pits or holes form, usually in a vertical direction.

Pixel: Picture Element — smallest element of an image, such as a dot on a computer monitor screen.

pK value: A measure of the strength of an acid on a logarithmic scale. The pK value is given by log10 (1/Ka), where Ka is the acid dissociation constant pK values often are used to compare the strengths of different acids.

Plastic deformation: Permanent or nonrecoverable deformation, accompanied by permanent atomic displacement.

Plasticizer: A low molecular weight polymer additive that enhances flexibility and workability and reduces stiffness and brittleness.

Point defect: A crystalline defect associated with one or several atomic sites.

Poisson's ratio (í): For elastic deformation, the negative ratio of lateral and axial strains that result from an applied axial stress.

Poisson's ratio: A bar of an isotropic, elastic material ordinarily shrinks laterally when it is stretched longitudinally. The lateral contracting strain divided by the applied tensile strain is Poisson's ratio, which varies from material to material.

Polarization (P): The total electric dipole moment per unit volume of dielectric material.

Potentiometric Device: Monitors the voltage between a sensing electrode and a reference electrode. A high input impedance voltmeter is used to minimize current flow. The voltage typically is proportional to the logarithm of the analyte concentration.

Power [W]: Product of voltage and current in a component; also, refers to the field of electric energy supply.

Precision: The degree of reproducibility among several independent measurements of the same true value under specified conditions.

Presentation semaphore: In medical nanorobotics, a mechanical device used to display specific antigens, chemical ligands, or other molecular objects to the external environment, with the purpose of selectively modifying the chemical or other surface characteristics of a nanorobot exterior.

Printed circuit board: PCB — selectively metallized insulating sheet for supporting and interconnecting circuit components.

Probability density function: Consider an uncertain physical property and a corresponding space describing the range of values that the property can have (e.g., the configuration of a thermally excited N particle system and the corresponding $3N$ dimensional configuration space). The probability density function associated with a property is defined over the corresponding space; its value at a particular point is the probability per unit volume that the property has a value in an infinitesimal region around that point.

Protein Design, Protein Engineering: The design and construction of new proteins; an enabling technology for nanotechnology.

Protein Folding: "The process by which proteins acquire their functional, preordained, three-dimensional structure after they emerge, as linear polymers of amino acids, from the ribosome." [The Scientist]

Protein: Living cells contain many molecules that consist of amino acid polymers folded to form more-or-less definite three-dimensional structures; these are termed proteins. Short polymers lacking definite three-dimensional structures are termed peptides. Many proteins incorporate structures other than amino acids, either as covalently attached side chains or as bound ligands. Molecular objects made of protein form much of the molecular machinery of living cells.

Proteomics: The study of the full expression of proteins in an organism The term proteome refers to all the proteins expressed by a genome, and thus proteomics involves the identification of proteins in the body and the determination of their role in physiological and pathophysiological functions. ... Ultimately it is believed that through proteomics new disease markers and drug targets can be identified that will help design products to prevent, diagnose and treat disease. [e-proteomics.net]

Proximal probes: A family of devices capable of fine positional control and sensing, including scanning tunneling and atomic force microscopes; an enabling technology for nanotechnology.

p-Type semiconductor: A semiconductor for which the predominant charge carriers responsible for electrical conduction are holes.

Pyroelectricity: The property of certain crystals, such as tourmaline, of acquiring opposite electrical charges on opposite faces when heated.

Q factor: A rating, applied to coils, capacitors, and resonant circuits, equal to the reactance divided by the resistance. The ratio of energy stored to energy dissipated per cycle in an electrical or mechanical system.

Quantization: The concept that energy can occur only in discrete units called quanta.

Quantum Computer: A computer that exploits quantum mechanical phenomena such as superposition and entanglement.

Quantum Confined Atoms (QCA): atoms caged inside nanocrystals. May find uses in clear-glass sunglasses, bio-sensors, and optical computing.

Quantum corral: Structures where electrons are corralled (confined) by atoms placed on the surface of a metal.

Quantum Cryptography: A system based on quantum- mechanical principles. Eavesdroppers alter the quantum state of the system and so are detected. Developed by Brassard and Bennett, only small laboratory demonstrations have been made.

Quaternary Structure: Three-dimensional relationship of the different polypeptide chains in a protein complex.

Qubit: The quantum computing analog to a bit. Qubits exhibit superposition. Thus, unlike normal bits, qubits can be both 1 and 0 at the same time.

Radiation damage: Chemical changes (bond breakage, ionization) caused by high-energy radiation (e.g., x-rays, gamma rays, high-speed electrons, protons, etc.).

Representative point: The point in a configuration space that represents the geometry of a system.

Reproducibility: A property of an experiment or process where one tends to receive consistent results from following a specific procedure.

Residual stress: A persisting stress in a material free of external forces or temperature gradients.

Resist: A material or coating that can protect a surface from chemical reactions.

Resistance [Ω-ohm]: Characteristic of a resistor: in a 1-ohm resistance a current of 1 A produces a voltage drop of 1 V.

Resistivity (ñ): The reciprocal of electrical conductivity, and a measure of a material's resistance to passing electric current.

Resistor: Energy dissipative element consisting of a poor conductor in series with connecting wires.

Resolution: The minimum distance between two objects that can be distinguished in microscopy or the minimum spacing between two features that can be fabricated with lithography. The smallest measurable change in input that will produce a small but noticeable change in the output. In the context of chemical separations, defines the completeness of separation.

Resonant frequency: The frequency at which a moving member or a circuit has a maximum output for a given input.

Respirocrit: In medical nanorobotics, the volume-fraction or bloodstream concentration of respirocyte nanorobots, expressed as a percentage.

Respirocyte: In medical nanorobotics, a theorized bloodborne spherical 1-micron (nanorobotic) device having a 1000-atm pressure vessel with active pumping powered by endogenous serum glucose, that serves as a mechanical artificial red blood cell.

Response time: The time it takes for the sensor's output to reach its final value. A measure of how quickly the sensor will respond to changes in the environment. In general, this parameter is a measure of the speed of the sensor and must be compared with the speed of the process.

Restriction Enzyme: An enzyme that cuts DNA at a specific site, allowing biologists to insert or delete genetic material.

Reverse bias: The insulating bias for a p-n junction rectifier; electrons flow into the p side of the junction.

RF: Radio Frequency — refers to alternating voltages and currents having frequencies between 9 kHz and 3 MHz.

Ribonuclease: An enzyme that cuts RNA molecules into smaller pieces.

SAMFET: (self assembled monolayer field effect transistor). Where a few molecules act as FETs, exhibiting both very strong gain, and extraordinarily rapid response. [Mark Ratner & MT 5(2) p. 20]

Sanguinatation: In medical nanorobotics, locomotion (especially swimming by a nanorobot) through the bloodstream.

Sarcomere: Repeating unit of a myofibril in a muscle cell, composed of an array of overlapping thick (myosin) and thin (actin) filaments between two adjacent Z discs.

Sarcoplasmic reticulum: Network of internal membranes in the cytoplasm of a muscle cell that contains high concentrations of sequestered Ca2+ : that is released into the cytosol during muscle excitation.

Satellite DNA: Regions of highly repetitive DNA from a eucaryotic chromosome, usually identifiable by its unusual nucleotide composition. Satellite DNA is not transcribed and has no known function.

Saturated molecule: Molecule containing carbon-carbon bonds that has only single covalent bonds.

Saturated: An organic molecule is described as saturated if it is a closed shell species lacking double or triple bonds; forming a new bond to a saturated molecule requires the cleavage of an existing bond.

Scanning Capacitance Microscopy (SCM): A variation of scanning force microscopy that maps the localized capacitance of a surface. A method for mapping the local capacitance of a surface. [NTN]

Scanning electron microscope (SEM): A microscope producing an image by using reflected electron beams that scan the surface of a specimen.

Scanning Electron Microscopy (SEM): A method of producing images of a surface by scanning an electron beam over the sample and measuring the electronic interactions with the interface.

Scanning Force Microscope (SFM): An instrument able to image surfaces to molecular accuracy by mechanically probing their surface contours. A kind of proximal probe. .... A device in which the deflection of a sharp stylus mounted on a soft spring is monitored as the stylus is moved across a surface. If the deflection is kept constant by moving the surface up and down by measured increments, the result (under favorable conditions) is an atomic-resolution topographic map of the surface. Also termed an atomic force microscope.

Scanning Near Field Optical Microscopy: A method for observing local optical properties of a surface that can be smaller than the wavelength of the light used. A type of scanning probe microscopy that maps the near-field optical properties of an interface.

Science Court: A name (originally applied by the media) for a government-conducted fact forum.

Science: The process of developing a systematized knowledge of the world through the variation and testing of hypotheses. The pursuit of knowledge and understanding, from the Latin term *scientia*, which means 'knowledge'.

Sealed assembler lab: A general-purpose assembler system in a container permitting only energy and information to be exchanged with the environment.

Sealed Assembler Laboratory: A work space, containing assemblers, encapsulated in a way that allows information to flow in and out but does not allow the escape of assemblers or their products.

Second messenger: Small molecule that is formed in or released into the cytosol in response to an extracellular signal and helps to relay the signal to the interior of the cell. Examples include cAMP, IP3 and Ca2+.

Secondary structure: Regular local folding pattern of a polymeric molecule; in proteins, a helices and b-pleated sheets.

Secretory vesicle: Membrane-bounded organelle in which molecules destined for secretion are stored prior to release. Sometimes called secretory granule because darkly staining contents make the organelle visible as a small solid object.

Selectivity: The ability of a sensor to measure only one metric or, in the case of a chemical sensor, to measure only a single chemical species.

Self-Assembled Monolayer (SAM): A two-dimensional film, one molecule thick, covalently assembled at an interface.

Self-Assembler: A specific type of assembler that makes use of self-assembly such that the assembly process would theoretically not require external energy or information input.

Serine/threonine kinase: Protein kinase that phosphorylates serines or threonines on its target protein.

Sertoli cell: Supporting cell of the mammalian testis that surrounds and nourishes developing sperm cells.

Sex chromosome: Chromosome that may be present or absent, or present in a variable number of copies, according to the sex of the individual; in mammals, the X and Y chromosomes.

Shape Memory Alloys: (SMA's) are a unique class of alloys which are able to "remember" their shape and are able to return to that shape even after being bent. The ability is known as the shape memory effect. ... This property has lead to many uses of SMA from orthodontics and coffee makers to methods of controlling aircraft and protecting buildings from earthquake damage. ... The first SMA to be discovered and the most commonly used is called Nitinol.

Shear modulus: Shear stress divided by shear strain; has the units of force per unit area.

Shear: A shear deformation is one that displaces successive layers of a material transversely with respect to one another, like a crooked stack of cards. Shear is a dimensionless quantity measured by the ratio of the transverse displacement to the thickness over which it occurs.

Species: In chemistry, a distinct kind of molecule, ion, or other structure.

Specific heat: The heat capacity per unit mass of material.

Specific Modulus (specific stiffness): The ratio of elastic modulus to specific gravity for a material.

Specific strength: The ration of tensile strength to specific gravity for a material.

Star Trek scenario: Someone builds potentially dangerous self-replicating devices that spread disastrously.

Steady-state diffusion: The diffusion condition for which there is no net accumulation or depletion of diffusing species. The diffusion flux is independent of time.

Step response: The response of a system to an instantaneous jump in the input signal.

Steric hindrance: Slowing of the rate of a chemical reaction owing to the presence of structures on the reagents that mechanically interfere with the motions associated with the reaction, typically by obstructing the reaction site.

Superconductor: An object or substance that conducts electricity with zero resistance.

Superintelligence: An intellect that is much smarter than the best human brains in practically every field, including scientific creativity, general wisdom and social skills. This definition leaves open how the superintelligence is implemented: it could be a digital computer, an ensemble of networked computers, cultured cortical tissue or what have you. It also leaves open whether the superintelligence is conscious and has subjective experiences. [Bostrom]

Superlattice Nanowire Pattern (transfer): [SNAP] a technique for producing "Ultra High Density Nanowire Lattices and Circuits".

Superlattice: Nanowire interwoven bundles of nanowires using substances with different compositions and properties.

Surfactant: Short for 'SURFace ACTive AgeNT' - a surfactant is a molecule that lowers suface tension.

Synapse: A structure that transmits signals from a neuron to an adjacent neuron (or other cell).

Synthesis: The production of a specific molecular structure by a series of chemical reactions.

Synthespian: An artificial actor, for example a 3D model animated by motion capture from a real actor or a computer program. [AS]

System: In scientific usage, usually equivalent to "a collection of matter and energy being analyzed as a unit." In engineering usage, usually equivalent to "a set of components working together to serve a set of purposes."

T lymphocyte (T cell): Type of lymphocyte responsible for cell-mediated immunity; includes both cytotoxic T cells and helper T cells.

TAB bonding: Tape automated bonding; semiconductor packaging technique that uses a tiny lead-frame to connect circuitry on the surface of the chip to a substrate instead of wire bonds.

TATA box: Consensus sequence in the promoter region of many eucaryotic genes that binds a general transcription factor and hence specifies the position where transcription is initiated.

Technocyte: A nanoscale artificial device (especially a nanite) in the human bloodstream used for repairs, cancer protection, as an artificial immune system or for other uses. [AS 1995]

Technofobics: Those who have a phobia to technology, and/or to advances in technology.

Technology: The application of knowledge and understanding in order to control processes and fabricate products, from the Greek root *techni*, meaning 'art'.

Telomere: End of a chromosome, associated with a characteristic DNA sequence that is replicated in a special way. Counteracts the tendency of the chromosome otherwise to shorten with each round of replication. (From Greek *telos*, end.)

Telophase: Final stage of mitosis in which the two sets of separated chromosomes decondense and become enclosed by nuclear envelopes.

Temperature sensitive (ts) mutant: Organism or cell carrying a genetically altered protein (or RNA molecule) that performs normally at one temperature but is abnormal at another (usually higher) temperature.

Temperature: A system in which internal vibrational modes have equilibrated with one another can be said to have a particular temperature. Two systems A and B are said to be at different temperatures if, when brought into contact, heat flows from (say) A to B, increasing the thermal energy of B at the expense of the thermal energy of A.

Template: An external source of information or structure often used in manufacturing and mass production. As a nanoscale example, biological systems are known to use nucleic acids as a templates for protein synthesis.

Templating: The directed assembly of materials through interaction with a template.

Tertiary structure: Complex three-dimensional form of a macromolecule, especially a protein.

Tesla [T]: Unit of magnetic induction: 1T = 1 weber/m^2 (also, 1T = 10^4 gauss).

Tetrahymena: Genus of ciliated protozoa used in studies of ciliary axonemes, self-splicing RNA, and telomere reproduction.

TFT: Thin film transistor.

The Atom: Atoms are the building blocks of matter. All things are made of arrangements of atoms. There are 118 (with some very short lived ones) different elements in the periodic table, which means that we know of 118 different kinds of atoms.

Thermal conductivity (ê): For steady-state heat flow, the proportionality constant between the heat flux and the temperature gradient. Also, a parameter characterizing the ability of a material to conducting heat.

Thermal energy: The internal energy present in a system as a result of the energy of thermally equilibrated vibrational modes and other motions (including both kinetic energy and molecular potential energy). The mean thermal energy of a classical harmonic oscillator is kT.

Vegetal pole: The end at which most of the yolk is located in an animal egg. The end opposite the animal pole.

Virtual Nanomedicine: Using VR to perform surgery and other functions inside the body.

Virtual reality system: A combination of computer and interface devices (goggles, gloves, etc.) that presents a user with the illusion of being in a three dimensional world of computer-generated objects.

Virus: A parasite (consisting primarily of genetic material) that invades cells and takes over their molecular machinery in order to copy itself. A small replicator consisting of little but a package of DNA or RNA which, when injected into a host cell, can direct the cell's molecular machinery to make more viruses. Particle consisting of nucleic acid (RNA or DNA) enclosed in a protein coat and capable of replicating within a host cell and spreading from cell to cell. Often the cause of disease.

Watt (W): Unit of power. One watt is the power that, in 1 second, gives rise to an energy of 1 joule.

Wave function: In quantum mechanics, a complex function extending over the configuration space of a system; its complex conjugate yields the probability density function, and other mathematical operations yield other physical quantities.

Weber: Unit of magnetic flux. One weber is a magnetic flux that, linking a circuit of 1 turn, would produce in it an electromotive force of 1 volt if it were reduced to zero at a uniform rate in 1 second.

Weight percent (wt%): Concentration specification on the basis of weight (or mass) of a particular element relative to the total alloy weight (or mass).

Western blotting: Technique by which proteins are separated and immobilized on a paper sheet and then analyzed, usually by means of a labeled antibody.

Bibliography

Baker, R.T.K. Synthesis, properties and applications of graphite nanofibers. In *R&D status and trends*, ed. Siegel et al.

Bate, R., Frazier, G., Frensley, W., Reed., M., "An Overview of Nanoelectronics," *Texas Instruments Technical Journal*, July-August 1989, pp. 13-20.

Braun, P.V., P. Osenar, and S.I. Stupp. 1996. *Nature*. 368: 2.

Bueno, O.: 2004, 'Von Neumann, Self-Reproduction and the Constitution of Nanophenomena', in: D. Baird, A. Nordmann & J. Schummer (eds.), *Discovering the Nanoscale*, Amsterdam: IOS Press, pp. 101-115.

Carroll, J.S.: 2001, 'Social Science Research Methods for Assessing Societal Implications of Nanotechnology', in: M.C. Roco and W.S. Bainbridge (eds.), *Societal Implications of Nanoscience and Nanotechnology*, Dordrecht: Kluwer, pp. 188-192.

Chen, C-Y., S.L. Burkett, H.-X. Li, and M.E. Davis. 1993. *Microporous Mater*. 2:27.

Chianelli, R.R. 1998. Synthesis, fundamental properties and applications of nanocrystals, sheets, and fullerenes based on layered transition metal chalcogenides. In *R&D status and trends,* ed. Siegel et al.

Chianelli, R.R., M. Daage, and M.J. Ledoux. 1994. *Advances in Catalysis* 40:177.

Dresselhaus, M., and G. Dresselhaus. 1995. *Ann. Rev. Mat. Sci*. 25:487.

Dresselhaus, M.S. 1998. Carbon-based nanostructures. In *R&D status and trends*, ed. Siegel et al.

Dresselhaus, M.S., G. Dresselhaus, and P. Eklund. 1996. *Science of fullerenes and carbon nanotubes*. San Diego: Academic Press.

Drexler, K.E.: 2004, 'Nanotechnology: From Feynman to Funding', *Bulletin of Science, Technology & Society*, 24, 21-27.

Dupuy, J.-P.: 2004, 'Complexity and Uncertainty: A Prudential Approach To Nanotechnology', in: European Commission (Community Health and Consumer Protection): *Nanotechnologies: A Preliminary Risk Analysis on the Basis of a Workshop, Brussels, 1-2 March 2004*, pp. 71-93 (www.europa.eu.int/comm/health/ph_risk/documents/ev_20040301_en.pdf).

Dupuy, J.-P.: 2004, 'Pour une évaluation normative du programme nanotechnologique', *Annales des Mines*, (February 2004), 27-32.

Einsiedel, E.F.; Goldenberg, L.: 2004, 'Dwarfing the Social? Nanotechnology Lessons from the Biotechnology Front', *Bulletin of Science, Technology & Society*, 24, 28-33.

Etzkowitz, H., 2001, 'Nano-Science and Society: Finding a Social Basis for Science Policy', in: M.C. Roco and W.S. Bainbridge (eds.), *Societal Implications of Nanoscience and Nanotechnology*, Dordrecht: Kluwer, pp. 121-128.

European Commission: 2004, *Converging Technologies: Shaping the Future of European Societies (Report of the High Level Expert Group "Foresighting the New Technology Wave")*, Brussels: European Commission Research.

F. Buot, "Mesoscopic Physics and Nanoelectronics: Nanoscience and Nanotechnology," *Physics Reports*, pp.73-174, 1993.

F. Capasso and S. Datta, "Quantum Electron Devices," *Physics Today*, pp.74-82, February 1990.

Feynman, R., "There's Plenty of Room at the Bottom: An invitation to Enter a New Field of Physics," Talk at the Annual Meeting of the American Physical Society, 29 December 1959. Reprinted in Appendix B of Crandall and Lewis. See citation above.

Fielder, F.A.; Reynolds, G.H.: 1994, 'Legal Problems of Nanotechnology: An Overview', *Southern California Interdisciplinary Law Journal*, 3, 593-629.

Fleischer, Th.; Decker, M. & Fiedeler, U. (eds.): 2004, *Große Aufmerksamkeit für kleine Welten ' Nanotechnologie und ihre Folgen*, special issue of *Technikfolgenabschätzung ' Theorie und Praxis*, 13 (2), 5-85 [www.itas.fzk.de/tatup/042/inhalt.htm].

Fogelberg, H. & Glimell, H.: 2003, 'Molecular Matters: In Search of the Real Stuff', in: H. Fogelberg & H. Glimell, *Bringing Visibility to the Invisible: Towards A Social Understanding of Nanotechnology*, Göteborg: Göteborg University, pp. 5-32 [www.sts.gu.se/publications/STS_report_6.pdf].

Glimell, H.: 2001, 'Dynamics of the Emerging Field of Nanoscience', in: M.C. Roco and W.S. Bainbridge (eds.), *Societal Implications of Nanoscience and Nanotechnology*, Dordrecht: Kluwer, pp. 156-160.

Glimell, H.: 2003, 'A Nano Narrative: The Mircopolitics of a New Generic Enabling Technology', in: H. Fogelberg & H. Glimell, *Bringing Visibility to the Invisible: Towards A Social Understanding of Nanotechnology*, Göteborg: Göteborg University, pp. 55-77 [www.sts.gu.se/publications/STS_report_6.pdf].

Glimell, H.: 2003, 'Dynamics of the Emerging Field of Nanoscience', in: H. Fogelberg & H. Glimell, *Bringing Visibility to the Invisible: Towards A Social Understanding of Nanotechnology*, Göteborg: Göteborg University, pp. 79-85 [www.sts.gu.se/publications/STS_report_6.pdf].

Glimell, H.: 2004, 'Grand Visions and Lilliput Politics: Staging the Exploration of the 'Endless Frontier'', in: D. Baird, A. Nordmann & J. Schummer (eds.), *Discovering the Nanoscale*, Amsterdam: IOS Press, pp. 231-246.

Goldhaber-Gordon, D., Montemerlo, M.S., Love, J.C., Opiteck, G.J., and Ellenbogen, J.C., "Overview of Nanoelectronic Devices," submitted to the *Proceedings of the IEEE*, February 1997. For more information, please send e-mail to nanotech@mitre.org.

Gorman, M.E.; Groves, J.F. & Shrager, J.: 2004, 'Societal Dimensions of Nanotechnology as a Trading Zone: Results from a Pilot Project', in: D. Baird, A. Nordmann & J. Schummer (eds.), *Discovering the Nanoscale*, Amsterdam: IOS Press, pp. 63-73.

Grinbaum, A. & Dupuy, J.-P.: 2004, 'Living with Uncertainty: Toward the Ongoing Normative Assessment of Nanotechnology', *Techne: Research in Philosophy and Technology*, 8(3) (forthcoming).

Grinbaum, A.: 2004, 'La condition de l'homme moderne et les nanotechnologies', in: G. Nivat (ed.), *Les limites de l'humain. 39èmes Rencontres Internationales de Genève*, L'Age d'Homme, Genève, p. 141.

Grunwald, A.: 2004, 'Ethische Aspekte der Nanotechnologie. Eine Felderkundung', *Technikfolgenabschätzung ' Theorie und Praxis*, 13 (2), 71-78.

Gupta, V.K. & Pangannaya, N.B.: 2000, 'Carbon nanotubes: bibliometric analysis of patents', *World Patent Information*, 22, 185-189.

Hansson, S.O.: 2004, 'Great Uncertainty about Small Things', *Techne: Research in Philosophy and Technology*, 8(3) (forthcoming).

Haruta, M. 1997. *Catalyst surveys of Japan* 1:61 and references therein.

Hennig, J.: 2004, 'Changes in the Design of Scanning Tunneling Microscopic Images from 1980 to 1990', *Techne: Research in Philosophy and Technology*, 8(3) (forthcoming).

Hessenbruch, A.: 2004, 'Nanotechnology and the Negotiation of Novelty', in: D. Baird, A. Nordmann & J. Schummer (eds.), *Discovering the Nanoscale*, Amsterdam: IOS Press, pp. 135-144.

Hullmann, A. & Meyer, M.: 2003, 'Publications and Patents in Nanotechnology. An overview of previous studies and the state of the art', *Scientometrics*, 58 (3) 507-527.

Huo, Q., D.I. Margolese, U. Ciesla, P. Feng, T.E. Gier, P. Sieger, R. Leon, P.M. Petroff, F. Schuth, and G.D. Stucky. 1994. *Nature* 368:317.

Jena, P., S.N. Khanna, and B.K. Rao. 1996. In *Science and technology of atomically engineered materials*, ed. P. Jena. River Edge, NJ: World Scientific.

Johansson, M.: 2003, 'Plenty of room at the bottom: Towards an anthropology of nanoscience', *Anthropology Today*, 19 (No. 6), 3-6.

Johnson, A.: 2004, 'The End of Pure Science: Science Policy from Bayh-Dole to the NNI', in: D. Baird, A. Nordmann & J. Schummer (eds.), *Discovering the Nanoscale*, Amsterdam: IOS Press, pp. 217-230.

Jones, C.W., and W.J. Koros. 1994. *Carbon* 32:1419.

Jounet, C., W.K. Maser, P. Bernier, A. Loiseau, M. Lamy de la Chapelle, S. Lefrant, P. Deniard, R. Lee, and J.E. Fischer. 1997. *Nature* 388:756.

K.E. Drexler, *Nanosystems: Molecular Machinery, Manufacturing, and Computation*, Wiley, New York, 1992.

Karger, J., and D.M. Ruthven. 1992. *Diffusion in zeolites*. New York: J. Wiley.

Keiper, A.: 2003, 'The Nanotechnology Revolution', *The New Atlantis*, 2, 17-34.

Khushf, G.: 2004, 'A Hierarchical Architecture for Nano-scale Science and Technology: Taking Stock of the Claims About Science Made By Advocates of NBIC Convergence', in: D. Baird, A. Nordmann & J. Schummer (eds.), *Discovering the Nanoscale*, Amsterdam: IOS Press, pp. 21-33.

Khushf, G.: 2004, 'Systems Theory and the Ethics of Human Enhancement: A Framework for NBIC Convergence', *Annals of the New York Academy of Sciences*, 1013, 124-149.

Krätschmer, W., L.D. Lamb, K. Fostiropoulos, and D.R. Huffman. 1990. *Nature* 347:354.

Kresge, C.T., M.E. Leonowicz, W.J. Roth, J.C. Vartuli, and J.S. Beck. 1992. *Nature* 359:710.

Kupperman, A., S. Nadimi, S. Oliver, G. Ozin, J. Garcés, and M. Olken. 1993. *Nature* 365:239.

Kuusi, O.; Meyer, M.: 2002, 'Technological generalizations and leitbilder ' the anticipation of technological opportunities', *Technological Forecasting & Social Change*, 69, 625-639.

Landon, B.: 2004, 'Less is More: Much Less is Much More: The Insistent Allure of Nanotechnology Narratives in Science Fiction', in: N.K. Hayles (ed.), *Nanoculture: Implications of the New Technoscience*, Bristol, UK: Intellect Books, pp. 131-146.

Lasžlo, P.: 2004, 'Is There Life After Partington?', *Hyle: International Journal for Philosophy of Chemistry*, 10(2), 169-178.

Lenhard, J.: 2004, 'Nanoscience and the Janus-Faced Character of Simulations', in: D. Baird, A. Nordmann & J. Schummer (eds.), *Discovering the Nanoscale*, Amsterdam: IOS Press, pp. 93-100.

Lent, C.S., Tougaw, P.D., Porod, W., Bernstein, G.H., "Quantum Cellular Automata," *Nanotechnology*, Vol. 4, p. 49, 1993.

Lewak, S.E.: 2004, 'What's the Buzz? Tell Me What's A-Happening: Wonder, Nanotechnology, and Alice's Adventures in Wonderland', in: N.K. Hayles (ed.), *Nanoculture: Implications of the New Technoscience*, Bristol, UK: Intellect Books, pp. 201-.

Lide, D.R., ed. 1993-1994. *CRC Handbook of Chemistry and Physics*, 74th ed.

Lin-Easton, P.C.: 2001, 'It's Time for Environmentalists to Think Small ' Real Small: A Call for the Involvement of Environmental Lawyers in Developing Precautionary Policies Molecular Nanotechnology', *Georgetown International Law Review*, 14, 106-134.

López, J.: 2004, 'Bridging the Gaps: Science Fiction in Nanotechnology', *Hyle: International Journal for Philosophy of Chemistry*, 10(2), 129-152.

Lösch, A.: 2004, 'Nanomedicine and Space: Discursive Orders of Mediating Innovations', in: D. Baird, A. Nordmann & J. Schummer (eds.), *Discovering the Nanoscale*, Amsterdam: IOS Press, pp. 193-202.

Marshall, K.: 2004, 'Atomizing the Risk Technology', in: N.K. Hayles (ed.), *Nanoculture: Implications of the New Technoscience*, Bristol, UK: Intellect Books, pp. 147-160.

Martin, T.P., N. Malinowski, U. Zimmerman, U. Naher, and H. Schaber. 1993. *J. Chem. Phys.* 99:4210.

Martin, T.P., U. Naher, H. Schaber, U. Zimmerman. 1993. *Phys. Rev. Lett.* 70:3079.

Mayer, S.: 2002, 'From genetic modification to nanotechnology: the dangers of 'sound science", in: T. Gilland (ed.), *Science: Can We Trust the Experts?*, London: Hodder and Stoughton, pp. 1-15.

Mehta, M.D.: 2002, 'Nanoscience and Nanotechnology: Assessing the Nature of Innovation in These Fields', *Bulletin of Science, Technology & Society*, 22(4), 269-273.

Mehta, M.D.: 2004, 'From Biotechnology to Nanotechnology: What Can We Learn from Earlier Technologies?, *Bulletin of Science, Technology & Society*, 24, 34-39.

Meyer, M. & Kuusi, O.: 2004, 'Nanotechnology: Generalizations in an Interdisciplinary

Field of Science and Technology', *Hyle: International Journal for Philosophy of Chemistry*, 10(2), 153-168.

Meyer, M.: 2000a, 'Does science push technology? Patents citing scientific literature', *Research Policy*, 29, 409-434.

Meyer, M.: 2000b, 'Patent citations in a novel field of technology: What can they tell about interactions of emerging communities of science and technology?', *Scientometrics*, 48, 151-178.

Meyer, M.: 2001a, 'Patent citations in a novel field of technology: An exploration of nano-science and nano-technology', *Scientometrics*, 51, 163-183.

Meyer, M.: 2001b, 'Socio-economic Research on Nanoscale Science and Technology: A European Overview and Illustration', in: M.C. Roco and W.S. Bainbridge (eds.), *Societal Implications of Nanoscience and Nanotechnology*, Dordrecht: Kluwer, pp. 217-241.

Milburn, C.: 2002, 'Nanotechnology in the age of post-human engineering: science fiction as science', *Configurations*, 10, 261-295 [reprinted in: N.K. Hayles (ed.), *Nanoculture: Implications of the New Technoscience*, Bristol, UK: Intellect Books, 2004, pp. 109-130].

Milburn, C.: 2004, 'Nano/Splatter: Disintegrating the Postbiological Body', *New Literary History*, 35 (in print).

Mnyusiwalla, A.; Abdallah, S.D.; Singer, P.A.: 2003, 'Mind the gap: science and ethics in nanotechnology', *Nanotechnology*, 14, R9-R13.

Mody, C.C.M.: 2004, 'How Probe Microscopists Became Nanotechnologists', in: D. Baird, A. Nordmann & J. Schummer (eds.), *Discovering the Nanoscale*, Amsterdam: IOS Press, pp. 119-133.

Mody, C.C.M.: 2004, 'Instruments in Training: The Growth of American Probe Microscopy in the 1980s', in: D. Kaiser (ed.), *Pedagogy and the Practice of Science: Producing Physical Scientists, 1800-2000*, Cambridge, MA: MIT Press (forthcoming).

Mody, C.C.M.: 2004, 'Small, but Determined: Technological Determinism in Nanoscience', *Hyle: International Journal for Philosophy of Chemistry*, 10(2), 99-128.

Mody, C.C.M.: 2004, *Crafting the Tools of Knowledge: The Invention, Spread, and Commercialization of Probe Microscopy, 1960-2000*, Ph.D. dissertation, Cornell University.

Montemerlo, M.S., Love, J.C., Opiteck, G.J., Goldhaber, D. J., and Ellenbogen, J.C., "Technologies and Designs for Electronic Nanocomputers," MITRE Technical Report 96W0000044, The MITRE Corporation, McLean, VA, July 1996. For more information, send e-mail to nanotech@mitre.org.

Moor, J.H. & Weckert, J.: 2004, 'Nanoethics: Assessing the Nanoscale From an Ethical Point of View', in: D. Baird, A. Nordmann & J. Schummer (eds.), *Discovering the Nanoscale*, Amsterdam: IOS Press, pp. 301-310.

Munn Sanchez, E.: 2004, 'The Expert's Role in Nanoscience and Technology', in: D. Baird, A. Nordmann & J. Schummer (eds.), *Discovering the Nanoscale*, Amsterdam: IOS Press, pp. 257-266.

Nordmann, A.: 2003, 'Shaping the World Atom by Atom: Eine nanowissenschaftliche WeltBildanalyse', in: A. Grunwald (ed.), *Technikgestaltung zwischen Wunsch und Wirklichkeit*, Berlin: Springer, pp. 191-199.

Nordmann, A.: 2004, 'Molecular Disjunctions: Staking Claims at the Nanoscale', in: D. Baird, A. Nordmann & J. Schummer (eds.), *Discovering the Nanoscale*, Amsterdam: IOS Press, pp. 51-62.

Nordmann, A.: 2004, 'Nanotechnology: Convergence and Integration', Presentation at *EuroNanoForum, Trieste, December 10, 2003* (Proceedings in preparation).

Nordmann, A.: 2004, 'Nanotechnology's WorldView: New Space for Old Cosmologies', *IEEE Technology and Society Magazine*, 23 (forthcoming).

Nordmann, A.: 2004, 'Social Imagination for Nanotechnology', in: European Commission (Community Health and Consumer Protection): *Nanotechnologies: A Preliminary Risk Analysis on the Basis of a Workshop, Brussels, 1-2 March 2004*, pp. 111-113 [www.europa.eu.int/comm/health/ph_risk/documents/ev_20040301_en.pdf].

Nordmann, A.: 2004, 'Was ist TechnoWissenschaft? ' Zum Wandel der Wissenschaftskultur am Beispiel von Nanoforschung und Bionik', in: T. Rossmann & C. Tropea (eds.), *Bionik ' Neue Forschungsergebnisse aus Natur-, Ingenieur- und Geisteswissenschaften*, Berlin: Springer, 2004 (forthcoming)

Paschen, H.; Coenen, C.; Fleischer, T.; Grünwald, R.; Oertel, D. & Revermann, C.: 2004, *Nanotechnologie: Forschung, Entwicklung, Anwendung*, Berlin: Springer (*Nanotechnologie, TAB-Arbeitsbericht 92*, Berlin: Büro für Technikfolgen-Abschätzung beim Deutschen Bundestag).

Pitt, J.C.: 2004, 'The Epistemology of the Very Small', in: D. Baird, A. Nordmann & J. Schummer (eds.), *Discovering the Nanoscale*, Amsterdam: IOS Press, pp. 157-163.

Pressman, J.: 2004, 'Nano Narrative: A Parable from Electronic Literature', in: N.K. Hayles (ed.), *Nanoculture: Implications of the New Technoscience*, Bristol, UK: Intellect Books, pp. 191-200.

Prigogine, I., and S. Rice. 1988. *Advances in chemical physics*, Vol. 70, Parts 1 & 2. New York: J. Wiley.

R. T. Bate, "Nanoelectronics," *Nanotechnology*, Vol. 1, pp. 1-7, 1990.

Rademann, K., B. Kaiser, U. Even, F. Hensel. 1987. *Phys. Rev. Lett.* 59:2319.

Rao, C.N.R., B.C. Satishkumar, and A. Govindaraj. 1997. *Chem. Commun.* 1581.

Rao, M.B., and S. Sircar. 1993. *Gas Separation and Purification* 7:279.

Reed, M.A., "Quantum Dots," *Scientific American,* January 1993, pp. 118-123.

Roberts, J.A.: 2004, 'Deciding the Future of Nanotechnologies: Legal Perspectives on Issues of Democracy and Technology', in: D. Baird, A. Nordmann & J. Schummer (eds.), *Discovering the Nanoscale*, Amsterdam: IOS Press, pp. 247-255.

Robinson, C.: 2004, 'Images in NanoScience/Technology', in: D. Baird, A. Nordmann & J. Schummer (eds.), *Discovering the Nanoscale*, Amsterdam: IOS Press, pp. 165-169.

Robison, W.L.: 2004, 'Nano-Ethics', in: D. Baird, A. Nordmann & J. Schummer (eds.), *Discovering the Nanoscale*, Amsterdam: IOS Press, pp. 285-300.

Roco, M.C. & Bainbridge, W.S. (eds.): 2001, *Societal implications of nanoscience and nanotechnology*, (Proceedings of a workshop organized by the National Science Foundation, September 28-29, 2000), Kluwer: Dordrecht [available online at http://itri.loyola.edu/nano/societalimpact/nanosi.pdf]

Roco, M.C. & Tomellini, R. (eds.): 2002, *Nanotechnology: Revolutionary Opportunities and Societal Implications, Workshop, Lecce (Italy), 31 January - 1 February 2002*, Luxemburg: Office for Official Publications of the European Communities, [200 pp.]

Roco, M.C.; Bainbridge, W.S. (eds.): 2002, *Converging Technologies for Improving Human Performance: Nanotechnology, Biotechnology, Information Technology and the Cognitive Science*, Arlington, VA: National Science Foundation.

Roher, H. 1993. *Jpn. J. Appl. Phys.* 32:1335.

Rohlfing, E.A., D.M. Cox, and A. Kaldor. 1984. *J. Chem. Phys.* 81:3846.

Rosen, A. 1998. A periodic table in three dimensions: A sightseeing tour in the nanometer world. In *Advances in quantum chemistry*. In press.

Ruthven, D.M., S. Farooq, K.S. Knaebel. 1994. *Pressure swing adsorption*. New York: VCH Publishers.

Sarewitz, D.; Woodhouse, E.: 2003, 'Small is Powerful', in: A. Lightman, D. Sarewitz & Chr. Desser, (eds.), *Living with the Genie: Essays on Technology and the Quest for Human Mastery*, Washington, DC: Island Press, pp. 63-83.

Schiemann, G.: 2004, 'Dissolution of the Nature-Technology Dichotomy? Perspectives on Nanotechnology from an Everyday Understanding of Nature', in: D. Baird, A. Nordmann & J. Schummer (eds.), *Discovering the Nanoscale*, Amsterdam: IOS Press, pp. 209-213.

Schmidt, J.C.: 2004, 'Unbounded Technologies: Working Through Technological Reductionism of Nanotechnology', in: D. Baird, A. Nordmann & J. Schummer (eds.), *Discovering the Nanoscale*, Amsterdam: IOS Press, pp. 35-50.

Schummer, J.: 2004, 'Why do Chemists Perform Experiments?', in: D. Sobczynska, P. Zeidler, E. Zielonacka-Lis (eds.), *Chemistry in the Philosophical Melting Pot*, Peter Lang (Frankfurt/M.), 2004, pp. 395-410.

Schummer, J.: 2005, 'Reading Nano: The Public Interest in Nanotechnology as Reflected in Book Purchase Patterns", *Public Understanding of Science*, 14 (2) (forthcoming).

Schummer, J.; Baird, D. (eds.): 2004-5, *Nanotech Challenges*, joint special issue of *Hyle: International Journal for Philosophy of Chemistry & Techne: Research in Philosophy and Technology* (forthcoming).

Stuckless, J.T, D.E. Starr, D.J. Bald, C.T. Campbell. 1997. *J. Chem. Phys.* 107:5547.

Suchman, M.: 2001, 'Envisioning Life on the Nano-Frontier', in: M.C. Roco and W.S. Bainbridge (eds.), *Societal Implications of Nanoscience and Nanotechnology*, Dordrecht: Kluwer, pp. 211-216.

Suchman, M.: 2002, 'Social Science and Nanotechnologies', M. Roco & R. Tomellini (eds.), *Nanotechnology ' Revolutionary Opportunities and Societal Implications*, Luxembourg: European Communities, pp. 95-99.

Sweeney, A.E.; Seal, S. & Vaidyanathan, P.: 2003, 'The promises and perils of nanoscience and nanotechnology: Exploring emerging social and ethical issues', *Bulletin of Science, Technology & Society*, 23 (4), 236-245.

Verbeek, A. & Callaert, J.: 2002, *Detailed analysis of the science-technology interaction in the field of Nanotechnology*, Luxemburg: Office for Official Publications of the European Communities (Linking Science to Technology. Bibliographic References in Patents, Vol. 9).

Index

A

Adleman, 5
Advanced technologies, 113
Aizawa, 7
Aksay, 2
Allara, 3
ATM, 7

B

Barbara Durrant, 38
Brus, 5

C

Cancer Diagnostics, 67
Cell Repair Machines, 34
Clear Springs Foods, 191
Consolidated Materials, 7
Control of Nanomedical Treatment, 56

D

Dave Fogel, 161
Diamondoid Receptor Design, 131
DNA, 2, 191
Donaldson, 49
Drexler, 46
Dynamic Pore Sizing and Gated Channels, 119

E

Electrostatic Bond, 127
Engines of Creation, 45
Ettinger, 48
European Research Infrastructures, 135

F

FDA, 43
Food and Drug Administration, 43

G

G.M. Fahy, 53
G.R. Taylor, 48
Gather, Darkness, 47
Gear Trains, 133
Gene Gun, 82
Germs, 113
Guarnieri, 6
Gunther Gross, 160

H

Halacy, 49
Hydraulics, 133

Hydrogen Bond, 127
Hydrophobic Forces, 128

I

Internal Transport Streams, 125
Isaac Asimov, 48

J

J. White, 48
Japan, 7
Jaworek, 8
Jeon, 48

K

Karak, 4
Kirschvink, 7
Krejchi, 1
Kumar, 3

L

Leonard Hayflick, 36
Ligand-Receptor Dynamics, 130
Ligand-Receptor Specificity, 129

M

Memory, 134
Microcosmic God, 47
MNT, 176
Molecular,
    Nanotechnology, 43, 52
    pumps, 123
    Science, 7
    sorting, 124
Monsanto Chemical, 161
Moore, 2

N

NanoCheck, 190
Nanodevice Versatility, 56
Nanomedicine, 43
Nanoparticles, 4
Nanosafety, 79
Nanotechnology, 1, 9
    Characterization Lab, 43
National Cancer Institute, 43
National Institute of Standards and Technology, 43
Naturophilia, 60
NCI, 43
NIST, 43
Nondegradation of Treatment Agents, 56
NRC, 2
NSOMs, 264
Nutraceutical, 311

O

Okazaki National Research Institutes, 7
Oncogene, 311

P

PCR, 5
Physical Forces in Molecular Recognition, 127
Population Pressure, 39
PTO, 256

R

Receptor-Based Transport, 126
Research Councils, 177
Richard Laing, 49
RNA, 2

S

San Francisco, 160
Seeman, 3
Seventh Biophysical Discussions, 6
Shrink sizes, 177
Slan, 47
Sligar, 7
Sorting Rotors, 123
Sosnowski, 3
Speed of Medical Treatment, 54
Star Trek scenario, 101
Stucky, 7
Superior Building Materials, 55
Svoboda, 8

T

Terrorism, 113
Theoretical Biophysics Group, 160
Tighten tolerances, 177
Transporter Pumps, 123

V

Virtual Hub, 176
virtual protein, 161

W

Whitesides, 3
WTO, 258

Y

Yin, 6

Z

Zhang, 3